BOSCH

Diesel-engine management

2nd
UPDATED AND
EXPANDED EDITION

SAE
INTERNATIONAL ®

Imprint

Published by:
© Robert Bosch GmbH, 1999
Postfach 30 02 20
D-70442 Stuttgart
Automotive Equipment Business Sector,
Product-Marketing software products (KH/PDI).

Editor-in-Chief:
Dipl.-Ing. (FH) Horst Bauer.

Editors:
Dipl.-Ing. Karl-Heinz Dietsche,
Dipl.-Ing. (BA) Jürgen Crepin,
Dipl.-Holzw. Folkhart Dinkler.

Layout:
Dipl.-Ing. (FH) Ulrich Adler,
Berthold Gauder, Leinfelden-Echterdingen.

Translation:
Peter Girling.

Technical graphics:
Bauer & Partner, Stuttgart.

Printed in Germany. Imprimé en Allemagne.
2nd edition, September 1999.
SAE Society of Automotive Engineers
400 Commonwealth Drive
Warrendale, PA 15096-0001 U.S.A.

(2.0 N)

ISBN 0-7680-0509-4

Authors

Diesel engine, diesel cycle, operation
Dr.-Ing. K.-O. Riesenberg,
Dipl.-Ing. (FH) W. Faupel.

Diesel fuels
Dr. rer. nat. B. Blaich.

Fuel management
Dipl.-Ing. E. Ungerer.

Exhaust-gas technology
Dr.-Ing. U. Pfeifer.

PE In-line fuel-injection pumps
Dipl.-Ing. (FH) E. Ritter, in cooperation with the responsible technical departments of Robert Bosch GmbH.

Governors for PE in-line fuel-injection pumps
Dipl.-Ing. (FH) E. Ritter, in cooperation with the responsible technical departments of Robert Bosch GmbH.

VE axial-piston distributor fuel-injection pump
Prof. Dr.-Ing. H. Tschöke, in cooperation with the responsible technical departments of Robert Bosch GmbH.

Electronic Diesel Control EDC
Dr. rer. nat. Jürgen Mössinger,
Dipl.-Ing. (FH) Frank Eichhorn.

PE-EDC in-line fuel-injection pumps
Dip.-Ing. F. Landhäußer.

VE-EDC axial-piston distributor fuel-injection pump
Prof. Dr.-Ing. H. Tschöke, in cooperation with the responsible technical departments of Robert Bosch GmbH.

VR radial-piston distributor fuel-injection pump
Dr.-Ing. U. Reuter,
Ing. (grad.) H. Nothdurft,
Dipl.-Ing. N. Rodriguez-Amaya,
Dipl.-Ing. (FH) K.-F. Rüsseler,
Dipl.-Ing. W. Vallon,
Dipl.-Ing. B. Veldten,
in cooperation with the responsible technical departments of Robert Bosch GmbH.

Common Rail accumulator fuel-injection system
Dipl.-Ing. R. Isenburg,
Dipl.-Ing. (FH) M. Münzenmay (STZ System- und Simulationstechnik, Prof. H. Kull, Esslingen),
in cooperation with the responsible technical departments of Robert Bosch GmbH.

Single-plunger fuel-injection pumps
Dr.-Ing. T. Stipek, Dipl.-Ing. T. Henze.

Innovative fuel-injection systems
Dipl.-Ing. S. Theobald.

Nozzles and nozzle holders
Dipl.-Ing. (FH) H. Reinauer.

Start-assist systems
Dipl.-Ing./Dipl.-Kfm. D. Franz.

Unless otherwise stated, the above are all employees of Robert Bosch GmbH, Stuttgart.

Foreword

All the manuals from the Bosch "Technical Instruction" publication range dealing with diesel technology were combined to form this reference book. It is intended to satisfy the thirst for knowledge of a large circle of readers.

The diesel engine and the diesel fuel-injection system form an inseparable unit, that is, one is unthinkable without the other. It was fuel-injection technology which first put the diesel engine on the road and enabled it to achieve its present-day significance as an indispensable power unit for vehicles and stationary units – from the subcompact-car engine, up to the large stationary diesel engine. The excellent interaction between the diesel engine and its injection system determines the engine's torque, power output, emissions, and noise.

Following its introduction, the diesel engine operated for many years without any form of electrical device or add-on unit. But on the diesel engine too, without electronics state-of-the-art injection techniques are inconceivable, and it would be impossible to comply with the increasingly stringent demands for low pollutant emissions and improved fuel economy. The innovative VR radial-piston distributor injection pump, and the "Common Rail" accumulator injection system, are examples of the latest developments in diesel fuel-injection technology.

This reference book provides comprehensive information on the state-of-the-art in diesel injection technology. It provides the reader who is interested in automotive-engineering technology with a wide range of easily understood descriptions of the most important components on and around the diesel engine.

The editorial staff

Contents

Combustion in the diesel engine

The diesel engine

Diesel combustion principle

The diesel engine is a compression-ignition (CI) engine which draws in air and compresses it to a very high level. With its overall efficiency figure, the diesel engine rates as the most efficient combustion engine (CE). Large, slow-running models can have efficiency figures of as much as 50% or even more.

The resulting low fuel consumption, coupled with the low level of pollutants in the exhaust gas, all serve to underline the diesel engine's significance.

The diesel engine can utilise either the 4- or 2-stroke principle. In automotive applications though, diesels are practically always of the 4-stroke type (Figs. 1 and 2).

Working cycle (4-stroke)

In the case of 4-stroke diesel engines, gas-exchange valves are used to control the gas exchange process by opening and closing the inlet and exhaust ports.

Induction stroke

During the first stroke, the downward movement of the piston draws in un-throttled air through the open intake valve.

Compression stroke

During the second stroke, the so-called compression stroke, the air trapped in the cylinder is compressed by the piston which is now moving upwards. Compression ratios are between 14:1 and 24:1. In the process, the air heats up to temperatures around 900°C. At the end of the compression stroke the nozzle injects fuel into the heated air at pressures of up to 2,000 bar.

Power stroke

Following the ignition delay, at the beginning of the third stroke the finely atomized fuel ignites as a result of auto-ignition and burns almost completely. The cylinder charge heats up even further and the cylinder pressure increases again. The energy released by the ignition is applied to the piston.

The piston is forced downwards and the combustion energy is transformed into mechanical energy.

Exhaust stroke

In the fourth stroke, the piston moves up again and drives out the burnt gases through the open exhaust valve.

A fresh charge of air is then drawn in again and the working cycle repeated.

Combustion chambers, turbocharging and supercharging

Both divided and undivided combustion chambers are used in diesel engines

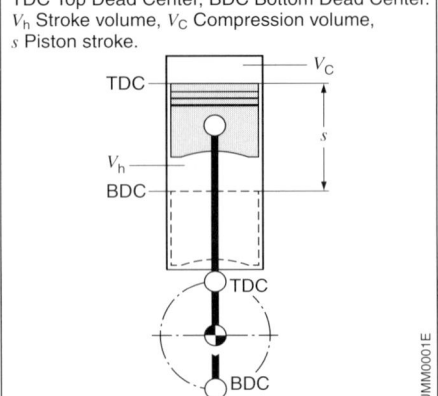

Fig. 1

Principle of the reciprocating piston engine

TDC Top Dead Center, BDC Bottom Dead Center.
V_h Stroke volume, V_C Compression volume,
s Piston stroke.

UMM0001E

(prechamber engines and direct-injection engines respectively).

Direct-injection (DI) engines are more efficient and more economical than their prechamber counterparts. For this reason, DI engines are used in all commercial-vehicles and trucks. On the other hand, due to their lower noise level, prechamber engines are fitted in passenger cars where comfort plays a more important role than it does in the commercial-vehicle sector. In addition, the prechamber diesel engine features considerably lower toxic emissions (HC and NO_X), and is less costly to produce than the DI engine. The fact though that the prechamber engine uses slightly more fuel than the DI engine (10...15%) is leading to the DI engine coming more and more to the forefront. Compared to the gasoline engine, both diesel versions are more economical especially in the part-load range.

Diesel engines are particularly suitable for use with exhaust-gas turbochargers or mechanical superchargers. Using an exhaust-gas turbocharger with the diesel engine increases not only the power yield, and with it the efficiency, but also reduces the combustion noise and the toxic content of the exhaust gas.

Diesel-engine exhaust emissions

A variety of different combustion deposits are formed when diesel fuel is burnt.
These reaction products are dependent upon engine design, engine power output, and working load.
The complete combustion of the fuel leads to major reductions in the formation of toxic substances. Complete combustion is supported by the careful matching of the air-fuel mixture, absolute precision in the injection process, and optimum air-fuel mixture turbulence. In the first place, water (H_2O) and carbon dioxide (CO_2) are generated. And in relatively low concentrations, the following substances are also produced:

– Carbon monoxide (CO),
– Unburnt hydrocarbons (HC),
– Nitrogen oxides (NO_X),
– Sulphur dioxide (SO_2) and sulphuric acid (H_2SO_4), as well as
– Soot particles.

When the engine is cold, the exhaust-gas constituents which are immediately noticeable are the non-oxidized or only partly oxidized hydrocarbons which are directly visible in the form of white or blue smoke, and the strongly smelling aldehydes.

Fig. 2

4-stroke diesel engine
1 Induction stroke, **2** Compression stroke, **3** Power stroke, **4** Exhaust stroke.

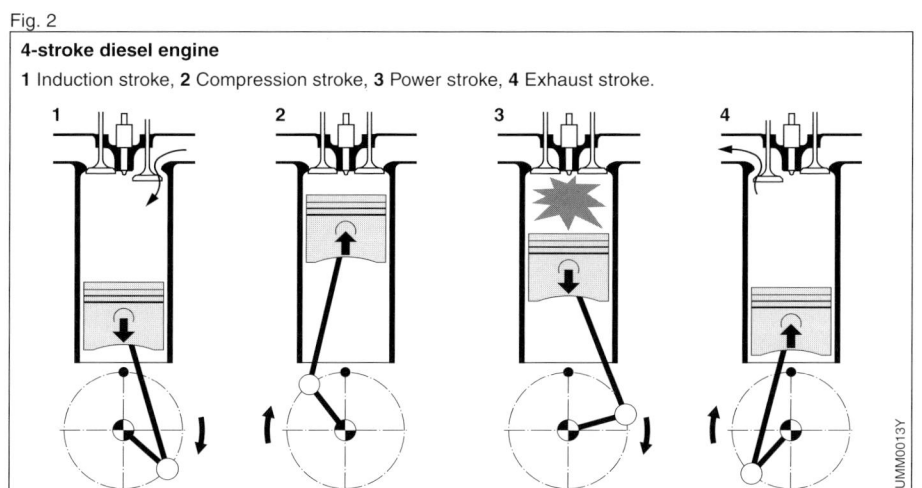

UMM0013Y

Diesel cycle and operation

Diesel combustion principle

Prechamber systems

In the prechamber system for passenger-car diesel engines, the fuel is injected into a hot prechamber (auxiliary chamber). Here, pre-combustion is initiated in order to achieve good mixture formation with reduced ignition lag for the main combustion process (Fig. 1).

The fuel is injected with a throttling pintle nozzle at a relatively low pressure (up to 300 bar). A specially designed baffle surface in the center of the chamber distributes the fuel jet which strikes it and mixes it intensively with air. Combustion starts and drives the partially-combusted air-fuel mixture through bores at the bottom end of the prechamber into the main combustion chamber above the piston, the mixture heating up even further in the process.

Here, intensive mixing takes place with the air in the main combustion chamber and combustion is continued and completed. A short ignition lag and controlled release of energy at a low overall pressure level in the main combustion chamber lead to "soft" combustion with low noise and less load on the engine.

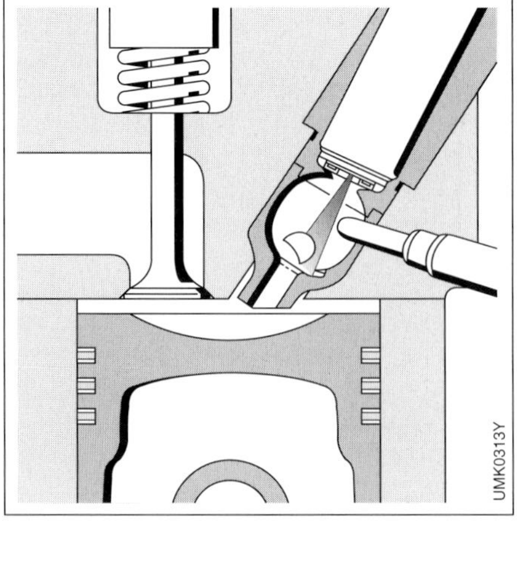

An optimized version of the prechamber permits combustion with an even lower toxic-substance content in the exhaust gas and an average of 40% less particulate emission. A modified prechamber shape, with evaporation recess and a changed shape and position of the baffle surface (ball pin), imparts a specific swirling action to the air after it flows out of the cylinder into the prechamber following compression. The fuel is injected at an angle of 5 degrees to the prechamber's axis (Figure 1). The glow plug is located downstream of the air flow to prevent it from interfering with the combustion process. Controlled post-glowing for up to 1 minute after cold starting (depending on the coolant temperature) contributes to improvement of the exhaust gas and reduction of noise in the warm-up period.

Whirl-chamber process

In this process used in passenger-car diesel engines, combustion is also initiated in an auxiliary chamber. The combustion process uses a ball-shaped or disk-shaped auxiliary combustion chamber (whirl chamber) with a throat area opening tangentially into the main combustion chamber (Figure 2).

A strong air vortex is generated during the compression stroke, and the fuel injected into this swirling air. The nozzle is positioned so that the fuel jet pene-

Fig. 1

Prechamber process

UMK0313Y

Fig. 2

Whirl-chamber process

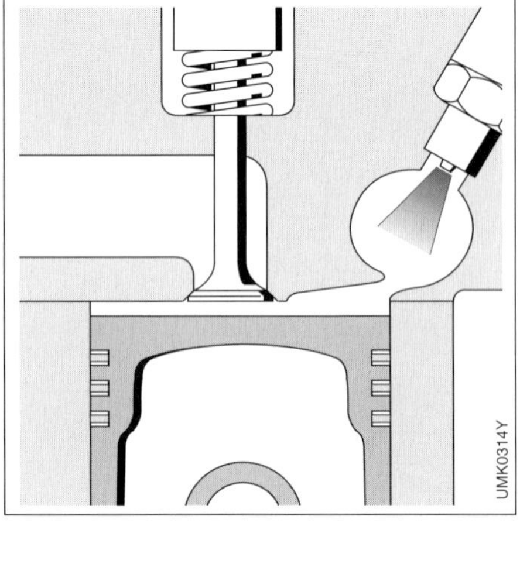

UMK0314Y

trates the swirling air perpendicularly to its axis and strikes the opposite chamber side in a hot wall zone.

At the start of combustion, the air-fuel mixture is forced into the main combustion chamber through the throat area and mixed with the residual combustion air. Compared with the prechamber process, the flow losses between the main combustion chamber and auxiliary chamber are lower for the whirl-chamber process because the flow cross-section is greater. This leads to lower charge-cycle work, with corresponding benefits to internal efficency and fuel consumption. At the same time, it is important that mixture formation takes place as completely as possible in the whirl chamber. The design of the whirl chamber, the arrangement and form of the nozzle jet and also the position of the glow plug must be carefully matched to ensure good mixture formation at all speed and load conditions.

An additional demand is rapid heating-up of the whirl chamber after cold starting. This reduces ignition lag and avoids the production of unburnt hydrocarbons (blue smoke) in the exhaust gas during warm-up.

Direct-injection process (DI)

In the direct-injection process, used mainly in commercial vehicle and stationary diesel engines of all sizes up to now, mixture formation in an auxiliary whirl chamber is dispensed with. The fuel is injected directly into the combustion chamber above the piston (Figure 3). The processes described up to now (fuel atomization, heating, evaporation and mixture with the air) must therefore occur in very rapid succession. High demands are placed both on the fuel injection and on air supply during induction. As in the whirl-chamber process, an air vortex is generated during the induction and compression strokes. The vortex is caused by the special shape of the intake port in the cylinder head. The design of the piston top with integrated combustion chamber contrib-

utes to the air movement at the end of the compression stroke, i.e. at the start of injection.

Of the combustion chamber shapes used during the course of development of the diesel engine, the cylindrical piston recess is widely used today, because it offers a compromise between economical manufacture and expedient air control.

In addition to good air turbulence, the fuel must also be distributed uniformly in order to ensure rapid mixing. Unlike the prechamber engine with its single-jet throttling pintle nozzle, a multihole nozzle is used for the direct injection (DI) process. Its spray position must be optimized in accordance with combustion-chamber design.

In practice, two methods are used for direct injection:
- Assistance of mixture formation by controlled air movement, and
- Assistance of mixture formation almost exclusively by fuel injection without controlled air movement.

In the second case, air turbulence involves no work. This becomes noticeable in the form of lower charge-cycle losses and better cylinder charge (filling). At the same time, considerably higher demands are placed on the fuel-injection equipment with respect to the position and number of nozzle holes, fineness of atomization by means of

Fig. 3

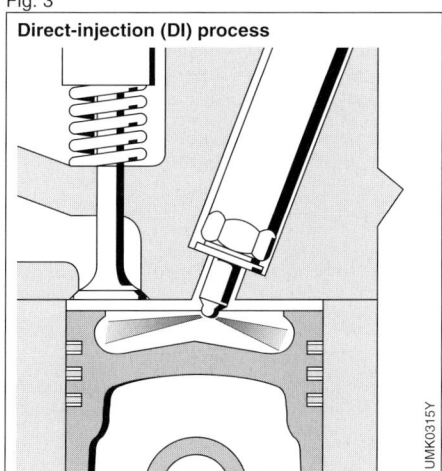

Direct-injection (DI) process

UMK0315Y

small spray-hole diameters, and the very high injection pressure necessary to reach the required short injection duration.

In the direct injection method described above, mixture formation is achieved by mixing and evaporating fuel particles with the air particles surrounding them (air-distribution method). In the method with wall distribution, on the other hand, the fuel is directed against the wall of the combustion chamber where it evaporates and is picked up by the air.

Direct-injection system with wall distribution (M system)

In this injection system for commercial and stationary diesel engines, the heat content of the piston-recess wall is used for evaporation of the fuel, and the air-fuel mixture is produced by suitable guidance of the combustion air (Figure 4). The process operates with a single-hole nozzle with a relatively low injection pressure. If the air movement in the combustion chamber is correctly adjusted, extremely homogeneous air-fuel mixtures can be obtained with long combustion duration, low pressure rise and thus quieter combustion. However, fuel consumption increases compared with the air-distribution processes.

Comparison of combustion processes

The disadvantages of the prechamber engines with respect to noise are most apparent during cold running, i.e. in the phase directly after cold starting. Inadequate mixture formation − caused not least by heat dissipation to the combustion-chamber walls − leads to relatively long ignition lags and to a knocking combustion noise. During the warm-up period, the whirl-chamber engine also tends towards higher combustion noise in the low load and low speed ranges. The prechamber method on the other hand, has advantages with respect to chamber temperature and ignition lag. The main advantage of the direct-injection system is a reduction in fuel con-

sumption of up to 20% compared with chamber engines.

Disadvantages of the direct-injection systems however, are combustion noise (particularly in the acceleration phase) and the restricted maximum speed. Basically speaking, the direct-injection system always requires higher injection pressures and thus a more complex fuel-injection system.

The advantages of the direct-injection system predominate for operating conditions where fuel consumption and thus economy are decisive and questions of comfort play a more subordinate role. Intensive development work with respect to mixture formation, which includes the fuel-injection installation, have already led to direct-injection systems being used in passenger cars.

Fig. 4

M-System

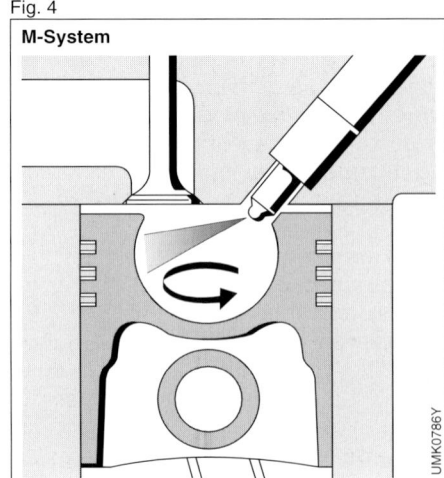

Supercharging processes

Supercharging as a means of increasing power has been in use for a long time for large diesel engines for stationary and marine applications, as well as for commercial-vehicle diesel engines. It is also being increasingly used for high-speed passenger-car diesel engines. In contrast to the naturally-aspirated engine, the air is supplied to the engine under pressure. This increases the air mass in the cylinder, thus permitting a higher power yield for the same swept volume with a correspondingly higher injected-fuel quantity. The lower the air temperature, the higher is the air mass forced into the cylinder (assuming other conditions remain the same). For this purpose, a charge-air cooler can be combined with the supercharging system. This also has the advantage that the thermal load on the cylinder chamber can be reduced.

Mechanical supercharging
In the mechanical supercharging process, the supercharger is driven directly by the engine. However, the drive power of the compressor reduces the engine's useful output power. Switching the supercharger on and off as required by means of a clutch improves the economy of the mechanically supercharged engine at partial loads, but also increases costs. The latest development is a spiral-piston charger which provides good volumetric efficiency over a wide speed range and which, particularly for small engines, can be seen as a possible alternative to exhaust-gas turbocharging.

Exhaust-gas turbocharging
A large amount of energy is lost with the combustion engine's exhaust gas. An obvious solution is to use this energy to generate pressure in the intake manifold through a flow compressor driven by an exhaust-gas turbine. Both flow machines together form the exhaust-gas turbocharger (Figure 5). The turbine and supercharger maps can be adjusted for favorable efficiency and thus a high level of turbocharging for stationary constant-speed operation. However, for vehicle engines, where high torque is expected particularly for low-speed acceleration, design is difficult. Low exhaust-gas temperatures, together with low exhaust-gas quantity and the mass acceleration of the turbocharger itself delay the pressure build-up in the compressor at the start of acceleration. This symptom is called the "turbohole" for turbocharged car engines.

Fig. 5

Exhaust-gas turbocharger with wastegate (schematic)

1 Air compressor,
2 Exhaust-gas turbine,
3 Bypass channel,
4 Wastegate.

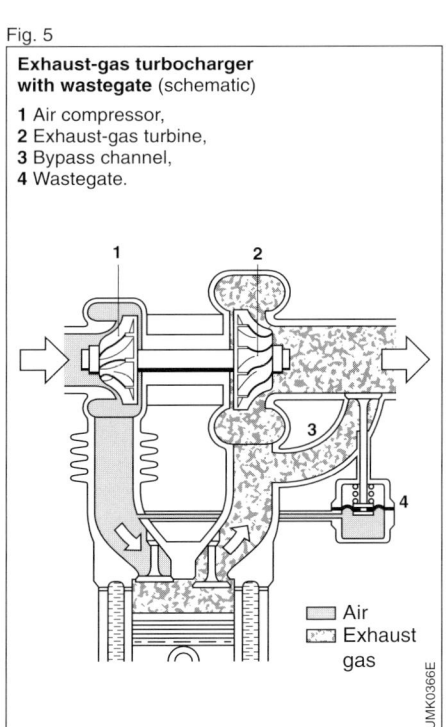

☐ Air
▨ Exhaust gas

UMK0366E

Turbochargers which thanks to their own low mass respond to even low exhaust-gas flows, were developed particularly for use in passenger cars and commercial vehicles. Driveability, particularly in the lower speed range, is considerably improved with such turbochargers.

In order to limit and adjust the charge-air pressure and to protect the turbocharger, the exhaust-gas flow to the turbine must be limited at high engine speeds and loads. Depending upon the maximum permissible charge-air pressure, a wastegate is opened and diverts part of the exhaust gas into the exhaust pipe through a bypass channel.

With regard to the adjustment of the charge-air pressure, in one version measures are taken to change the air flow approach at the turbocharger's blades and thus influence the charge-air pressure. Electronic closed-loop control can be applied to adapt this form of variable turbine geometry to the engine map.

Pressure-wave supercharging

One variant of supercharging is the pressure-wave supercharger known under the name "Comprex"® (Figure 6).

Table 1: Comparative data for spark-ignition and diesel engines

Engine type	Engine speed min⁻¹	Compression ratio ε	Mean pressure bar	Power output kW/l	Power/weight ratio kg/kW	Fuel consumption g/kWh	Torque increase %
Spark-ignition engine for cars							
Naturally-aspirated engine	4500...7500	8...12	8...11	35...65	3...1	350...250	15...25
with supercharging	5000...7000	7...9	11...15	50...100	3...1	380...280	10...30
Spark-ignition engine for trucks							
	2500...5000	7...9	8...10	20...30	6...3	380...270	15...25
Diesel engine for cars							
Naturally-aspirated engine	3500...5000	20....24	7...9	20...35	5...3	320...240	10...15
with supercharging	3500...4500	20...24	9...12	30...45	4...2	290...240	15...25
Diesel engine for trucks							
Naturally-aspirated engine	2000...4000	16...18	7...10	10...20	9...4	240...210	10...15
with supercharging	2000...3200	15...17	10...13	15...25	8...3	230...205	15...30
with LLK[1]	1800...2600	14...16	13...18	25...40	5...2.5	225...195	20...40

[1] LLK; charge-air cooling.

Fig. 6

Pressure-wave supercharger
(Principle)

1 Engine,
2 Cell rotor,
3 Belt drive,
4 High-pressure exhaust gas,
5 High-pressure air,
6 Low-pressure air inlet,
7 Low-pressure exhaust-gas outlet.

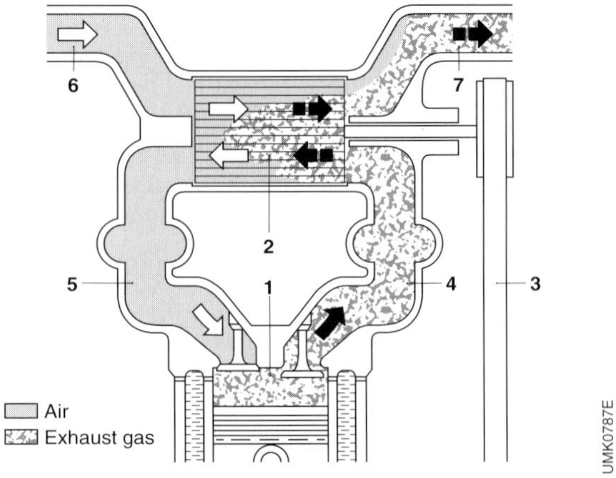

Air
Exhaust gas

UMK0787E

The exhaust-gas pressure waves drive a cell wheel whose specially shaped cells increase the pressure of the fresh-gas flow. The most important characteristic of pressure-wave supercharging is the direct energy exchange between the exhaust gas and the charge air without intermediate mechanical parts.

There are no disadvantages due to delayed turbocharger response. The pressure-wave supercharger reacts spontaneously to changes in load with an increase in charge-air pressure. Favorable torque characteristics, such as are not possible in the same way

with other supercharging processes can be achieved in the non-stationary area as well by clever design of the cell wheel (Figure 7).

A disadvantage, however, is the space required by the cell wheel and exhaust pipe on the engine (especially if the engine compartment is cramped) as well as the necessity to achieve suitable adjustment of the exhaust-gas oscillations for all loads and speeds.

Like mechanical superchargers, pressure-wave superchargers have a good response characteristic and ensure rapid torque pick-up during acceleration. However, with today's state-of-the-art an optimized exhaust-gas turbocharger probably offers the best compromise regarding function and cost.

Comparison of supercharging processes

Torque and power depend, among other things, on the mean pressure (mean piston or working pressure). The mean pressure values for small supercharged diesel ergines correspond to those of non-supercharged spark-ignition engines and in some cases even exceed them (refer to Table 1).

In larger commercial-vehicle engines, a further increase in the mean pressure is achieved by raising the supercharging level and reducing the compression, although restrictions in cold starting capability must be accepted as a result. With respect to power output per liter, diesel engines are less favorable than spark-ignition engines owing to their lower maximum engine speeds.

However, modern passenger-car diesel engines nevertheless achieve rated speeds up to 5,000 min^{-1}.

Operating conditions

The operating conditions of a diesel engine are based on a variety of different relationships which are typical for the processes involved:

In the diesel engine, the fuel is injected directly into the highly-compressed, hot air with the result that it auto-ignites.

Fig. 7

Comparison of different supercharging processes with the corresponding induction processes (stationary operation)

a Exhaust-gas turbocharger,
b Pressure-wave supercharger,
c Mechanical supercharger.
1 Supercharged engine,
2 Naturally-aspirated engine.

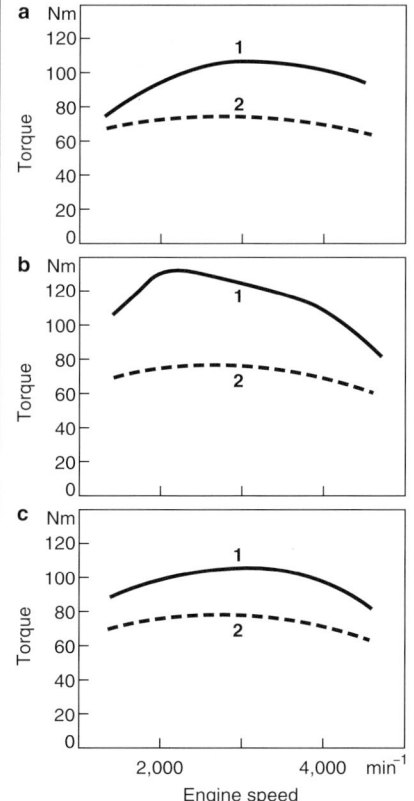

The diesel engine is thus not bound to ignition limits like the spark-ignition engine. Consequently, considering that the air quantity in the combustion chamber remains constant, only the fuel quantity needs to be regulated.

The fuel-injection system is thus of decisive importance for engine functioning.

At all speeds and loads, it is responsible for metering the fuel and for its uniform distribution throughout the whole charge. In addition, the pressure and temperature of the intake air must be taken into account.

Each engine operating point thus requires

– The correct injected fuel quantity,
– At the correct time,
– At the correct pressure,
– At the correct time sequence, and
– At the correct point in the combustion chamber.

In addition to the requirements for optimum mixture formation, other engine-specific and vehicle-specific operating limits must be taken into account for fuel metering, such as

– Smoke limit,
– Combustion-pressure limit,
– Exhaust-gas temperature limit,
– Engine-speed and torque limits, and
– Vehicle-specific and housing-specific loading limits.

Smoke limit

Since a considerable part of mixture formation takes place during combustion, local over-enrichment occurs and an increase in the emission of black smoke occurs even with moderate excess air. The air-fuel ratio which leads to smoke emissions bordering on the legal limit is a measure of how well the air is utilized. Prechamber engines operate at the smoke limit with an air excess of 10…25%, while direct-injection engines have an air excess of 40…50%.

Combustion-pressure limit

With diesel engines, because the vaporized fuel mixed with the air burns abruptly under extreme compression during the ignition process, we speak of "hard" or "noisy" combustion, and high peak pressures result which require a comparatively heavy engine. The forces which are generated during combustion cause periodically changing loads on the engine components and, due to their dimensions and service lives, these components limit the combustion pressure.

Exhaust-gas temperature limit

The exhaust-gas temperature limit of a diesel engine is determined by the high thermal stressing of the engine components surrounding the hot combustion chamber, the heat resistance of the exhaust-gas system and the dependence on temperature of the toxic substances in the exhaust gas.

Engine-speed limits

The excess air in the diesel engine and the regulation of the fuel quantity already dealt with mean that at constant speed the engine power depends solely on the injected fuel quantity. If fuel is supplied to the diesel engine without a corresponding torque being taken off, the engine speed increases. If the injected fuel quantity is not reduced before a critical engine speed is exceeded, the engine "races" and can destroy itself. Speed limitation or "governing" is therefore absolutely essential for diesel engines. When the diesel engine is used as a machine drive, a certain speed is stipulated which is kept constant or remains within permissible limits irrespective of the load. When the diesel engine is used for automotive applications, the driver must be able to use his accelerator pedal to select whatever speed he wants, whereby the engine speed must not fall below the idling limit to standstill when the pedal is released. Consequently, we distinguish between the variable-speed governors and minimum-maximum-speed governors as control systems.

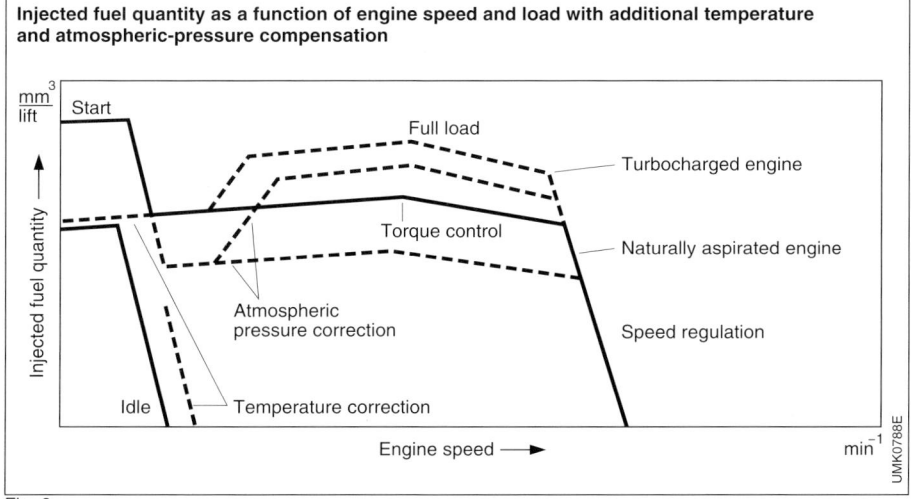

Injected fuel quantity as a function of engine speed and load with additional temperature and atmospheric-pressure compensation

Fig. 8

Taking into account all the specified requirements, a characteristic map can be defined for the operating range of an engine. This map (Figure 8) shows the injected fuel quantity as a function of speed and load as well as the required temperature and air-pressure compensation. The injected fuel quantity corresponds to

Fig. 9

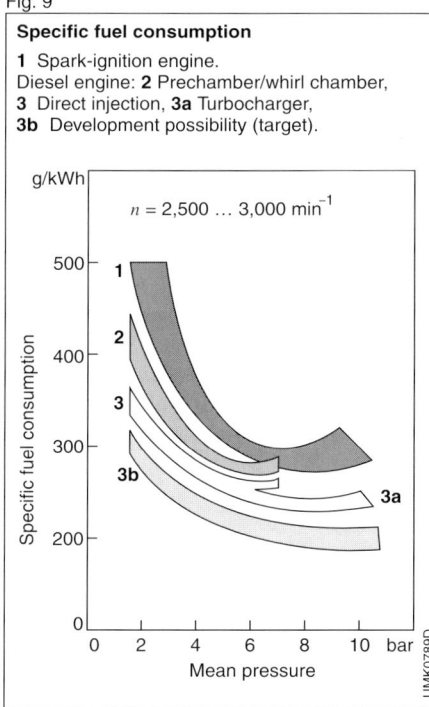

Specific fuel consumption

1 Spark-ignition engine.
Diesel engine: **2** Prechamber/whirl chamber,
3 Direct injection, **3a** Turbocharger,
3b Development possibility (target).

the mean requirement of all cylinders and the mean quantity at a specific speed. As the following example shows, the specified operating conditions place high demands on the accuracy of the fuel-injection system: The full-load fuel quantity for a four-cylinder, four-stroke engine with a power of 75 kW and a specific fuel consumption of 200 g/kWh necessitates an overall fuel requirement of 15 kg/h. This is equivalent to 288,000 injection strokes in one hour for a four-stroke engine operating at 2,400 min⁻¹. Converted to one injection stroke, this means a fuel quantity of 59 mm³ per injection stroke.

Compared with this, a raindrop has a volume of approx. 30 mm³. The fuel-injection system must perform this exact metering for one cylinder and for uniform distribution to the individual cylinders of a multiple-cylinder engine.

The theoretically determined injected fuel quantity applies as a guide value for designing a fuel-injection system. The full-load characteristic is limited by the engine's smoke limit in the lower speed range in particular, and by the permissible exhaust gas or component temperature in the upper speed range. The actually required fuel quantities are determined at the engine in accordance with empirical values. Systems are usually designed for mean sea level, i.e.

the power values are reduced to this level: if the engine is operated at altitudes above mean sea level, the fuel quantity must be corrected in accordance with the barometric altitude formula. An air density reduction of 7% per 1,000 m altitude applies as a guide value.

However, in contrast to specific fuel consumption, which is determined with a warm engine under constant test conditions (Figure 9), only driving consumption provides practically useful values. Cars in particular are mostly operated over short distances with frequent cold starting and in the low load range. The necessary cold-running enrichment leads to clear consumption differences (Figure 10).

Operating states

Start

Engine starting includes the process of ignition and acceleration up to sustained operation. The air heated in the compression stroke must ignite the injected fuel. The required ignition temperature for diesel fuel is approx. 220 °C. This temperature must be guaranteed with sufficient reliability at the minimum possible speed, and at low external temperatures with a cold engine. Several physical laws oppose these conditions: the lower the engine speed, the lower the final compression pressure and the lower the final compression temperature (Figure 11).

The causes of this behavior are leakage losses which occur because an oil film has not formed initially between the piston and cylinder wall. When the engine is cold, heat losses still occur during the compression stroke. In engines with divided combustion chambers, the heat losses are particularly high owing to the larger combustion-chamber surface area. In addition, the engine frictional forces at low temperatures are higher owing to reduced mechanical clearances of the engine components and higher engine-oil viscosity. In addition, the starter speed is particularly low owing to the

Fig. 10

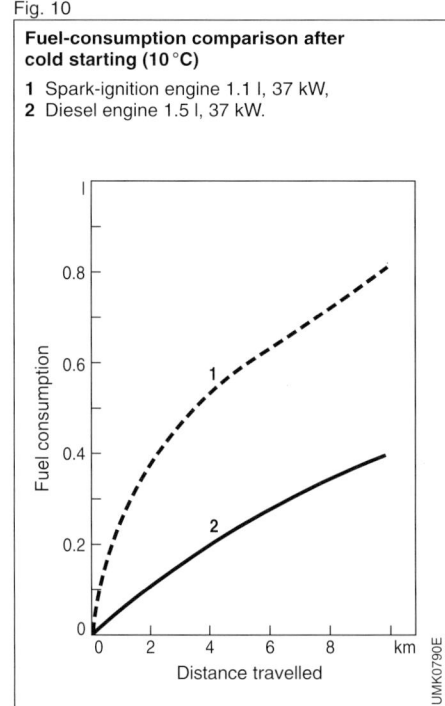

Fuel-consumption comparison after cold starting (10 °C)

1 Spark-ignition engine 1.1 l, 37 kW,
2 Diesel engine 1.5 l, 37 kW.

Fig. 11

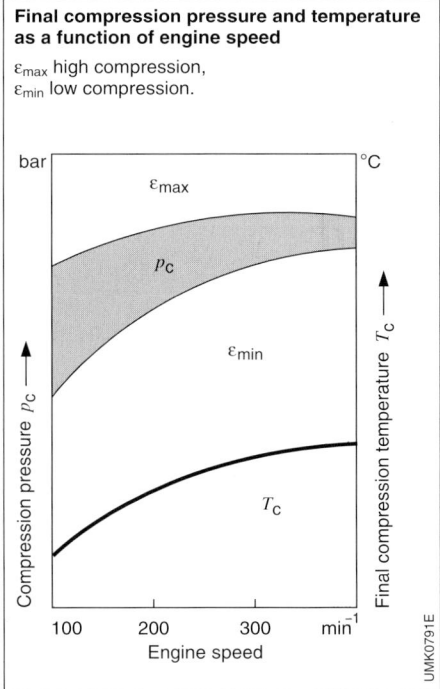

Final compression pressure and temperature as a function of engine speed

ε_{max} high compression,
ε_{min} low compression.

drop in battery voltage in cold conditions. There are several possibilities of counteracting these physical circumstances:

Fuel adaptation
Fuel problems which normally occur at low temperatures as a result of precipitation of paraffin crystals can be avoided by filter heating or direct fuel heating (Figure 12). Or, the flow properties can be improved by admixture of petroleum or regular-grade gasoline. Guide value: 10…30% can be added, depending on the temperature. The instructions of the vehicle manufacturer must be observed here. Diesel fuel is offered regionally which guarantees problem-free operation down to –23 °C.

Start-assist systems
In direct-injection engines, pre-heating of the intake air acts as a start-assist measure. In the case of prechamber engines, a glow plug in the whirl chamber acts as a starting aid. Modern glow plugs with a preheating time of only a few seconds permit rapid starting (Figure 13).

Both measures serve to improve fuel vaporization and mixture preparation and thus achieve reliable ignition of the air-fuel mixture.

Injection adaptation
One possible measure is the provision of excess fuel for starting to compensate for condensation and leakage losses, and in order to increase the engine torque in the run-up phase. A further measure is advancing the start of injection to compensate for the ignition lag and to ensure ignition in the TDC area, i.e. at the maximum final compression temperature.

The optimum start of injection must be achieved as exactly as possible within a narrow tolerance. If the fuel is injected too early, it is deposited on the cold cylinder walls and only a very small quantity evaporates because the mixture temperature is still too low at this time. If the fuel is injected too late, ignition takes place only in the combustion (or power) stroke and the piston is accelerated only slightly.

Fig. 12

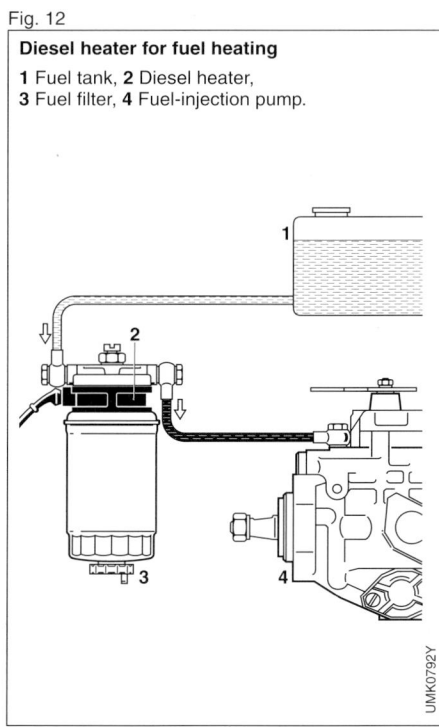

Diesel heater for fuel heating
1 Fuel tank, **2** Diesel heater,
3 Fuel filter, **4** Fuel-injection pump.

UMK0792Y

Fig. 13

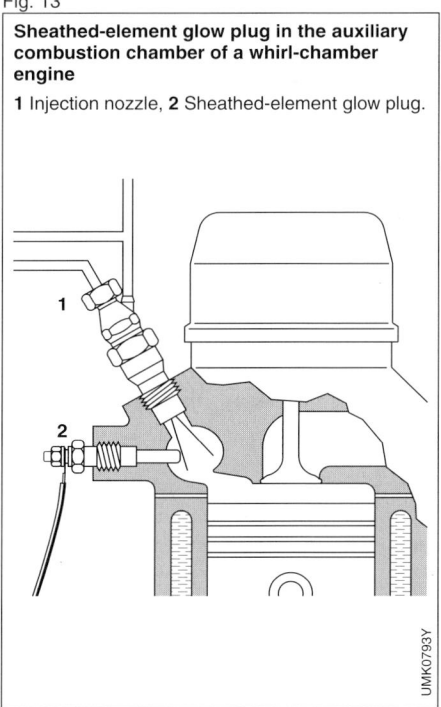

Sheathed-element glow plug in the auxiliary combustion chamber of a whirl-chamber engine
1 Injection nozzle, **2** Sheathed-element glow plug.

UMK0793Y

Figure 14 shows an example of a compression temperature curve during a single piston stroke referred to degrees crankshaft (°cks). By means of fuel distribution and mixture formation in the combustion chamber, the fuel-injection system (pump and nozzle) must ensure that the correct fuel-droplet size is available in the combustion chamber for the fastest possible air-fuel mixing.

Idle

Critical variables affecting diesel engines are the idle and low part-load states. Although the fuel-consumption values are extremely favorable compared with the spark-ignition engine in this operating range, noise and knocking pose a problem, particularly when the engine is cold. Ignition lag is one of the most important causes of idling noise.

As described for starting, the final compression temperature is lower for low speed and low load. This applies particularly at idle.

Compared with full-load operation, the combustion chamber is relatively cold in this operating range (even when the engine has reached operating temperature) because energy supply, and therefore temperature rise, are low. Heating of the combustion chamber occurs slowly and incompletely. Prechamber and whirl-chamber engines are particularly problematic in this respect, because the heat dissipation losses are specially high owing to the large surface area involved. One remedy is an increase in the engine compression ratio. However, the possibilities are limited in this case too, owing to the fuel-consumption disadvantages at full-load and the increase in mechanical noise. High demands are placed on the fuel-injection system regarding the accuracy of the start of injection, injected fuel quantity and the injection rates. As with the compression-pressure characteristic for starting, the maximum combustion temperature for idling is present only in a small piston-stroke range at TDC. The start of injection is adjusted to this extremely precisely.

Fig. 14

Compression temperature for cold starting as a function of crankshaft angle

a Ignition temperature range, diesel fuel,
t_a Outside temperature.

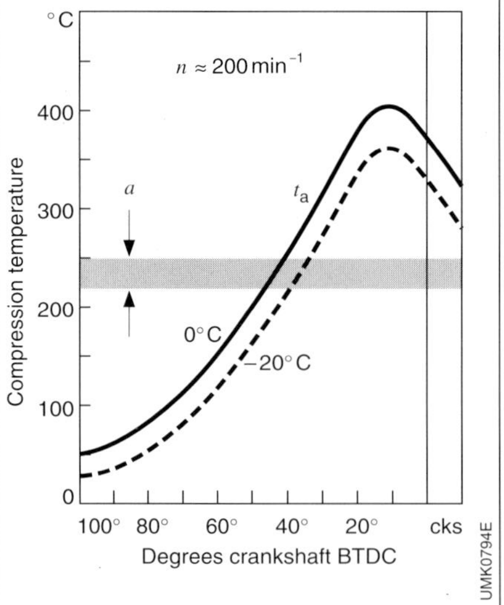

Compression temperature / °C

$n \approx 200 \, min^{-1}$

a t_a

$0\,°C$ $-20\,°C$

Degrees crankshaft BTDC — 100° 80° 60° 40° 20° cks

UMK0794E

Fig. 15

Excess-air factor characteristic

λ Excess-air factor at a stationary fuel droplet.

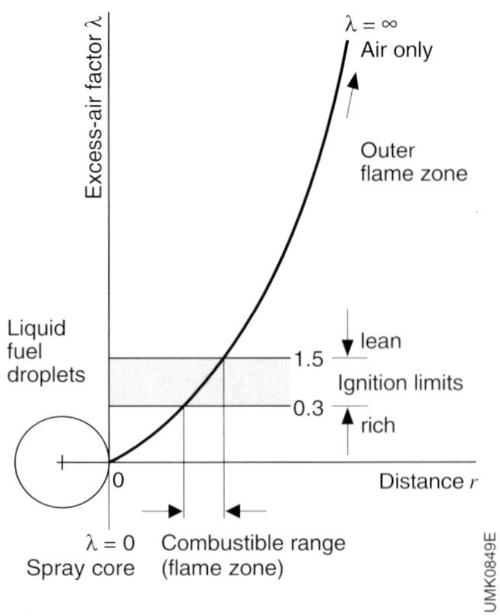

Excess-air factor λ

$\lambda = \infty$
Air only

Outer flame zone

Liquid fuel droplets

lean — 1.5
Ignition limits
0.3 — rich

Distance r

$\lambda = 0$ Combustible range
Spray core   (flame zone)

UMK0849E

With an idle delivery of 5...7 mm³ per injection, the degree of accuracy demanded from the fuel metering (0.5 mm³ per injection corresponds to 10%) immediately becomes apparent. Only a small amount of fuel must be injected during the injection-lag phase, because the fuel quantity in the combustion chamber at the moment of ignition is decisive for the sudden pressure rise in the cylinder. The noise produced depends directly on the pressure rise, whereby the steeper the rise, the more perceptible the "diesel knock" becomes (Figure 16). In addition to guaranteeing the exact start of injection and the exact delivery quantity, the fuel-injection system must therefore also ensure that the delivery quantity (0.25 mm³ per stroke and per degree crankshaft) is evenly distributed throughout 15...20°cks, and prepared uniformly in the combustion chamber. The fuel-injection pump is responsible for metering and control, while the injection nozzle is responsible for mixture formation.

Fig. 16

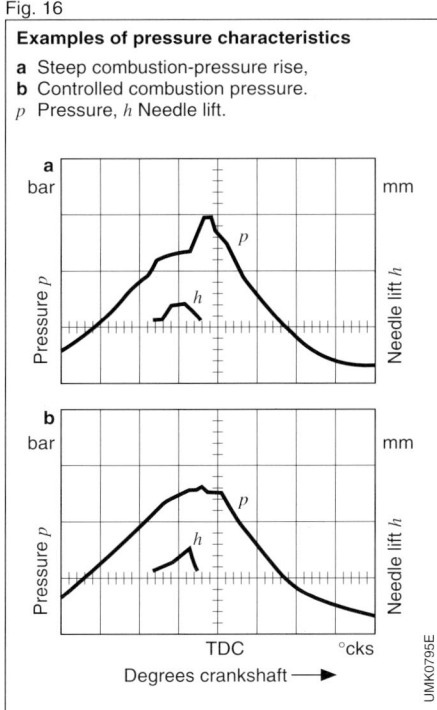

Examples of pressure characteristics

a Steep combustion-pressure rise,
b Controlled combustion pressure.
p Pressure, *h* Needle lift.

Full-load
Full-load designates the maximum torque which is permitted taking into account various parameter conditions. The torque characteristic – as a function of the speed – produces a torque maximum at approximately half rated speed.

The fuel-injection system must meet this demand. Mechanical, pneumatic and hydraulic adjustment possibilities are available for this purpose. Here, only the hydraulic measures to achieve the required full-load characteristic will be described. In the Chapter "Mixture Formation", it will be shown under "Duration of injection and rate-of-discharge curve" how the injection pressure and the rate-of-discharge curve change between the cam lift and the injection nozzle.

The "pre-delivery and post-delivery effect" is used for drawing-up the characteristic map. If the delivery quantity is considered for a piston pump, it is calculated by the piston area times effective stroke. In practical terms, delivery begins earlier and ends later.

The actual effective stroke is thus greater than the geometric effective stroke. This dynamic behavior is called the pre-delivery and post-delivery effect.

By varying the cross-section and flow speed, this effect can be changed as a function of engine speed to produce dynamically changed effective strokes and to permit implementation of a rising or falling fuel-delivery characteristic.

Diesel fuels

Diesel-fuel components

Diesel fuels consist of a large number of hydrocarbons which have boiling points in the range between about 180 °C and 370 °C. They are obtained by the step-wise distillation of crude oil.
Refineries are increasingly adding conversion products (crack components) to diesel fuels. These are obtained from heavy oils by using fission (cracking) of larger modules into smaller ones by the application of heat, pressure, and catalysts.

Characteristic values

The requirements placed on diesel fuels are defined in national standards. The EN 590 standard applies in Europe, and DIN 51 601 in Germany. The most important parameters as stipulated in these standards are given below:

Cetane number, ignition quality
Since the diesel engine operates without externally supplied ignition, following its injection into the hot, compressed air in the combustion chamber, the diesel fuel must ignite of its own accord with the minimum possible delay (ignition lag).
Ignition quality is defined as that property of the fuel which serves to initiate diesel-engine auto-ignition. Ignition quality is expressed by the cetane number (CN). The higher the cetane number, the easier it is for the fuel to ignite.
Cetane, which has very good ignition qualities is assigned the cetane number 100, whereas methyl naphtalene, which features poor ignition qualities, is given the cetane number 0. DIN 51 601 specifies a minimum cetane number of 45 for diesel fuels.

However, higher cetane numbers of around 50 are desirable for optimum operation of modern engines (quiet running, particulate emissions). High-quality diesel fuels contain a large proportion of paraffins with high CN numbers. In contrast, aromatics of the kind found in crack components impair ignition quality.

Cold behaviour, filterability
At low temperatures, the precipitation of paraffin crystals can cause clogging of the fuel filter, and thus result in interruption of the fuel supply. The start of paraffin precipitation can be as early as 0 °C in most unfavorable cases. Consequently, winter diesel fuels must be specially selected, or treated, in order to guarantee problem-free operation in cold weather.
Normally, "flow improvers" are added at the refinery. Although these do not prevent precipitation of the paraffins, they limit the crystal growth to a very high degree. The crystals that do form are so small that they can still pass through the filter pores. As a result, filterability can be extended down to lower temperatures. In accordance with DIN 51 601, filterability should be guaranteed down to at least −15 °C.
The resistance to cold can be improved even further by additives which prevent precipitation of paraffin crystals. The winter diesel fuels which are widely available today guarantee cold resistance down to at least −22 °C.
Additional measures are the addition of petroleum to the diesel fuel. Admixing of regular-grade gasoline can also delay crystal precipitation; however, the ignition quality is impaired and the flash point considerably reduced (gasoline has very low cetane numbers).
Today, if correct fuels are used as stipulated by standards, such additives are no longer necessary.

Flash point

The flash point is the temperature at which a combustible liquid gives off just enough vapor to the air surrounding it that the vapor-air mixture above the liquid can be ignited by an ignition source. For safety reasons (transport, storage), diesel fuels must be in Hazard Class A III, i.e. they have a flash point above 55 °C. For example, a gasoline content of less than 3% in the diesel fuel can reduce the flash point to such an extent that ignition is possible at room temperature.

Boiling range

The position of the boiling range influences parameters which are important for the operating behavior of the diesel fuel. Although extension of the range towards lower temperatures leads to a fuel which is suitable for cold operation, the cetane number is reduced at the same time, and this impairs the lubrication properties in particular. However, poorer lubrication properties increase the risk of fuel-injection-equipment wear.

If, on the other hand, the final boiling-point temperature is increased, which is desirable in order to achieve better utilization of crude oil, this can lead to increased soot production and nozzle coking (deposit of combustion residues).

Density

The diesel fuel's calorific value depends to a good approximation on its density, and increases with increasing density. Therefore, if fuels with greatly differing densities are used with the fuel-injection pump at the same setting – in this case, volume metering by the pump is constant – shifts occur in the mixture composition owing to the fluctuations in the calorific values, which in turn leads to increased soot emission for high densities, and a reduction in output power for low densities.

Sulphur

Diesel fuels contain sulphur in chemically bound form, depending on the crude-oil quality and the components used in the mixture. Crack components in particular have high sulphur contents, but these can be reduced by treatment with hydrogen at the refinery. Since sulphur is converted into sulphur dioxide (SO_2) during combustion in the engine (this substance is environmentally hostile owing to its acidic reaction), maximum permitted sulphur content is limited by law.

It has been reduced in several stages in recent years, and since 1 Oct 1996 must not exceed 0.05% by weight in Europe.

Among other things, the intention is to reduce the mass of emitted particulates which, in the case of catalytic exhaust-gas aftertreatment, also contain sulphates in addition to soot.

Additives

Quality improvement by the addition of additives, as has been common practice with gasoline for years now, has recently also become popular for diesel fuels ("super-grade" or "premium-grade" diesel fuel). Additive packages are mostly used which have a multiple effect:

- Ignition improvers raise the cetane number and are responsible in particular for quieter combustion.
- Detergents to prevent nozzle coking.
- Corrosion inhibitors to prevent corrosion of metal parts (in the event of water being entrained into the fuel system).
- Anti-foaming agents to facilitate tank filling.

The overall concentration of additives is generally below 0.1%, so that the physical characteristics of the fuel, such as density, viscosity, and boiling curve do not change.

Air supply

Air filters

By preventing air-borne dust being drawn into the engine with the intake air, the air filter helps inhibit internal engine wear.

On paved roads, the air's average dust content is about 1 mg/m^3; on unpaved roads however, and during construction work, it can range as high as 40 mg/m^3. This means that depending on roads and operating conditions, a medium-sized engine can draw in up to 50 mg every 1,000 km (600–700 miles).

Passenger-car air filters

Paper elements, contained in centrally located or fender-mounted housings serve as the air filters on passenger cars (Figs. 1 and 2). In addition to filtering the intake air, these units preheat and regulate its temperature, as well as attenuat-ing the intake noise. Intake-air tempera-ture regulation is important for smooth re-sponse, and also for exhaust-gas emis-sions. The temperatues for part and full-throttle operation can be different.

The hot air is taken off adjacent to the ex-haust system and directed by means of a flap mechanism into the air filter entrance so that it can mix with the cold intake air. Regulation is usually automatic, using either a pneumatic vacuum unit connect-ed to the intake manifold or expansion elements. Since it improves A/F mixture formation and distribution, the controlled (and thus constant) intake-air tempera-ture has positive effects upon engine power, fuel consumption, and exhaust emissions.

In addition, and this applies particularly at very low outside temperatures, the heating-up of the intake air reduces the duration of the warm-up phase after the engine has started.

Fig. 1

Central air filter for passenger cars

1 Fresh-air intake, **2** Warm-air intake, **3** Outlet for warm/fresh air mixture, **4** Vacuum unit.

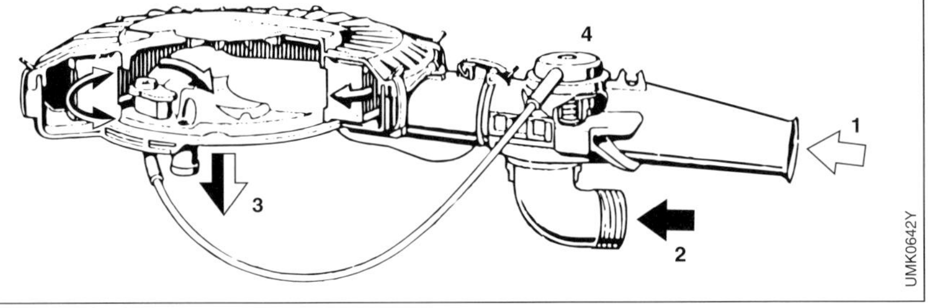

UMK0642Y

Fig. 2

Fender-mounted passenger-car air filter

1 Fresh-air intake, **2** Warm-air intake, **3** Outlet for warm/fresh air mixture.

UMK0639Y

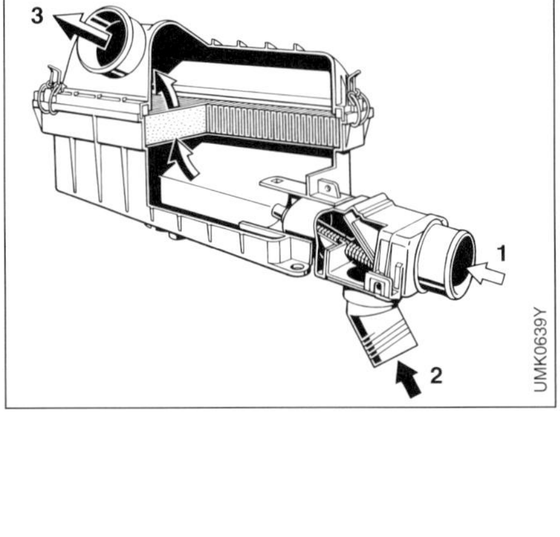

Fig. 3

Paper air filter with cyclone for commercial vehicles

1 Air intake,
2 Air outlet,
3 Cyclone air vanes,
4 Filter element,
5 Dust bowl.

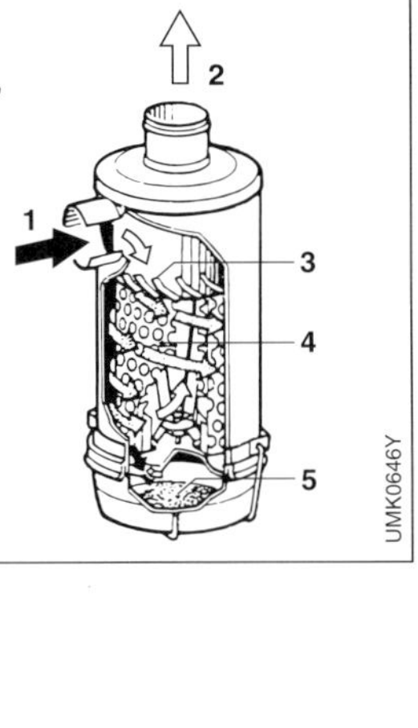

UMK0646Y

Passenger-car air filters are in the form of centrally located or fender-mounted filters with paper elements. Characteristic for these filters is their high filtering efficiency which is independent of the loading. It is a simple matter to change the paper cartridges at the intervals prescribed by the vehicle manufacturer. Passenger-car air filters must be carefully aligned to the particular engine in order to optimise power output, fuel consumption, intake-air temperature, and damping.

Commercial-vehicle air filters

Most of the air filters used in commercial vehicles are of the paper-element type, although oil-bath filters are found in isolated cases. Characteristic for paper filters is their high filtering efficiency in all load ranges and their higher level of flow resistance as levels of retained contaminants increase. The paper air filter can be supplemented by a cyclone prefilter directly incorporated in the housing to save space (Fig. 3). This is the preferred combination and is at present in widespread use. Service entails replacing the element and/or emptying the dust cup.

For paper air filters, the service interval is often indicated by a special display.

Regarding servicing, the vehicle manufacturer's instructions are to be complied with. For service simplification, specially aligned dust-removal valves can be incorporated depending upon the magnitude of the engine's air pulsation.

Cyclone's serve to extend the filter's useful life and, therefore, also the service intervals. Guide vanes in the cyclone cause the air to go into rotation, so that a large proportion of the dust is removed before the air reaches the downstream air filter. Cyclones are suitable for installation upstream of the paper air filter and/or the oil-bath air filter. Due to their inadequate filtration efficiency, cyclones are unsuitable for use as the only engine-air filter. Pertinent standard: DIN 71 459.

Intake-noise damping

In order for legal regulations concerning the vehicle's overall noise level to be complied with, it is necessary that the intake-air noise caused by passenger-car and commercial-vehicle air filters be damped. This damping is implemented almost exclusively by designing the air filter to act as a reflection sound absorber having the special shape of a Helmholtz resonator.

Acting as a suction resonator, the Helmholtz resonator has a damping effect in the area of its resonant frequency. When acting as a throughflow resonator it amplifies at resonant frequency, above which it features a wide damping range. Assuming that the air filter is of sufficient size (a good empirical value for 4-stroke engines is 15 to 20 times the piston displacement of 1 cylinder), intake noise can generally be damped by between 10 and 20 dB(A). In case noise is excessive at specific frequencies, special supplementary dampers must be used (Fig. 4).

Fig. 4

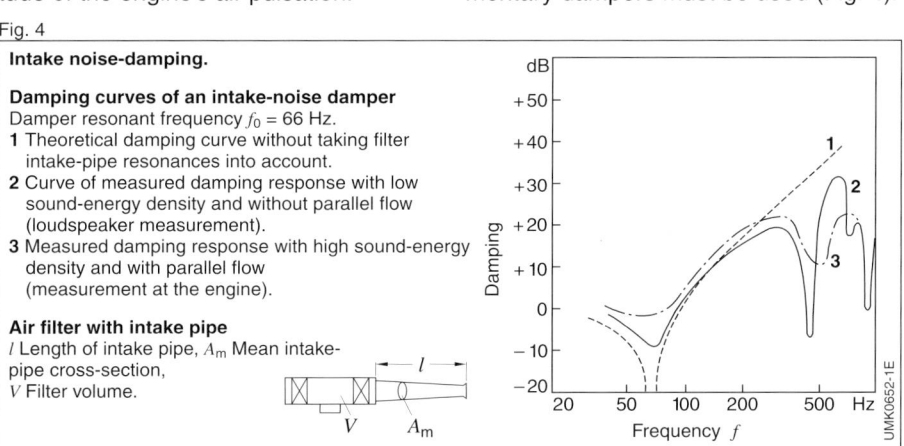

Intake noise-damping.

Damping curves of an intake-noise damper
Damper resonant frequency $f_0 = 66$ Hz.
1 Theoretical damping curve without taking filter intake-pipe resonances into account.
2 Curve of measured damping response with low sound-energy density and without parallel flow (loudspeaker measurement).
3 Measured damping response with high sound-energy density and with parallel flow (measurement at the engine).

Air filter with intake pipe
l Length of intake pipe, A_m Mean intake-pipe cross-section,
V Filter volume.

Fuel management

The air-fuel mixture formation considerably influences fuel consumption, exhaust-gas composition and the combustion noise of the diesel engine. The fuel-injection system makes an important contribution to mixture formation. Several parameters of the fuel-injection system influence mixture formation and the combustion process in the engine's combustion chamber:
- Start of delivery (port closing) and start of injection,
- Duration of injection and rate-of-discharge curve,
- Injection pressure,
- Injection direction and number of injection jets,
- Excess air.

The following sections describe the effects of these factors.

Start of delivery (port closing) and start of injection

The term "start of delivery" refers to the actual start of injection-pump delivery. Together with the start of delivery (FB), the actual start of injection (SB) is also of great importance for optimum engine response. Since the start of delivery (port closing) can be determined more simply than the actual start of injection for an engine at standstill, the timing of the diesel-injection pump to the engine takes place at the start of delivery.

This is possible because a defined relationship exists between the start of delivery and the start of injection.

Start of injection is defined by the crankshaft angle in the area of piston top dead center (TDC) at which the nozzle opens and fuel is injected into the combustion chamber. Start of fuel injection into the combustion chamber has a considerable influence upon the start of combustion of the air-fuel mixture. Maximum final compression temperature occurs at TDC. If combustion is initated well before TDC, combustion pressure increases steeply and brakes upward piston movement, thus impairing efficiency. The steep rise of combustion pressure also results in loud engine running. Combustion though must have been completed before the exhaust valve opens. The lowest fuel consumption is also achieved if combustion starts in the area of the TDC.

If the start of combustion is advanced, the temperature in the combustion chamber increases, thus also leading to a rise in NO_X emissions. If the start of injection is too late, this can lead to incomplete combustion and to emission of incompletely burnt hydrocarbons (Figure 1).

The instantaneous position of the piston influences the movement of air in the combustion chamber, its density and its temperature. Consequently, the speed of movement and the mixing quality of the air-fuel mixture depend on the start of injection. The start of injection thus also influences the emission of soot,

Fig. 1

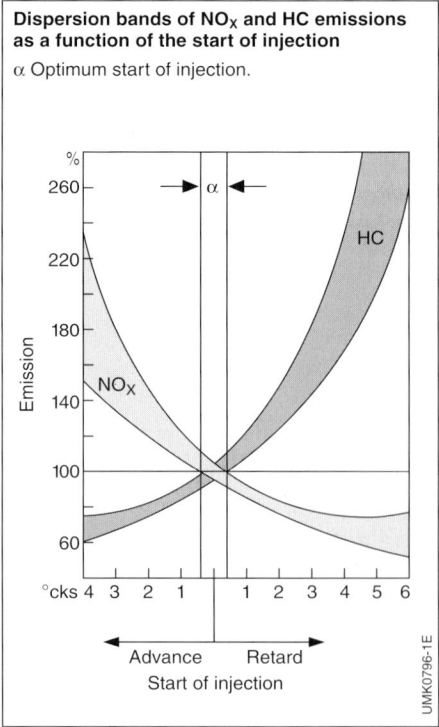

Dispersion bands of NO_X and HC emissions as a function of the start of injection

α Optimum start of injection.

a product of incomplete combustion. The opposing interdependencies of specific fuel consumption and hydrocarbon emissions on the one hand, and emission of black smoke and NO_X on the other, demand the minimum possible tolerances for the start of injection in order to achieve the respective optimum values. The different ignition lags at different temperatures necessitate temperature-dependent start of injection.

During delivery, the fuel's propagation time depends upon line length. At high speeds this results in an injection lag[1]). In addition, the higher the engine speed, the higher the ignition lag[2]). Both these factors must be compensated for, and this is why a fuel-injection system must incorporate a timing device for the speed-dependent advance of the start of injection. For noise and emission reasons, a different start-of-injection map for full-load is often required than for part-load. The start-of-injection map shows schematically the dependence of the start of injection on temperature, load and engine speed (Figure 2).

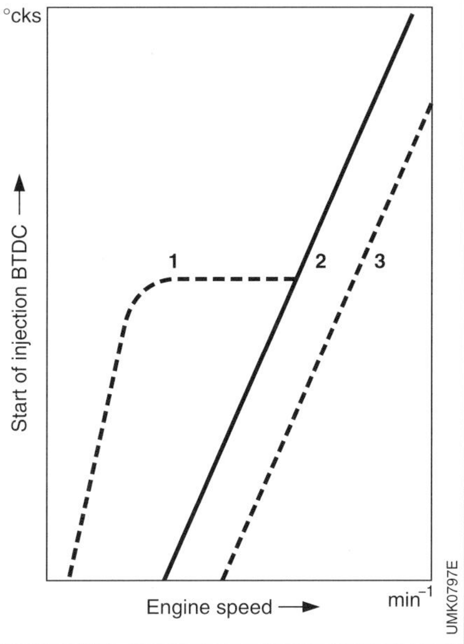

Duration of injection and rate-of-discharge curve

The term "rate of discharge" describes the characteristic curve of the fuel quantity injected into the combustion chamber as a function of crankshaft or camshaft angle (degrees crankshaft and degrees camshaft respectively). One of the main parameters affecting the rate-of-discharge curve is the duration of injection. This is measured in degrees crankshaft or degrees camshaft, or in milliseconds, and is the period during which the injection nozzle is open and fuel is injected into the combustion chamber.

Figure 3 shows how delivery of the injected fuel quantity is initiated by the pump camshaft and how fuel is injected at the nozzle (as a function of the camshaft angle). It can be seen that pressure characteristic and the rate-of-discharge curve change greatly between pumping element and nozzle, and that they are influenced by components which determine the injection (cam, pumping element, delivery valve, delivery line and nozzle). Various diesel combustion methods each require differing durations of injection:

The direct-injection engine requires approx. 25...30° crankshaft angle at rated speed and the prechamber engine 35...40° crankshaft angle. The injection duration of 30° crankshaft, corresponding to 15° camshaft, means a duration of injection of 1.25 ms for an injection-pump speed of 2,000 min^{-1}.

In order to keep fuel consumption and soot emission to a low level, the spray duration must be defined as a function of the operating point and timed to the start of injection (Figs. 3 and 5). At the start of injection, only a small quantity of fuel should flow, while a large quan-

Fig. 2

Start-of-injection map as a function of speed, cold-start temperature and load

1 Cold start, **2** Full load, **3** Part load.

°cks

Start of injection BTDC →

1   2   3

Engine speed → min^{-1}

UMK0797E

[1]) Time from start of delivery to start of injection.
[2]) Time from start of injection to start of ignition.

Fig. 3

Chain of actuating variables from the cam lift to the rate-of-discharge curve as a function of cam rotation angle

Example: Distributor injection pump.
t_L Injection lag.

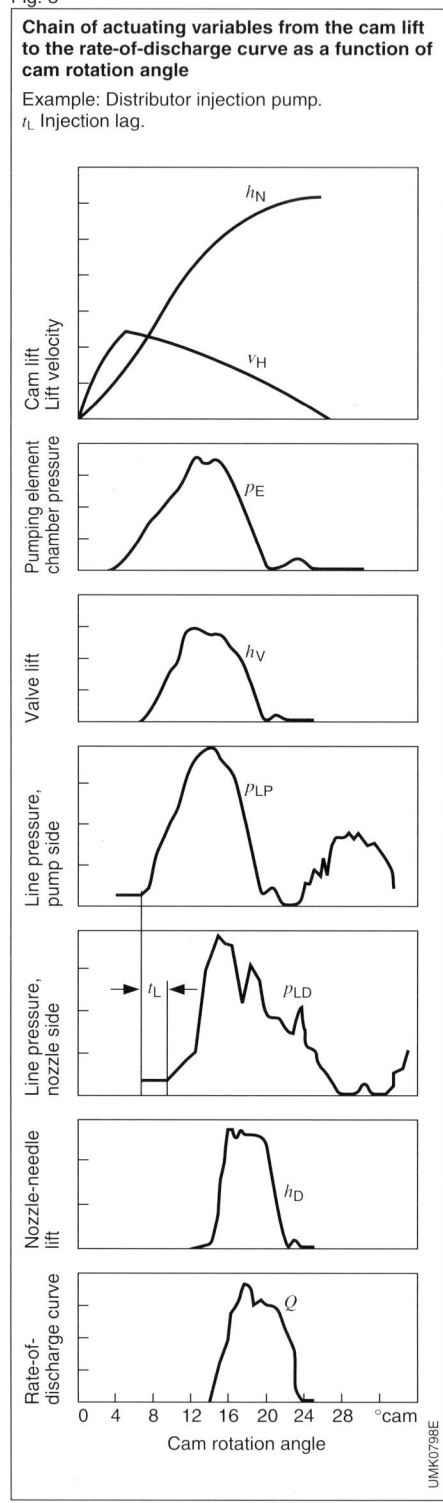

UMK0798E

tity of fuel is required at the end. The nozzle should then close as quickly and reliably as possible. Such a rate-of-discharge curve leads to a slowly rising combustion pressure. Combustion thus takes place quietly. In engines with direct injection, the combustion noise is reduced considerably if a small part of the fuel quantity is injected into the combustion chamber in finely atomized form before main injection.

Such a pilot-injection method is still very expensive. Throttling pintle nozzles are used in engines with divided combustion chamber (prechamber or whirl-chamber engines). These nozzles generate a single fuel jet and determine the rate-of-discharge curve. The nozzles control the discharge cross-section as a function of the stroke of the injection valve.

Secondary injection (or so-called "dribble") is particularly unfavorable, and results from the nozzle opening briefly again after having closed once and injecting poorly prepared fuel later on in the combustion process. This fuel

Fig. 4

Influence of nozzle design on hydrocarbon emissions

a Nozzle without blind hole,
b Nozzle with miniature blind hole.
1 Engine with 1.3 l/cyl.,
2 Engine with 2 l/cyl.

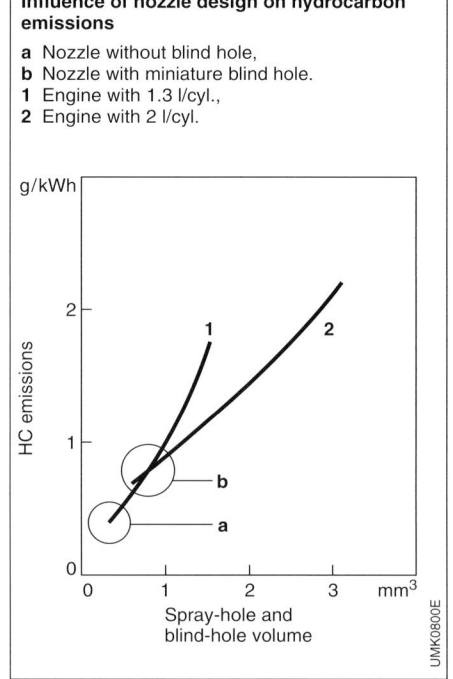

UMK0800E

burns incompletely or not at all and enters the exhaust as unburnt hydrocarbon.

Rapidly closing nozzles prevent such "dribble". A "dead volume" in the nozzle downstream of the seat has a similar effect as dribble. The fuel vapor stored in this volume is discharged into the combustion chamber after completion of combustion and also flows into the exhaust where it increases the emissions of unburnt hydrocarbon. The smallest "dead volume" is achieved with hole-type perforated-seat nozzles in which the spray holes are bored in the seat (Figure 4).

Fig. 5

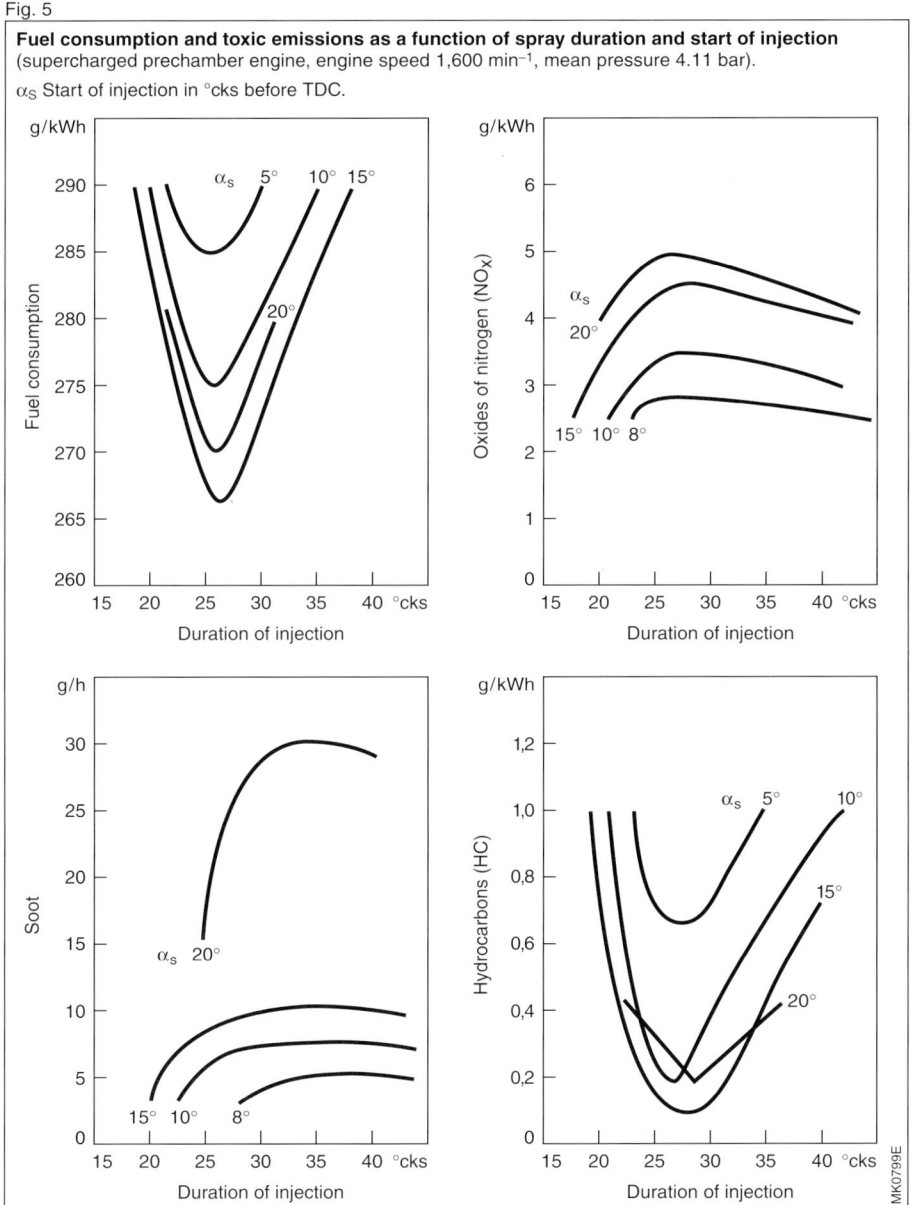

Fuel consumption and toxic emissions as a function of spray duration and start of injection
(supercharged prechamber engine, engine speed 1,600 min^{-1}, mean pressure 4.11 bar).
α_S Start of injection in °cks before TDC.

UMK0799E

Injection pressure

The higher the relative speed between the fuel and the air, and the higher the density of the air in the combustion chamber, the finer the atomization of the diesel fuel. High fuel pressure leads to high fuel speed. Diesel engines with divided combustion chamber operate with high air speed in the whirl chamber, or in the auxiliary combustion chamber, or in the connection channel joining the whirl and main combustion chambers. Here, no advantages are obtained by using pressures over approx. 350 bar. For direct-injection diesel engines, the air speed in the combustion chamber is relatively low and mixing is normal.

Mixing is considerably improved if the fuel is injected into the combustion chamber at high pressure (Figure 6). Soot emission can be greatly reduced, particularly at low engine speeds, by using injection pressures of up to approx. 1,000 bar. Higher injection pressures significantly increase the fuel consumption, among other things because the drive power for the injection pump increases.

Injection direction

Prechamber or whirl-chamber diesel engines operate with only one injection jet whose direction is matched to the respective combustion chamber. Deviations lead to poorer utilization of the combustion air and thus to a rise in black smoke and hydrocarbon emissions.

Direct-injection diesel engines generally operate with 4 to 6 spray holes whose injection direction is adapted very precisely to the respective combustion chamber. Deviations around 2° from the optimum injection direction lead to a measurable increase in black smoke emission and fuel consumption.

Excess air and exhaust-gas behavior

Diesel engines generally operate without throttling of the intake air. If there is a large air excess, the fuel burns "cleanly" in the combustion chamber.

Exhaust-gas components such as carbon monoxide and soot are formed in very low concentrations. The air excess in the combustion chamber decreases with increasing injected-fuel quantity. Taking into consideration low engine weight and engine cost, there is a given engine swept volume for maximum possible power. The engine must therefore operate with a small air excess at high load. If the air excess is small though, the emissions must be limited, i.e. the fuel quantity must be exactly metered for the available air quantity and as a function of the engine speed. Low air pressure (e.g. at high altitudes) requires adaptation of the injected fuel quantity to the reduced quantity of air available.

Fig. 6

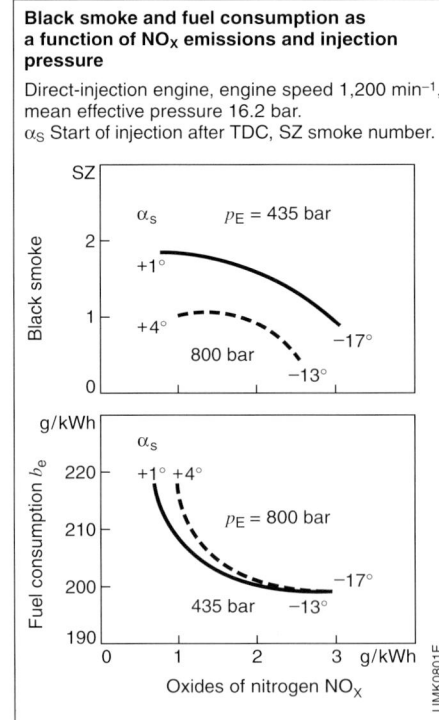

Black smoke and fuel consumption as a function of NO_X emissions and injection pressure

Direct-injection engine, engine speed 1,200 min^{-1}, mean effective pressure 16.2 bar.
α_S Start of injection after TDC, SZ smoke number.

Turbocharging

In the case of turbocharged engines, the injected fuel quantity is limited as a function of the pressure in the engine's intake manifold.

Exhaust-gas recirculation (EGR)

In engines with EGR, exhaust air can be mixed with the intake air at part-load operation to reduce NO_X emissions. This measure reduces the oxygen concentration of the charge, and in addition exhaust gas has a higher specific heat than air. Both influences reduce the combustion temperature (and with it the formation of NO_X). An increasing EGR rate reduces the fresh-air throughput of the engine and thus the quantity of excess air. Consequently, the emissions of hydrocarbons and soot in the exhaust gas increase if the charge contains an excessive proportion of exhaust gas (Figures 7 and 8).

Attempts at greatly reducing NO_X emissions by exhaust-gas recirculation also require precise adjustment of the injected-fuel quantity to the available air quantity during part-load operation.
In other words, the recirculated exhaust gas must be limited so that there is sufficient oxygen available for combustion of the injected fuel in the combustion chamber.

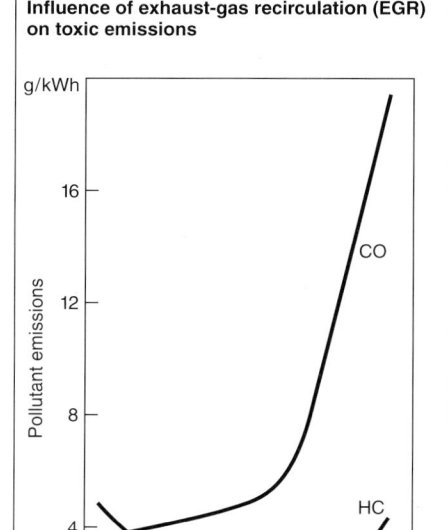

Influence of exhaust-gas recirculation (EGR) on toxic emissions

Fig. 7

Fig. 8

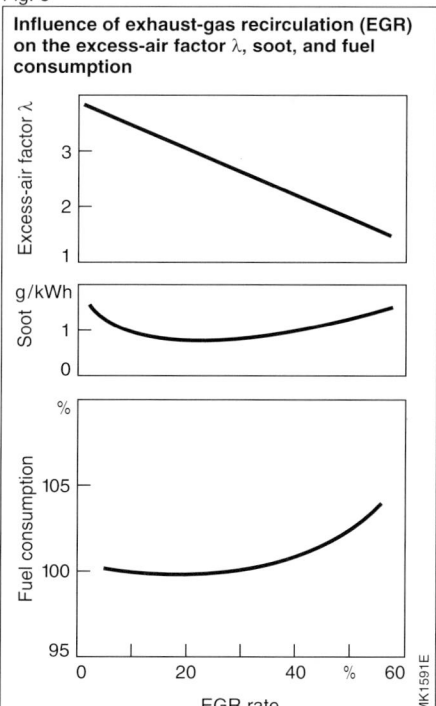

Influence of exhaust-gas recirculation (EGR) on the excess-air factor λ, soot, and fuel consumption

Toxic substances in exhaust gas

Combustion

Diesel engines operate with "internal" mixture formation, whereby combustion takes place during and after fuel injection. Combustion is greatly influenced by the extremely extensive mixture-formation process.

Fuel is injected into the highly compressed and correspondingly heated air in the combustion chamber shortly before the top dead center point. The fuel auto-ignites after a certain ignition lag. Since a considerable part of mixture formation occurs during combustion, we speak of a diffusion flame (diffusion: mutual permeation).

Start of injection, injection characteristic, and fuel atomization affect the exhaust-gas emissions. The start of injection has considerable influence upon start of ignition. If injection takes place too late, hydrocarbon emissions increase. A deviation of the start of injection by 1° crankshaft can lead to increases in NO_X, particulate, and HC emissions of up to 15%.

The injected fuel has only a very short time to mix with the air supplied for combustion (a matter of milliseconds); consequently, a non-uniform mixture is produced with zones having a low fuel content (lean) and zones having a high fuel content (rich). The utilization of the combustion air is thus not optimum, which is why the diesel process must occur with a high excess-air factor[1] ($\lambda > 1.2$). Compared with the spark-ignition engine, this results in a lower mean pressure (mean piston or effective pressure) and in some cases a clear reduction in the emission of gaseous toxic substances. The figures compare with those attainable for a spark-ignition engine with catalytic converter (Figure 1).

The diesel engine operates with auto ignition. This means that it can be designed with a high compression ratio ($\varepsilon \leq 24$). In turn, this necessitates a design with a higher weight, but also permits more favorable thermodynamic efficiency and thus a lower specific fuel consumption.

Fig. 1

Comparison of the toxic emissions in the European test (4-cylinder engines, engine swept volume 1.7 l, European series 1997)

☐ Spark-ignition engine (with catalytic converter),
☑ Diesel engine.

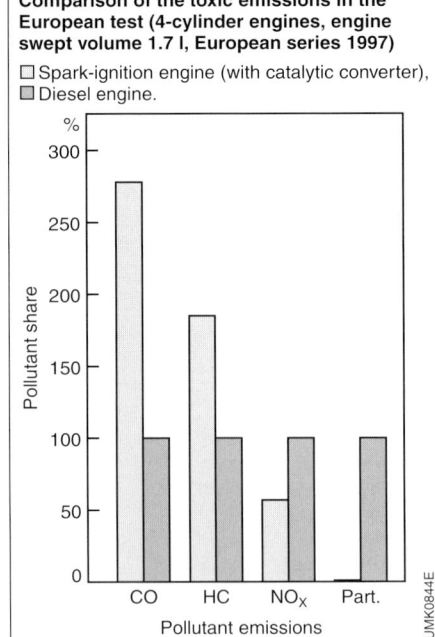

Pollutant emissions: CO, HC, NO_X, Part.
Pollutant share (%)

UMK0844E

---

[1] The excess-air factor λ specifies by how much the actually present air-fuel mixture deviates from the mass ratio theoretically necessary for complete combustion: λ = supplied air mass/theoretical air requirement.

Formation of toxic substances

Since the reaction partners of air and fuel only mix partially during reaction when combustion takes place in a diesel engine, mixture formation, ignition and combustion do not occur independently but mutually influence each other. Unlike a spark-ignition engine, different fuel concentrations and air conditions therefore exist in the combustion chamber of a diesel engine. The first reactions take place in the rich areas of the air-fuel mixture which has heated-up as a result of the compression. These reactions though only take place in the vapor jacket of the fuel droplets. Free carbon is produced. If combustion of these carbon particles is prevented during continuation of the reaction, e.g. owing to poor mixture, local lack of oxygen or extinguishing of the flame at cold points, the particles leave the exhaust pipe in the form of soot.

These highly complex processes in the combustion chamber show that the variables affecting mixture formation — these are influenced by the fuel-injection process and by the local movement of air in the combustion chamber — must be matched to each other very carefully to keep particulate emission as low as possible. However, this circumstance requires a compromise in engine optimization. Measures which have a positive effect on soot and particulate emission mostly have a negative influence on fuel consumption, NO_X, and noise emission. With regard to its toxic emissions, the diesel engine features long-term stability which deteriorates only insignificantly throughout its service life.

Characteristics of toxic substances

Gaseous toxic substances
Even the most basic version of the diesel engine produces low levels of toxic emissions. This results from the high excess-air factors (λ) for carbon monoxide and for the hydrocarbons (CO, HC), together with the low process temperatures for oxides of nitrogen (NO_X). In order to comply with the currently applicable emissions limits though, it is nevertheless necessary to resort to specific measures on both the engine and the fuel-injection system.

Particulate emissions
Particulate emission is a special characteristic of the diesel engine, and is considerably higher than for the gasoline engine. Depending upon the combustion process and engine operating state, such emissions are largely comprised of carbon particles (soot). The rest is accounted for by hydrocarbon compounds (bound to the soot), aerosols (solid or liquid substances dispersed in gases), and sulphates. The sulphur content of the diesel fuel is responsible for the latter.

The soot particles represent chains of carbon particles with a very large specific surface area on which uncombusted or partially combusted hydrocarbons are deposited. These are mostly aldehydes (compounds with high molecular number) with a powerful odor. Particulate emission from diesel engines can be seen as environmental pollution because of the effects of dirt and smell, and the impairment of vision. In addition, there is the frequently expressed opinion that health is endangered (lung cancer) by certain aromatic compounds (generic term for aromatic hydrocarbons) in the soot. Investigations that have taken place in this respect have not led to conclusive results. Owing to their diameter of only a few ten-thousandths of a millimeter, the soot particles emitted into the atmosphere float there for a long time. They can enter the lungs of human beings, but they are also breathed out again to a great extent for the same reason.

Exhaust-gas treatment

In 1994, in Germany the proportion of road-traffic emissions in the overall pollution of the air due to particle emissions was 8.5%. The black, blue, and white smoke emitted by the diesel engine are directly perceptible emissions and, like the smell of exhaust gas, are considered to be particularly disturbing. Modern diesel engines which comply with the Euro-II standards (valid since 1996) feature a very low level of particulate and smoke emission. These engines hardly emit any visible smoke at all, no matter whether the engine is cold or at operating temperature.

In order to comply with the Euro-IV standards (as from around 2005), the use of particle filters is being discussed for passenger cars and commercial vehicles (provided exhaust-gas recirculation (EGR) is used in the latter). As their name implies, these filters are intended to remove the particles from the stream of exhaust gas emitted by the diesel engine and dispose of them without imposing a load on the environment. To a great extent, the toxic CO and HC emissions can be burnt in an oxidation-type catalytic converter downstream of the engine. The reduction of NO_X has proved to be the most difficult assignment. A reduction agent must be added to the exhaust-gas flow in order achieve NO_X reduction in a downstream catalytic converter.

Soot burn-off filter

The diesel engine operates permanently with excess air. This means that the exhaust gas contains so much oxygen that at temperatures of above about 550 °C, the soot that has been deposited and collected in the filter burns off of its own accord in a soot burn-off filter, with the effect that the filter cleans itself. However, local peak temperatures of up to 1,200 °C during soot burn-off necessitate particularly demanding material specifications. For this reason, different designs of almost exclusively ceramic filter materials have been developed for this purpose.

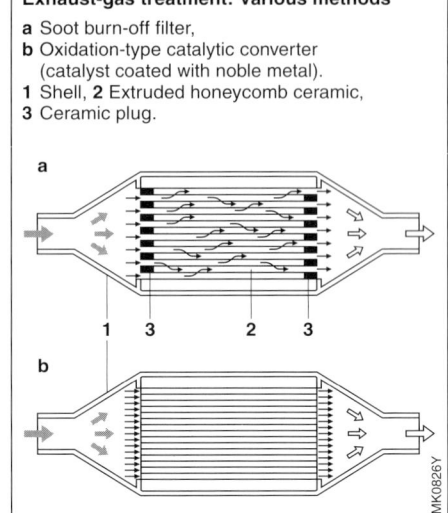

Exhaust-gas treatment: Various methods
a Soot burn-off filter,
b Oxidation-type catalytic converter
  (catalyst coated with noble metal).
1 Shell, 2 Extruded honeycomb ceramic,
3 Ceramic plug.

Fig. 1

The extruded honeycomb ceramic element (Figure 1) is similar in design and material to the catalyst substrate used with spark-ignition engines. However, the ends of honeycombs are sealed off alternately with ceramic plugs. Consequently, exhaust gas flowing into an open channel can flow through the porous ceramic walls into the neighboring channels leading to the tailpipe. The ceramic walls are less than 0.5 mm thick.

So-called deep-bed filters are being developed as an alternative to the honeycomb ceramic filters. These have a considerably larger pore size, and separation takes place only at sufficient filter depth (wall thickness). Filter "candles" consisting of wound ceramic fibers are used here.

In order to exclude excessive counterpressures, and thus the risk of blockage, regeneration aids must be provided. The ignition temperatures can be reduced to 200...250 °C by addition of metallo-organic substances to the diesel fuel. Burn-off is then still sufficient even with the filter system located underneath the vehicle body. Alternatively, forced regeneration of the filter can be implemented by the supply of external heat energy via an electric heater or fuel burner.

Catalytic-converter techniques

Oxidation-type catalytic converter
With the diesel engine, the oxidation-type catalytic converter (Fig. 1b) produces a marked reduction in carbon monoxide and hydrocarbon emissions. Since hydrocarbon emission contributes to particle emission, this means that to a limited extent, the oxidation-type catalytic converter also contributes to a reduction in particle emission. The use of low-sulphur fuel (EU Regulations: ($\leq 0.005\%$ S as from 2005), means that the catalytic converter remains effective over a very long period.

SCR method
At present, the highest levels of NO_X conversion are obtained using the SCR-method (SCR = Selective Catalytic Reduction). Here, a dosing system injects a reduction agent into the multi-chamber catalytic converter as a function of a map. The most common reduction agent at present in use is a carbamide-water solution (Fig. 2). Ammonia, the actual reduction agent, is generated from this solution in the hydrolysis stage. The NO_X reduction then takes place in the SCR stage. Usually, the catalytic converter is completed by means of an oxidation stage which oxidises the unburnt portions of CO and HC. Alternative SCR concepts use diesel fuel for instance as the reduction agent, although the reduction rates are comparatively low.

CRT method
This is a continually regenerating particulate-filter system in which the CO, HC, and NO toxic emissions are first of all oxidized in an oxidation-type catalytic converter. The NO_2 which is generated as a result combines in a downstream soot filter with the hydrocarbon C of the particles which have been deposited there and continually burns these to form CO_2. The NO_2 is thus reduced again to NO. If complete NO_X reduction is required, this must subsequently take place using the SCR method. This system avoids temperature peaks in the soot filter and therefore increases the filter's useful life. In order to ensure that optimum reduction remains permanent, the engine must be run with non-sulphur diesel fuel ($< 0.001\%$ S).

Fig. 2

Diesel-engine exhaust-gas treatment using the SCR method

1 Metering ECU, **2** Compressed-air reservoir, **3** Air compressor, **4** Throttle valve, **5** 2/2 directional-control valve, **6** Pressure-control valve, **7** Carbamide reservoir, **8** Carbamide pump, **9** Injection valve, **10** Air-assisted injection head, **11** Untreated exhaust gas, **12** Hydrolysis catalytic converter, **13** SCR catalytic converter, **14** Oxidation-type catalytic converter, **15** Muffler, **16** Clean exhaust gas. CAN **C**ontroller **A**rea **N**etwork (data bus in the vehicle).

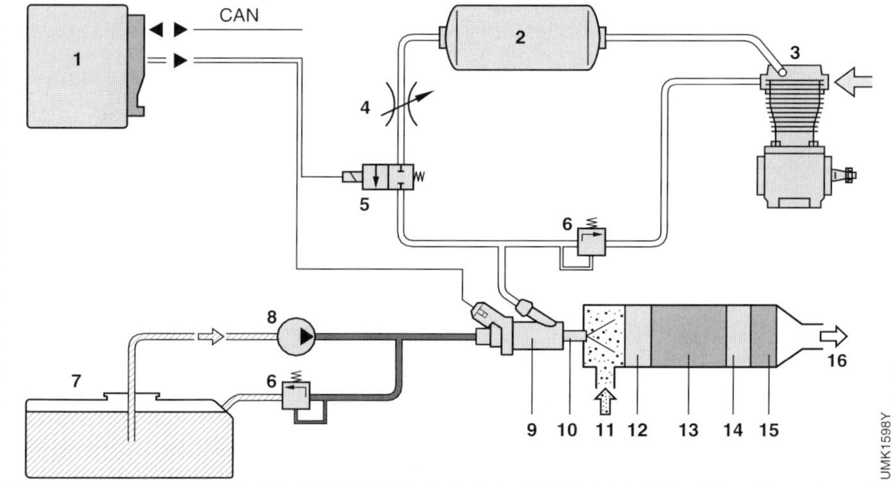

UMK1598Y

Exhaust-gas analysis techniques

Emission-control legislation

Many countries restrict the pollutant emissions of automotive and stationary diesel engines by means of legislation. The regulations contain test methods for engines and/or vehicles, measurement techniques, and limits. These are applied uniformly in a number of countries, while in other countries their application differs to a greater or lesser degree owing to ecological, economic, climatic, or political considerations. Limits which must not be exceeded apply to the following exhaust-gas components:
- Unburned or partially burned hydrocarbons (HC),
- Carbon monoxide (CO),
- Nitrogen oxides (NO_X),
- Particulate matter, and
- Smoke (vision-impairing component of particulate matter).

The toxic-substance emissions result from:
- Combustion in the engine (gases, sulphur compounds, particulate matter, odors),
- Crankcase-ventilation emissions (gases, sulphur compounds, odors), and
- Evaporative fuel emissions (from the fuel system).

Toxic emissions from the diesel-engine crankcase are practically negligible. Only pure air is compressed during the compression stroke, and the blowby gases entering the crankcase during the working stroke account for only approx. 10% of the toxic-substance mass generated by spark-ignition engines. Nevertheless, in the meantime, closed crankcase-ventilation systems have also been stipulated by law for the diesel engine. Unlike the SI engine, testing of evaporative emissions is not necessary for diesel engines, because the fuel system is closed and the diesel fuel does not contain any highly-volatile components. Sulphur compounds in the exhaust gas are the result of the sulphur content in the fuel. Worldwide, this has resulted in stricter limits for diesel-fuel sulphur content being introduced (EU: $\leq 0.05\%$ since 1996, $\leq 0.035\%$ as from 2000, and $\leq 0.005\%$ as from 2005).

The problem of diesel odor has not yet been solved; attempts to clarify the background behind processes in the diesel engine and behind emissions which cause odor are still in the initial stages. There is no generally recognized measuring method. Most countries have introduced legislation to restrict particle emissions and/or smoke, or are at least planning to do so. The continual sharpening of the exhaust-gas limits necessitates continuous engine development aimed at reducing the emissions output and at further refinement of the exhaust-gas measuring techniques.

Test procedures and classification

Following the USA, the countries of the EC and Japan have also developed their own test procedures for motor-vehicle exhaust-gas testing. Other countries have taken over these methods in the same or modified form. In this context, the USA has assumed a leading role.

Three test procedures laid down by the lawmakers are used, depending on the vehicle class and purpose of the test:
- Type homologation to obtain the General Certification for a vehicle,
- Series testing as a random-sample check of production by the acceptance autority, and
- Field monitoring to check specific exhaust-gas components during actual vehicle operation (OBD systems, periodic technical checks by the responsible authorities).

The greatest amount of test work is necessitated by the General Certification test. Highly-simplified methods are used for field monitoring.

Generally speaking, in all countries with motor-vehicle exhaust-gas testing, vehicles are sub-divided into three classes which are standard apart from a few minor overlaps:

- Passenger cars: Testing takes place on a roller-type test stand.
- Light commercial vehicles: Depending on the national laws, the upper limit of the vehicle weight is around 3.5...3.8 t. Testing is performed on a roller-type test stand.
- Heavy commercial vehicles: Vehicle weight over 3.5...3.8 t. Testing takes place on an engine test bench; no vehicle test takes place.

Type homologation
(Type approval/TA)
Successful completion of the exhaust-gas tests is a prerequisite for the award of the General Certification for a given vehicle and/or engine type. These tests specify the completion of test cycles which are run through under defined conditions, and the compliance with specific emission-limit values. These test cycles, and in particular the emission-limit values, are specific to the country concerned and are continually getting tighter and tighter. In the USA (State of California), referred to a vehicle manufacturer's fleet average, the emissions of NMOG (Non-Methanous Organic Gases) are limited. The vehicle manufacturer can apply a variety of different vehicle concepts which, depending upon their values for NMOG, CO, NO_X, and particle emissions are assigned to the following categories:
- Transitional Low-Emissions Vehicle (TLEV),
- Low-Emissions Vehicle (LEV),
- Ultra-Low-Emissions Vehicle (ULEV),
- Super Ultra-Low-Emissions Vehicle (SULEV), and
- Zero-Emissions Vehicle (ZEV).

USA type homologation
(Type approval/TA)
In order for a diesel-powered vehicle to be approved, the manufacturer must verify that toxic substances in the form of hydrocarbons (HC) or NMOG, carbon monoxide (CO), oxides of nitrogen (NO_X), particulate matter (solid matter), and smoke emissions (impairment of vision/opacity), do not exceed the limit emission values over a distance of 50,000 miles (approx.

80,000 km), and in some cases double this distance. The manufacturer must provide two vehicle fleets from series production for this type test:
1. One fleet for the endurance test which allows the deterioration factors of the individual components to be determined in continuous-duty operation. In this test, the vehicles are driven for 50,000 (or 100,000) miles in accordance with a certain drive program whereby exhaust emissions are measured at intervals of 5,000 miles. Service and maintenance must be performed only at the prescribed intervals (currently 12,500 miles).
2. One fleet from series production, whereby each vehicle must have been driven 4,000 miles before the test. These vehicles are also used for determination of the emission data. For reasons of simplification, some of the users of the USA driving schedule (e.g. Switzerland), permit defined deterioration factors to be applied.

EC type homologation
(Type approval/TA)
This follows a similar procedure to that in the USA, but with the following deviations: The toxic emissions HC, CO, NO_X, particulate matter (PM), and impairment of vision/opacity are measured. Before starting measurements, the test vehicle must have been run-in for 3,000 km. A deterioration factor as stipulated by law for each toxic component, is applied to the test results. Alternatively, following a stipulated endurance run of 80,000 km/50,000 miles (as from 2005: 100,000 km/62,500 miles) the vehicle manufacturer can prove that a lower deterioration factor applies.

Series test
Normally, the manufacturer performs the series test himself during production as part of quality control. The approval authority can order follow-up tests as often as it wishes. To a large extent, EC Regulations and ECE Directives take fabrication spread into account by sample measurements performed on between 3 and a maximum of 32 vehicles. The most stringent legislation is in effect in the USA where the lawmakers demand

practically 100 % quality monitoring from automotive manufacturers. This applies in particular to California.

Field monitoring

In most countries with obligatory field monitoring, testing of diesel vehicles has up to now been restricted to the opacity (opacity or sight impairment) caused by diesel smoke. For this purpose, the smoke emission is determined at free acceleration [1]. The maximum measured for each vehicle type in the type homologation test must be within certain tolerances. Regular exhaust-gas tests for diesel vehicles came into effect in Germany as from 1993. In the USA, far more extensive regulations apply and are aimed at maintaining exhaust-gas quality throughout the whole of the vehicle's service life.

---

[1] Full acceleration from low idle without gear engaged. This acts against the engine's flywheel mass.

Test methods

An exhaust-gas test should permit a quantitative statement to be made about the exhaust emissions to be expected in road traffic under normal operating conditions, without the necessity of performing measurements with the vehicle actually on the road. For this reason, for a passenger car, the operating conditions in road traffic are simulated on a roller-type test stand (dynamometer) or engine test bench. It is assumed that emissions during "test-stand driving", and when actually driving on the road, are the same when the speed and forces acting on the vehicle agree with respect to their time characteristic on both the test stand and on the road. For each vehicle, suitable braking loads and flywheel masses simulate the inertia forces (frictional resistance to rolling motion

Fig. 1

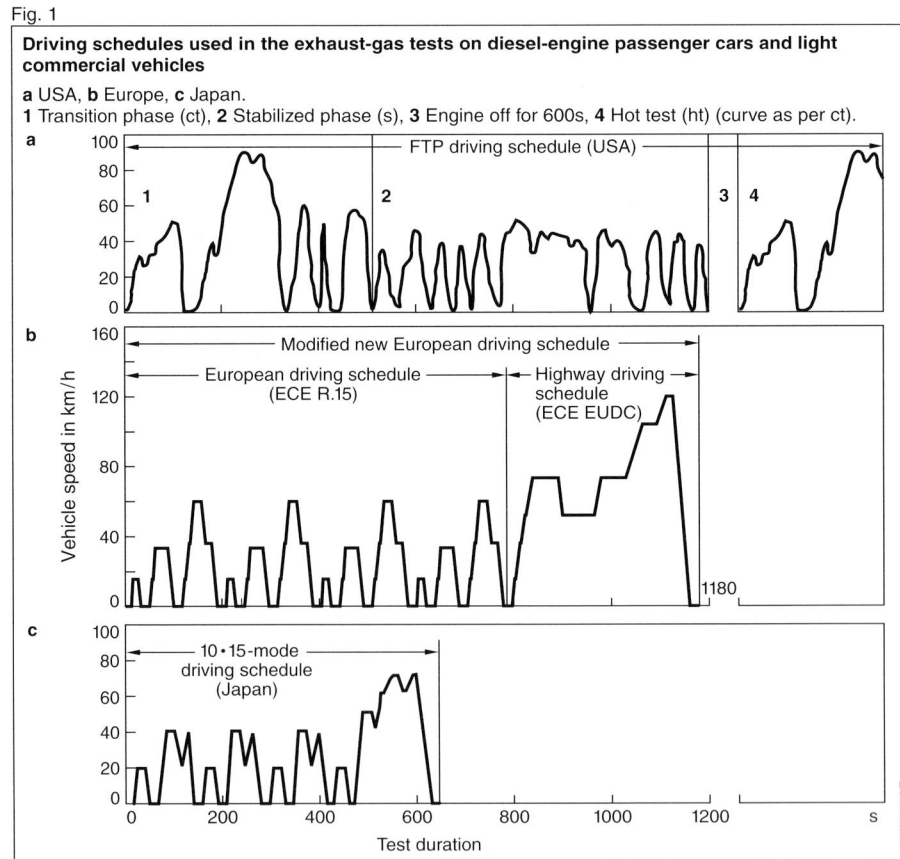

Driving schedules used in the exhaust-gas tests on diesel-engine passenger cars and light commercial vehicles

a USA, **b** Europe, **c** Japan.
1 Transition phase (ct), **2** Stabilized phase (s), **3** Engine off for 600s, **4** Hot test (ht) (curve as per ct).

and air resistance) on the test stand, so that it is sufficient to specify the same speed for roller stand and road.

In all countries with statutory emissions control, the relevant test specifications contain driving curves. Currently, several driving schedules, differing with respect to speed characteristic and duration, are prescribed for passenger cars and light commercial vehicles (Figs. 1 and 2). It is possible to distinguish between two types here in accordance with the manner in which they are developed:

- Driving curve from recordings of actual road journeys, e.g. US FTP driving cycle (FTP: Federal Test Procedure) and
- Driving curves constructed from sections with constant acceleration and speed, e.g. European and Japanese driving cycles.

The exhaust-gas testing of heavy-duty commercial-vehicle engines takes place on an engine test bench. Here, in order in particular to be able to take into account the effects of acceleration and deceleration on the exhaust-gas behavior, precisely specified engine operating conditions are being stipulated to an increasing degree. This test procedure necessitates cost-intensive investments in test and measuring equipment. Fully dynamic (continuous-action), electronic control devices for the simulation of the operating conditions based on the stipulated driving schedule and for the complex processing of the exhaust-gas samples are needed.

On the other hand, the exhaust-gas tests based on specified stationary (non-dynamic) load and speed values are not performed until the particular engine operating point has stabilized. This means that the required control, testing, and measuring equipment is far simpler. These exhaust-gas tests though are in the process of being phased-out.

Fig. 2

CVS test method for passenger cars and light commercial vehicles

1 Blower, **2** Dynamometer, **3** Air, **4** Filter, **5** Pump, **6** Dilution tunnel, **7** Flow meter, **8** Gas counter, **9** Heat exchanger/heater, **10** Burner, **11** Airbag, **12** Calibrating gas, **13** Zero gas, **14** Exhaust-gas sample bag, **15** Heated line, **16** Roots-blower, **17** Extraction, **18** Integrator, **19** Computer, **20** Recorder.

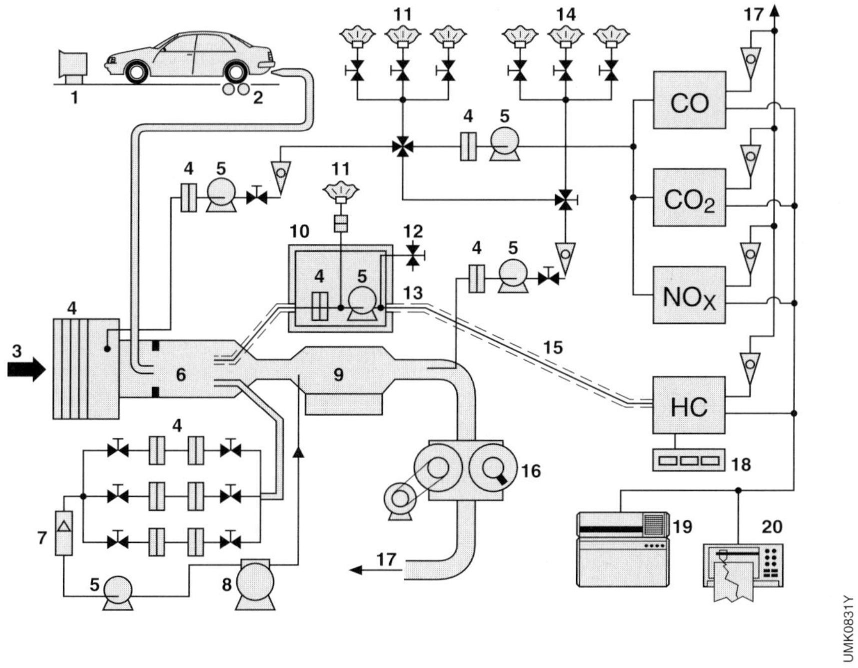

Fig. 3

Transient driving schedules for exhaust-gas testing of heavy-duty commercial-vehicle engines

The standardized engine speed n^*, as well as the standardized torque M^* are Table values as stipulated by legislation.
a Specific speed and torque characteristic; USA,
b Specific speed and torque characteristic; Europe (as from 2000/2005): ETC test cycle.

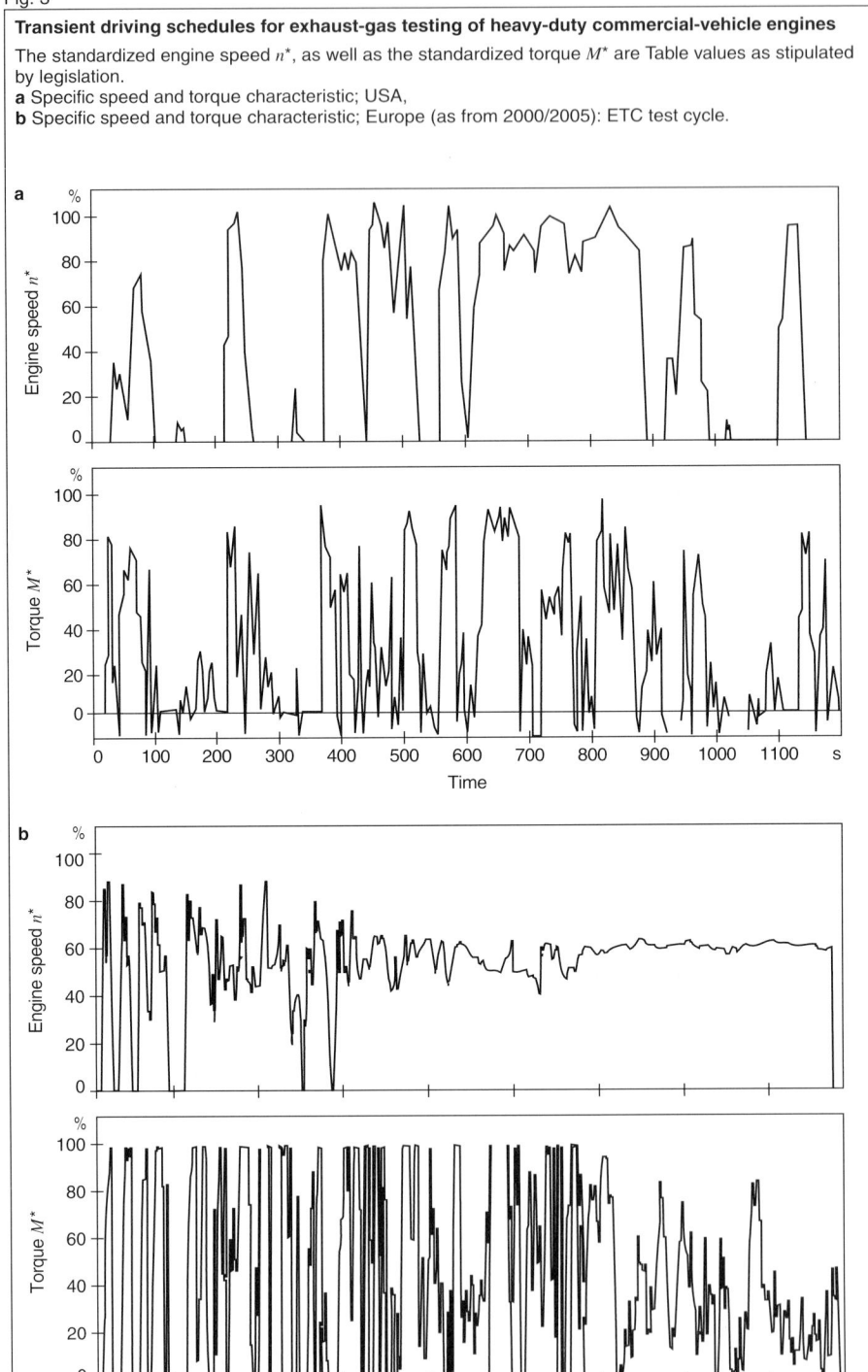

CVS test method (USA)

The CVS test method (CVS: Constant Volume Sampling) was introduced in the USA for passenger cars and for light diesel-powered commercial vehicles in 1972, and in the meantime has been improved in a number of steps. Its principle is based on using pure air from the surroundings to dilute the (varying) exhaust-gas volume in order to generate a constant exhaust-gas/air mixture volume. This in turn is used for calculation of the mean pollutant content and its magnitude.

This method is characterized by:
- Consideration of the exhaust-gas volume actually emitted by the engine in question during the test,
- Real registration of all the steady-state and transient vehicle conditions,
- Avoidance of the condensation of water-vapor and nonburned hydrocarbons, and
- From the point of view of measurement techniques, precise definition of particle emissions.

In the USA, all diesel-powered vehicles have been included in the CVS test method since 1975. For this purpose, it was necessary to modify the sampling and analysis systems for measurement of hydrocarbons. In order to avoid condensation of those hydrocarbons in the exhaust gas with a high boiling point, and to ensure evaporation of hydrocarbons in the mixture volume already condensed from the diesel exhaust gas, it is necessary to heat up the complete sampling system to approx. 190 °C.

The CVS test method was modified by inclusion of particulate limit values in the US exhaust-gas legislation. For this purpose a "dilution tunnel" with high internal turbulence (Reynolds' number > 40,000) was integrated in the measuring system and supplemented by corresponding filter measuring points for collection of particles. Owing to dilution with a mean ratio of 1:10, the measured pollutant concentrations are very low. Consequently, highly-sensitive analyzers must be used.

All countries which have included the CVS test method in their emission-control legislation also use uniform measuring principles for analysis of the exhaust-gas and of the individual toxic constituents:
- Measuring the CO and CO_2 concentrations with NDIR infrared absorption analyzers (Non-Dispersive Infra-Red),
- Measuring the NO_X concentration with devices operating according to the chemi-luminescence principle CLD (Chemi-luminescence Detector),
- Gravimetric determination of the particulate emissions (conditioning and weighing of the particulate-matter filter before and after operation).

Transient test method (USA, Europe)

The above-described CVS test method is also applied in the transient test method (Fig. 3) prescribed in the USA as from model year 1986, and in Europe as from 2000/2005, for emission control of diesel engines in heavy-duty commercial vehicles as from 8,500 lbs (approx. 3.8 ton), or above 3.5 ton gross vehicle weight.

Presuming that the same dilution conditions are to apply as for the CVS measuring method, the size of such truck engines necessitates a test system which features a considerably higher throughput capacity than that used on passenger cars and light commercial vehicles. The double dilution (using a secondary tunnel) as permitted by legislation contributes to keeping the equipment outlay down to a reasonable level.

The diluted exhaust-gas volume can be measured either with a calibrated Roots blower (revolution counter), or with a venturi tube (critical pressure area).

13-step test ECE R.49 (Europe)

This test method for the emissions test on heavy commercial vehicles remains valid until the end of 1999 in the EU and a number of other European countries.

The weighting factors for the individual test points take into account the characteristic operating conditions which prevail in European road traffic.

The overall result is determined from weighted part-results of the individual steps, and referred to an average engine power calculated in accordance with the same method.

The frequency distribution of load and speed during operation in normal road traffic is taken into account by different assessment factors for the individual steps (Fig. 5). With the introduction of particulate-emission limits, it has become necessary to include the CVS measuring method in the 13-step test. In order to limit the equipment outlay, the lawmaker regards partial-flow dilution (only a defined portion of the exhaust-gas volume is analyzed and from it the total exhaust-gas emission is computed) as being of equal standing with the full-flow dilution method (Fig. 4).

13-stage test ESC (Europe)

The ESC test cycle supersedes the 13-stage test ECE R.49 for heavy-duty commercial vehicles which is being phased-out. It will come into force in 2000, and will apply generally only for heavy-duty commercial-vehicle engines without exhaust-

gas aftertreatment. It is expected to remain in force only until 2005, after which the dynamic ETC (European Transient Cycle) test for all tests performed on heavy-duty commercial-vehicle engines becomes binding.

With this ESC test cycle, the emissions test also takes place under 13 defined test conditions which are run through from point to point as shown in Fig. 6. The engine operating points though, and the weighting of the resulting measurement figures for engine power, fuel consumption, and toxic emissions differ considerably.

10 · 15-step test (Japan)

This test is valid in Japan for passenger cars and light commercial vehicles. It comprises an emissions test of gaseous and particulate pollutants emitted by the vehicle when operated on a chassis dynamometer in accordance with a special driving schedule.

This corresponds to typical driving conditions in Tokyo and, similar to the European driving schedule, has been extended by

Fig. 4

Sampling and measuring system for the 13-mode test schedule (Europe)

1 Exhaust gas, **2** Air, **3** Zero gas, **4** Calibrating gas, **5** Fuel, **6** Air backflush, **7** Outlet, **8** Heating line/housing, **9** Filter, **10** Pump, **11** Cooler, **12** Water separator, **13** Flow meter, **14** Prefilter, **15** Oxygen.

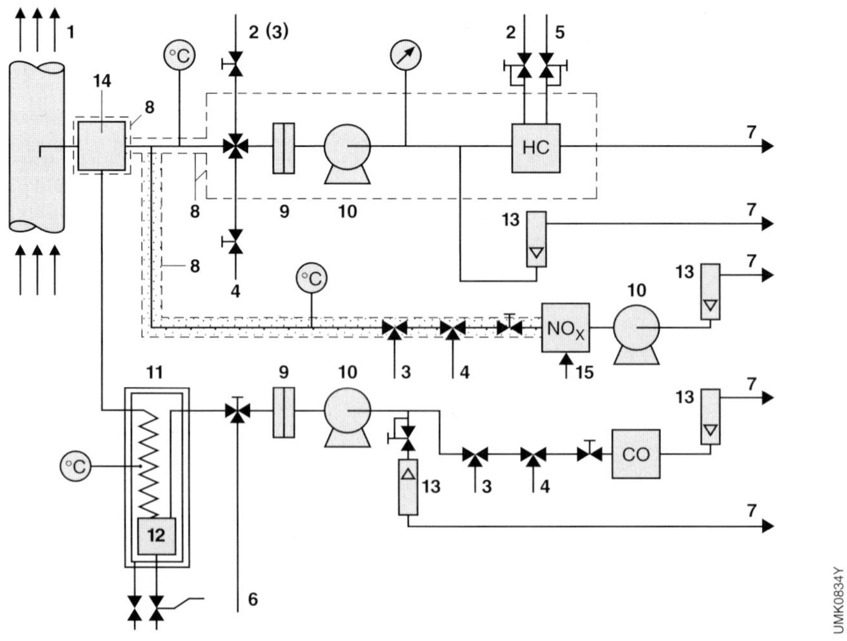

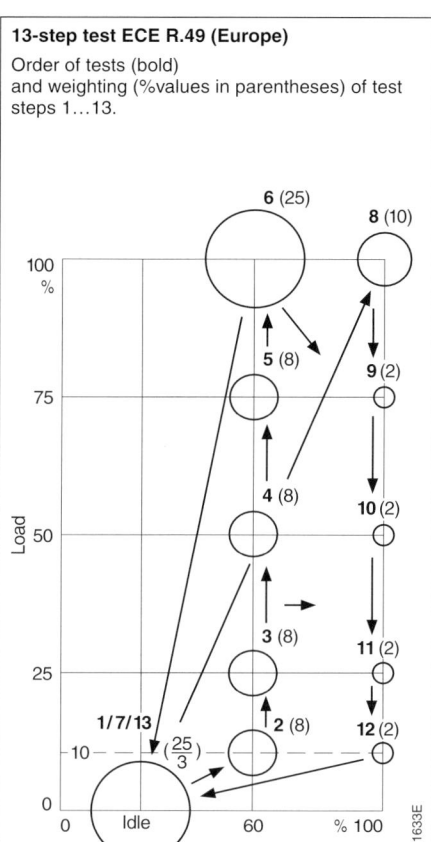

13-step test ECE R.49 (Europe)

Order of tests (bold)
and weighting (%values in parentheses) of test
steps 1...13.

Fig. 5

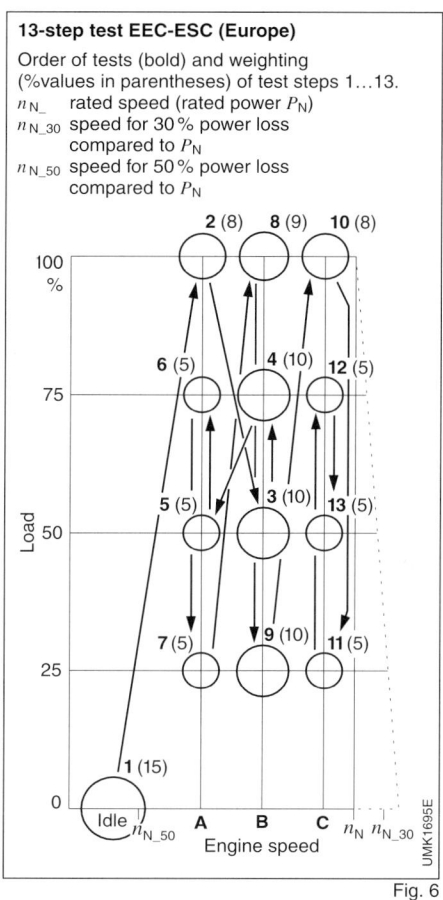

13-step test EEC-ESC (Europe)

Order of tests (bold) and weighting
(%values in parentheses) of test steps 1...13.
n_N — rated speed (rated power P_N)
n_{N_30} speed for 30% power loss
compared to P_N
n_{N_50} speed for 50% power loss
compared to P_N

Fig. 6

the addition of a high-speed portion. The measurement method is identical to the mass-emission technology employed for the CVS measurement procedure in the USA.

13-step test (Japan)

This test method is valid in Japan for heavy-duty commercial vehicles with a gross vehicle weight exceeding 2500 kg. The operating points and their respective weighting (Fig. 7), differ considerably from those in the European 13-step test. The introduction of particulate-emission limits in Japan has meant that the CVS measurement method has become necessary here too.

Test cycles for smoke measurement

Specific driving schedules are stipulated in the USA, Japan, and the EC for measuring the smoke emission from diesel engines. These are run on a dynamometer. The type of curve used defines the operating conditions or points at which the maximum smoke emission can be expected, and does not simulate driving in road traffic.

Smoke tests

Long before legislation was introduced for the monitoring of gaseous pollutants, separate laws were already in place governing the monitoring of diesel-engine smoke emissions. This smoke-test legislation is still valid in almost unchanged form. The countries with statutory diesel-smoke control do not use uniform test methods, and the results of all existing smoke tests depend to a high degree upon the measuring instruments and methods used.

Smoke test USA (Federal Smoke Test)
This test covers a defined sequence of driving conditions which are run through immediately following engine warm-up. It can only be performed on a dynamometer. The braking load as defined before the test is derived from the driving specification which stipulates that a 6-step driving schedule is run through 3 times in succession, whereby the smoke values are to be ascertained for each of the three steps given below (Fig. 8):

– Step 2: Within a defined time window, braking-load-controlled WOT* acceleration of the test unit with linear speed increase from approx. 30% to approx. 90% of the rated speed. Essentially, this is unthrottled full-load acceleration against the intertia masses of the engine and the unloaded dynamometer.
*WOT = Wide Open Throttle

– Step 4: Within a tightly-toleranced broad time window, the braking-load-controlled WOT acceleration starting at the engine speed for maximum torque, or 60% of the rated engine speed, whichever is the higher of the two.

– Step 6: Braking-load-controlled WOT deceleration (by increasing the braking load) with linear speed reduction from approx. 100% of rated speed to the engine speed for maximum torque, or to 60% of the rated engine speed, whichever is the higher of the two.

The arithmetic mean for each of the above weighting steps is derived from a stipulated number of repeat measurements. The smoke values in the form of turbidity values can only be measured with a light-absorption meter.

Smoke test Japan (3-step test)
In Japan, this test is obligatory for the homologation of all diesel-engine vehicles. It is performed on a dynamometer, and under steady-state conditions comprises the measurement with filter-type measuring instruments of diesel smoke at WOT, and at 3 different engine speeds (Fig. 9).

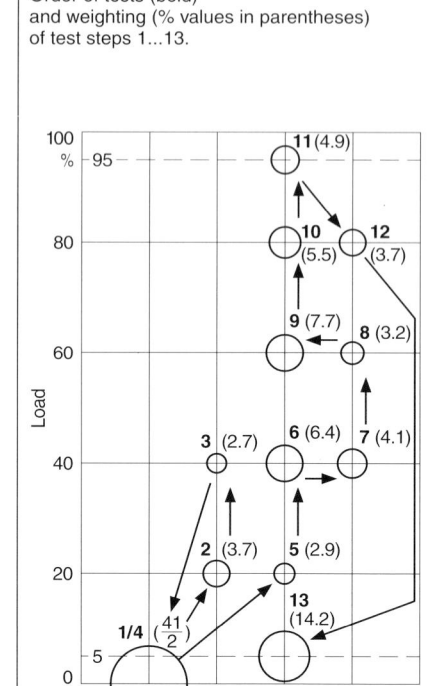

13-step test for commercial vehicles (Japan)
Order of tests (bold) and weighting (% values in parentheses) of test steps 1...13.

UMK1188E

Fig. 7

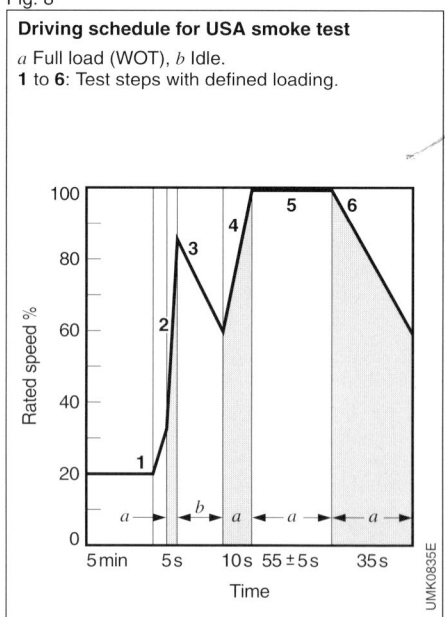

Driving schedule for USA smoke test
a Full load (WOT), *b* Idle.
1 to **6**: Test steps with defined loading.

UMK0835E

Fig. 8

EC smoke test (ECE R.24)

The full-load method is used for diesel-engine smoke testing as part of the stipulations to be satisfied in order to obtain a General Certification for a vehicle. The run-in engine must be operated under full load at constant speed with the exhaust system scheduled for series installation and with observation of the scheduled operating temperatures for lube-oil and water. The full-load opacity must be determined for a number of speeds distributed uniformly over the speed range between 100% and 45% of the rated speed, but not less than 1,000 min^{-1} (Figure 10).

A light-absorption meter is prescribed for the measurements. The absorption coefficient *k, whose limit-value curve is defined, serves as a comparison value for opacity. This full-load method guarantees clear and easily reproducible characterization/definition of the full-load smoke behavior only when the respective thermal equilibrium has been established. If diesel smoke also contains oil mist or water vapor in addition to soot, false readings must be expected, the absorption meter erroneously indicating higher diesel smoke emissions than is actually the case. Part 2 of ECE R.24 regulates smoke measurement under conditions of "free acceleration" (the accelerator pedal is pushed rapidly to the floor). It is performed immediately after the full-load smoke test, and is used to define the puff of smoke emitted when the engine is accelerated freely. The result is applied as a comparison figure for monitoring performance in the field. In the acceleration test, the unloaded engine, which must be at normal operating temperature, is accelerated from idle to maximum speed by pressing the accelerator pedal down quickly (kick-down). Total full-load (WOT) power is then utilized to accelerate the rotating masses. The test duration is therefore limited to between 2 and 5 seconds.

EC smoke test (EEC-ELR)

This smoke test (Fig. 11) must be successfully completed by heavy-duty commercial-vehicle engines following type approval (TA) in line with ESC or ETC test methods. The test is in the form of a "multi-layer" engine loading test at each of the 3 engine speeds "A", "B", and "C" as applied in the ESC test method. An additional, fourth, engine speed "D" is located between speeds "A" and "C".

Fig. 9

Driving schedule for the Japan smoke test

The particular test step (full load) starts after reaching the steady state, and terminates upon completion of the smoke measurement.
1) or 1,000 min^{-1} for vehicles, on which 0.4 n_{max} is less than 1,000 min^{-1}.

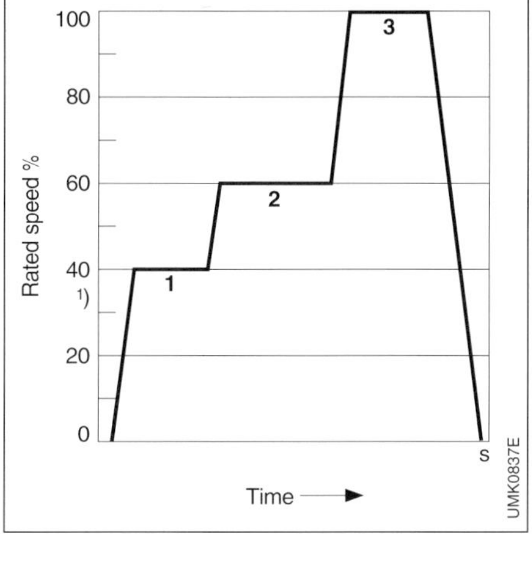

Fig. 10

Driving schedule for the ECE R.24 smoke test

The particular test step (full load) terminates upon completion of the smoke measurement.
1) or 1,000 min^{-1} for vehicles, on which 0.45 n_{max} is less than 1,000 min^{-1}.

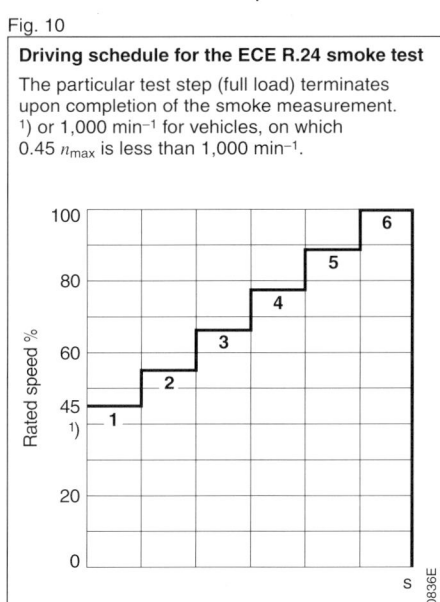

At each of the three engine speeds "A", "B", and "C", the injected fuel quantity is abruptly increased to the full-load level which is appropriate to the speed concerned. In each case, the starting point is a torque corresponding to 10% of the maximum torque which can be developed at the speed in question. In the process, the load is regulated so that engine speed remains practically constant. The same test procedure is used at speed "D", the only difference being that the starting-point torque can be specified as $\geq$ 10% of the corresponding full-load torque. Three identical test runs are to be performed one after the other for each of the engine speeds. Test speed "D" and the torque of the corresponding starting point are stipulated by the official testing authorities (Fig. 11).

Similar to the EC smoke test ECE R.24, the smoke emission must be measured with a light-absorption measuring instrument. Regarding the validity of the final results, specific regulations are in place for their evaluation and approval. The resulting absorption factor represents the puff of smoke emitted by the engine in case of sudden loading in the lower and middle speed ranges, and as such takes into account the critical conditions actually encountered in everyday operation.

Measuring instruments

The following measuring principles are used all over the world for the stipulated test procedures:

Hydrocarbon analysis

The overall hydrocarbon content in diesel-engine exhaust gas is determined using a flame ionization detector analyzer (FID). The FID measuring principle is based on the generation of ions when hydrocarbons are burnt in a hydrogen flame (Fig. 12). The flow of ions between 2 electrodes is proportional to the atomic C-fraction of the hydrocarbon compound in question.

The exhaust gas contains a large number of different hydrocarbon compounds from the engine (consisting individually of unburnt, split, and partially oxidized compounds) which occur in different ratios depending upon the type of fuel and the engine operating mode.

Fig. 11

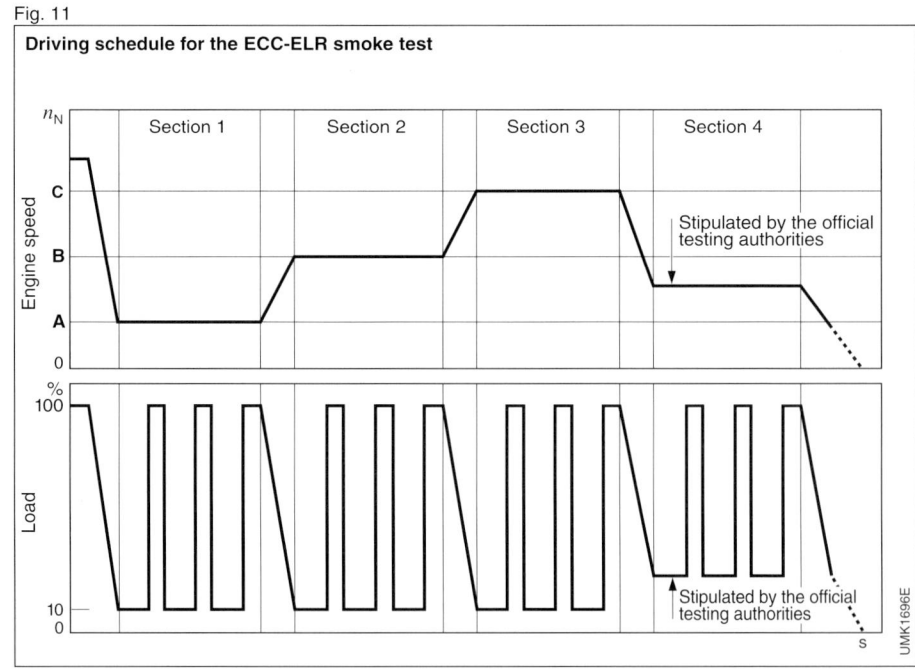

Driving schedule for the ECC-ELR smoke test

The preparation of the test-gas samples is of decisive importance, particularly for hydrocarbon analysis of diesel exhaust emissions. The considerable differences in the condensation temperatures of the individual exhaust-gas components mean that in contrast to the gasoline engine, the diesel test-gas system must be heated continually from the point where the test-gas is extracted right up to entry into the FID burner. The wall temperature of the sampling line system must be maintained at 190 °±10 °C.

Carbon-monoxide and carbon-dioxide analysis (Fig. 13)

The non-dispersive infrared (NDIR) analyzer process is applied for determining

Fig. 12

FID measurement method for HC analysis

1 Display, 2 Burner, 3 Outlet, 4 Hydrogen, 5 Air without HC, 6 Calibrating gas, 7 Exhaust gas.

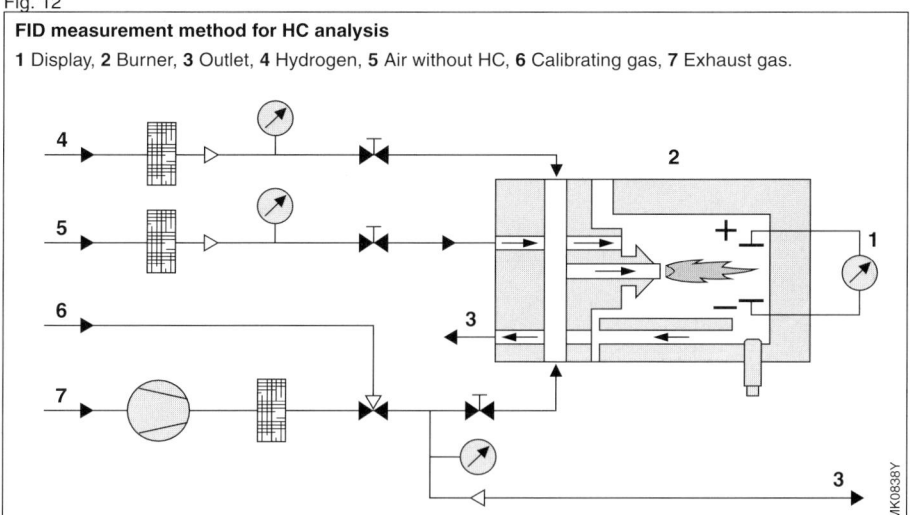

Fig. 13

NDIR measurement method for CO/CO₂ analysis

1 DC power supply, 2 Amplifier, 3 Mains stabilization, 4 Receiving chamber, 5 Metallic diaphragm, 6 Beam trimmer, 7 Exhaust gas, 8 Measuring cell, 9 Reference cell, 10 Filter cells, 11 Motor-driven chopper, 12 Light source.

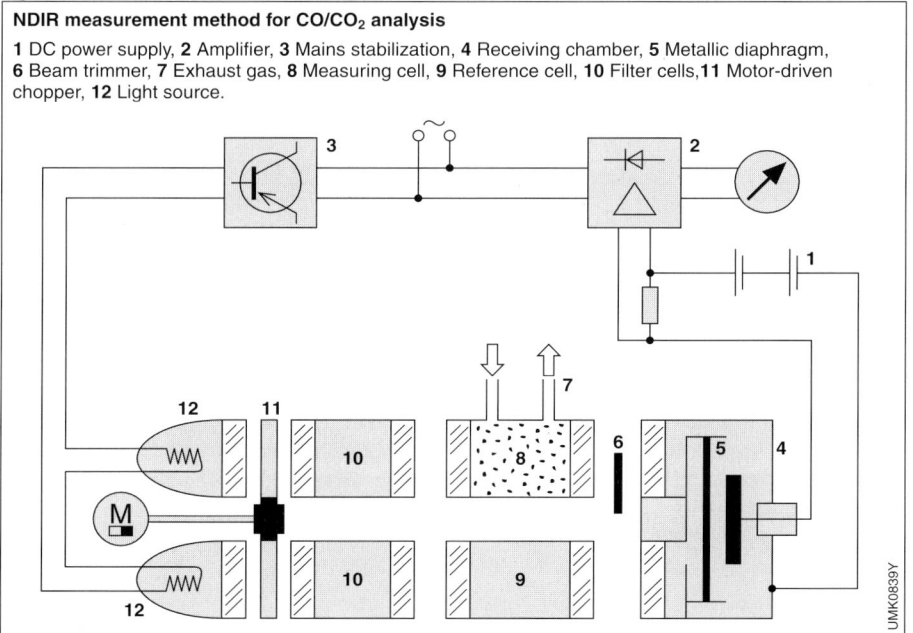

the concentration of these two gaseous components. This exploits the fact all polyatomic, non-elementary gases absorb infrared radiation in distinct bands specific to each gas. The sample gas is directed through a measuring cell located in the measuring-beam path. A reference cell located in the second beam path is charged with a gas which does not absorb radiation in the wavelengths concerned. A chopper wheel diverts the infrared radiation alternately through one cell and then through the other, so that the radiation finally enters one of the two receiving chambers. These are filled with the gas components which are to be analyzed, and separated from each other by a metal diaphragm in the form of a condensor plate. The radiation entering the chambers is absorbed only in the specific absorption bands of the gas in the respective chamber, that is, selectively. A difference in the amount of absorbed energy leads to a difference in temperature and pressure between the two chambers. This is converted into a voltage which is proportional to the concentration of the measured components.

Nitrogen-oxide analysis

This measuring principle makes use of chemiluminescence (chemically generated optical phenomenon) which occurs in the band between 590 and 3,000 nm after reaction of nitrogen oxide (NO) with ozone (O_3) (Figure 14).

The gas sample not only contains the nitrogen oxide formed by combustion in the engine, but also products arising from the partial oxidation of NO with the residual oxygen in the exhaust gas. For instance, such nitrogen oxides as NO_2, N_2O, and N_2O_4 are generated in the process. In addition to NO as the major component, NO_2 can also reach a significant concentration, whereas the other nitrogen-oxide compounds lie only slightly above the basic values of the surrounding air. If the total of all nitrogen-oxide compounds (NO_X) in the test gas is to be ascertained, these must be reduced in the "converter" to NO, whereby either thermal or thermo-catalytic methods are used (this process has no effect upon the NO component already present in the test gas). This means that the CLD reaction chamber is only filled with a concentration of pure NO which represents the various nitrogen oxides which have been reduced in the "converter".

Fig. 14

CLD chemiluminiscence detector for NO_X analysis

1 High-vacuum pump, **2** Molecular sieve, **3** Reference line, **4** Quantity regulator, **5** Filter, **6** Air, **7** Oxygen, **8** O_3 generator, **9** Capillary tube, **10** Reaction chamber, **11** Optical filter, **12** Photo-electric multiplier, **13** Amplifier, **14** Indicator device, **15** Exhaust gas, **16** NO_X/NO converter.

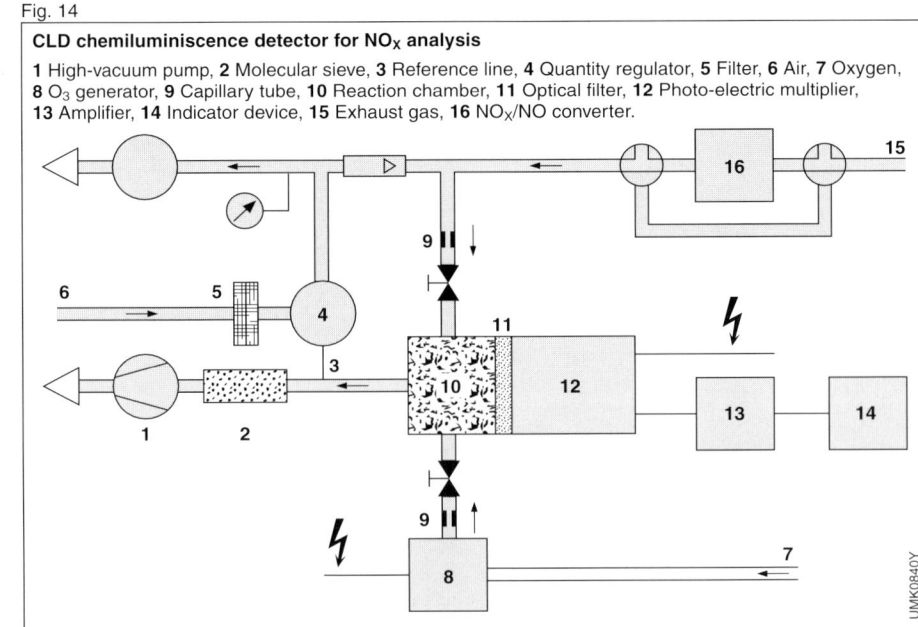

The chemiluminescence generated by the introduction of O_3 in the reaction chamber thus corresponds to the overall nitrogen oxide content. In order to eliminate disturbing luminiscence caused by other molecules contained in the gas mixture, only the radiation band between 600 and 660 nm is taken into account by using an optical filter.

Thanks to this selection process and a very low verification limit, the chemiluminescence measuring principle (CLD) is suitable for NO measurement in diluted or undiluted diesel-engine exhaust-gas. Since NO_2 dissolves in water, the test line must be heated to approx. $150 \pm 50\,°C$ in order to prevent the condensation of water vapor.

Particulate-matter measurement

Particulates are defined as those exhaust-gas components which at a test-gas temperature of $52\,°C$ are precipitated on standardized fluoride-carbon-coated glass-fiber filters. Mass measurement takes place using differential weighting methods (empty and loaded filters) at constant levels of dampness and temperature. A precision weighing instrument is used. The above definition was drawn up for the first time in US legislation. In the meantime, this test method has become the only generally accepted method in all countries which specify a particulate-emission limit.

Determination of soot emission

The filter and absorption methods are currently listed in emission-control legislation for measurement of the soot content in diesel exhaust gas. A reciprocal relationship exists between the measured results of both measuring methods if, for the absorption (opacity) measurement, the exhaust gas contains neither water vapor nor fuel mist. Both measuring methods provide measured values which increase logarithmically with the increasing soot concentration. A higher display accuracy than 5% can hardly be archieved using optical devices.

In the case of the filter method, the blackening of a filter paper is used as a measure for the quantity of soot deposited on it (Figure 15). In some countries (e.g. Switzerland), the filter device is prescribed for measurement of the smoke emission during free acceleration as a criterion for field monitoring. For this purpose, the length of time during which a test-gas sample is taken must be adap-

Fig. 15

Filter method for soot-emission measurement

a Bosch smokemeter, **b** Evaluation unit.
1 Exhaust gas, **2** Soot-coated paper filter disc, **3** Plunger position before measurement, **4** Sample volume, **5** Plunger position after measurement, **6** Tensioning spring, **7** Battery, **8** Light source, **9** Indicator, **10** Light receiver.

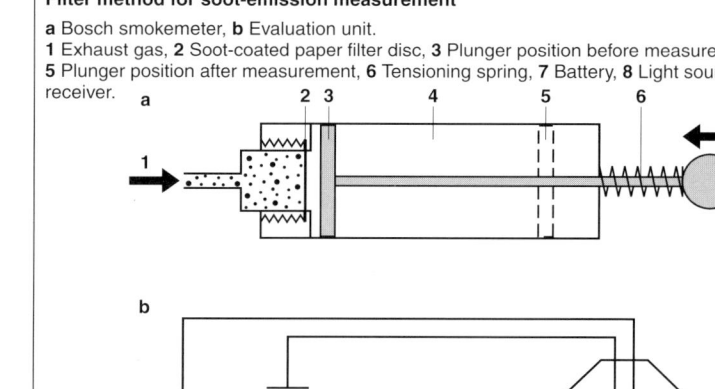

UMK0841Y

ted to the time taken for the engine to accelerate to the required speed for emitting the puff of smoke. Generally, 6 seconds is stipulated in order that all smoke can be trapped by the filter paper. Evaluation is performed by means of a photocell (Figure 15b), or using the "Bacharach grey scale".

The smokemeter (light-absorption or opacity measurement) uses the attenuation of a light beam as a measure of the soot concentration. During the measurement, some of the exhaust gas is drawn by a pump through the exhaust-sample probe and the hose and into the measuring chamber. This process above all avoids the exhaust-gas pressure, and its fluctuations, negatively affecting the measurement results.

A beam of light passes through the exhaust gas drawn into the measuring chamber. The reduction in light intensity is measured photo-electrically and displayed in %-opacity T or as the absorption coefficient $*k$. High display accuracy and reproducibility demand that the measuring-chamber length is precisely defined and that the measuring-chamber window is kept free of soot by applying thermal cleaning methods (Figures 16, 17, and 18).

When tests are performed under load, measurement and display are continuous. In the case of free acceleration, the complete measuring curve is stored digitally. The tester itself automatically processes the peaks, from which it then calculates the mean value resulting from a number of acceleration emissions.

Assessment

All exhaust-gas measurements contain both random (stochastic) and systematic errors. The random errors can be reduced by repeated measurement. The systematic errors are greatest if only one test apparatus is available. This error component can be reduced only by use of additional measuring equipment (e.g. second test bench). Only the mean value from the results of many measurements can provide a satisfactory assessment of exhaust-gas emissions.

Smokemeter to measure opacity T

1 Exhaust-sample probe, **2** Purge-air changeover valve, **3** Measuring chamber, **4** Measuring distance, **5** Lamp, **6** Receiver, **7** Pump.

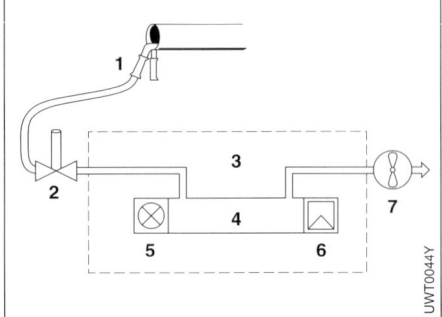

Fig. 16

Fig. 17

Smoke measurement as a function of time

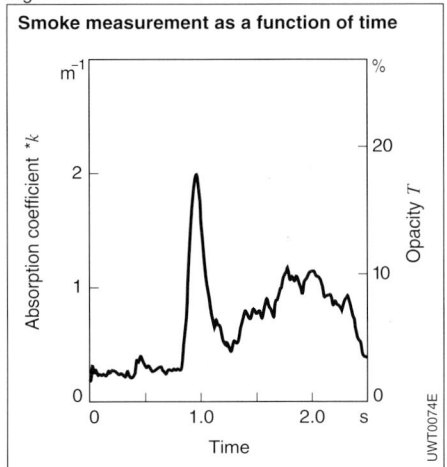

Time

Fig. 18

Limit-value curve for diesel smoke ECE R.24

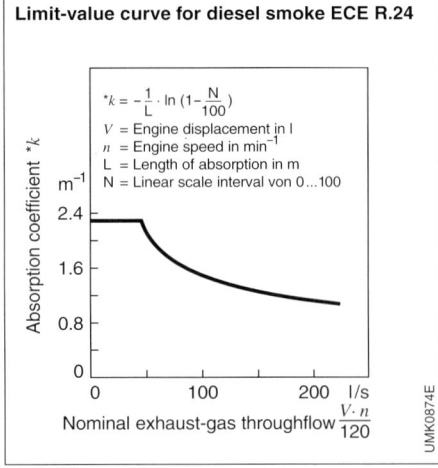

$$*k = -\frac{1}{L} \cdot \ln\left(1 - \frac{N}{100}\right)$$

V = Engine displacement in l
n = Engine speed in min^{-1}
L = Length of absorption in m
N = Linear scale interval von $0...100$

Nominal exhaust-gas throughflow $\frac{V \cdot n}{120}$

Exhaust-gas limits: Europe (status 1998)

Table 1
Limit values for passenger cars in the EC. Gross weight rating (GWR) $\leq$ 2.5 t and seating positions for $\leq$ 6 persons.

Regulation	Test	Date as from	Engine type	HC+NO$_X$ g/km [5]	CO g/km [5]	Particulates g/km [5]
91/441/EWG [1] Stage 1 (Euro I)	Type homologation/ First registration	1.7.92 1.1.93	IDI [3]	0.97 1.13	2.72 3.16	0.14 0.18
94/12/EU [2] Stage 2 (Euro II)	Type homologation/ First registration	1.7.96 1.1.97	IDI [3] DI [4]	0.7 0.9	1.0 1.0	0.08 0.1
Proposal EP [2] Stage 3 (Euro III)	Type homologation/ First registration	1.1.2000	IDI [3] & DI [4]	0.56 (0.50) [7]	0.64	0.05
Proposal EP [2] Stage 4 (Euro IV)	Type homologation/ First registration	1.1.2005	IDI [3] & DI [4]	0.3 (0.25) [7]	0.5 [6]	0.025 [6]

[1] Special regulation for $V_H < 1.4$ l. [2] No series-tolerance bonus (type homologation limit value identical to series limit value). [3] Prechamber engines. [4] Direct-injection (DI) engines. [5] New European driving schedule (NEFZ), start of sampling after 40 s. [6] Modified NEFZ, without 40 s delay. [7] NO$_X$ on its own must not esceed the above values in parentheses.

Table 2
Limit values for heavy-duty vehicles (HDV) in Europe. Gross weight rating (GWR) > 3.5 t (engine power output > 85 kW).

Regulation	Test	Date as from	HC g/kWh	NO$_X$ g/kWh	CO g/kWh	Particulates g/kWh
91/542/EWG 1.Stufe (Euro I)	Type homologation/ First registration	1.7.92 1.10.93	1.1 [1] 1.23 [1]	8.0 [1] 9.0 [1]	4.5 [1] 4.9 [1]	0.36 [1] 0.4 [1]
91/542/EWG 2.Stufe (Euro II)	Type homologation/ First registration	1.10.95 1.10.96	1.1 [1]	7.0 [1]	4.0 [1]	0.15 [1]
Vorschlag 3.Stufe (Euro III)	Type homologation/ First registration	1.10.2000	0.66 [2] 0.78 [3][4]	5.0 [2] 5.1 [3]	2.1 [2] 5.45 [3]	0.1 [2] 0.16 [3]
Vorschlag 4.Stufe (Euro IV)	Type homologation/ First registration	approx. 2005	0.33 [2] 0.39 [3][4]	2.5 [2] 2.5 [3]	1.05 [2] 2.73 [3]	0.05 [2] 0.08 [3]

In addition, the smoke limit values as per ECE R.24 (k$_{lim}$) and EEC-ELR (k = 0.8 m^{-1}).
[1] European 13-step test (as per ECE R.49). [2] New European 13-step test (EEC-ESC). [3] New European transient test EEC-ETC). [4] NMHC.

Table 3
Limit values for light commercial vehicles in Europe. Gross weight rating (GWR) < 3,5 t (corresponding to 93/59/EWG). New European driving schedule (NEFZ).

Standard regulation (in addition, special regulation D1 is also in force for vehicle classes M and N1)						Proposals	
Reference mass kg	Toxic components	Type homologation				Type homologation	
		10.94 [1] DI g/km	1.97 [1] IDI g/km	1.97 [1] DI g/km	10.99 [1] DI g/km	1.2000 [2] IDI/DI g/km	2005 [2] IDI/DI g/km
$\leq$ 1,250	HC+NO$_X$ CO Particulates	0.97 2.72 0.14	0.7 1.0 0.08	0.9 1.0 0.1	0.7 1.0 0.08	0.56 (0.50 [3]) 0.64 0.05	0.3 (0.25 [3]) 0.5 0.025
$\leq$ 1,700	HC+NO$_X$ CO Particulates	1.4 5.17 0.19	1.0 1.25 0.12	1.3 [4] 1.25 [4] 0.14 [4]	1.0 1.25 0.12	0.72 (0.65 [3]) 0.8 0.07	0.39 (0.33 [3]) 0.63 0.04
> 1,700	HC+NO$_X$ CO Particulates	1.7 6.9 0.25	1.2 1.5 0.17	1.6 [4] 1.5 [4] 0.2 [4]	1.2 1.5 0.17	0.86 (0.78 [3]) 0.95 0.1	0.46 (0.39 [3]) 0.74 0.06

[1] New European driving schedule (NEFZ), sampling takes place after 40 s. [2] Modified NEFZ, without 40 s delay. [3] NO$_X$ on its own must not exceed the above values in parentheses. [4] Applies since 01.98.

Exhaust-gas limits USA (status 1998)

Table 4
Limits for passenger cars [g/mile].
Error control: OBD II (On-board-diagnosis of the system components which are decisive for emissions).
Age/Mileage: 10 years old or 160,000 km (100,000 miles), whichever occurs first.

Model year	1993	1995	1996	1996	1998	1998	2001	2004	2003	− [5]
Validity [1]	Cal.	Cal.	Fed.	Fed.	Cal.	Cal.	Fed.	Fed.	Cal.	Fed.
Standard	Impr.[2]	TLEF [2]	Tier1 [3]	CFV1 [3]	LEF [2]	ULEF [2]	CFV2 [3]	Tier2 [3]	ZEF [2]	ILEF [3]
Driving schedule [4]	FTP-75	FTP-75	FTP-75	FTP-75	FTP-75	FTP-75	SFTP	SFTP	SFTP	SFTP
NMHC [5]	0.31	0.31	0.31	–	–	–	–	0.125	–	− [6]
HMOG [5]	–	0.156	–	0.156	0.09	0.055	0.09	–	0	0.075
HCHO [5]	–	0.018	–	0.018	0.018	0.011	0.015	–	0	− [6]
NO$_X$	1.0	0.6	1.25	0.6	0.3	0.3	0.3	0.2	0	0.20
CO	4.2	4.2	4.2	4.2	4.2	2.1	4.2	1.7	0	3.4
Particulates	0.08	0.08	0.1	0.08	0.08	0.04	0.08	0.1	0	− [6]

[1] Cal.: California, Fed.: 49 States. [2] Limit-value classifications for Cal.: Impr.: for basic development; LEV = Low-Emissions Vehicle; ULEV = Ultra-Low-Emissions Vehicle; ZEV = Zero-Emissions Vehicle. [3] Limit-value classifications for Fed.: Tier1, Tier2 for standard fuels; CFV1, CFV2 for "clean" fuels; ILEV = Inherently-Low-Emissions Vehicle. [4] Driving schedule FTP = Federal Test Procedure; Driving schedu SFTP = Supplemental Federal Test Procedure. [5] Hydrocarbons: NMHC = Hydrocarbon emissions (methane content ignored); NMOG = Choice of oxygen-free and oxygenic hydrocarbons; NCHO = Formaldehyd. [6] Not yet stipulated.

Table 5
NMOG [1] limit values [g/mile] for fleet average (only applies to California).

Model year	1994	1995	1996	1997	1998	1999	2000	2001	2002	2003	2004	2005
NMOG [1]	0.25	0.231	0.225	0.202	0.157	0.113	0.073	0.07	0.068	0.062	0.053	0.049

[1] NMOG: Choice of oxygen-free or oxygenic hydrocarbons.

Table 6
Limit values for heavy-duty vehicles (HDV)

Model year	1994	1994	1996	1996	1998	1998	1998
Validity [1]	Fed.	Cal.	Fed.	Cal.	Fed.	Fed.	Cal.
Standard [2]	HDV	HDV	UB	UB	HDV	UB	HDV
Exhaust-gas test: Test cycle HDTTC (Heavy-Duty Transient Test Cycle) [g/(hp · h)] [3].							
THC [4]	1.3	1.3 [5]	1.3	1.3	1.3	1.3	1.3 [5]
NMHC [4]	–	1.2 [5]	–	1.2	–	–	1.2 [5]
NO$_X$	5.0	5.0	5.0	4.0	4.0	4.0	4.0
CO	15.5	15.5	15.5	15.5	15.5	15.5	15.5
Particulates	0.1	0.1	0.05	0.05	0.1	0.05	0.1
Smoke test: Test cycle FST (Federal Smoke Test) [%].							
A [6]	20	20	20	20	20	20	20
B [6]	15	15	15	15	15	15	15
C [6]	50	50	50	50	50	50	50

[1] Cal.: California, Fed.: 49 States. [2] HDV: Heavy-Duty Vehicle; UB: Urban Bus [3] hp · h = horse power · hour (power · time). [4] Hydrocarbons: THC = Total of all emitted hydrocarbons; NMHC: Hydrocarbon emissions (methane content ignored). [5] Alternatively THC or NMHC. [6] A = Acceleration; B = Full-load deceleration; C = Max puff of smoke.

Exhaust-gas limits: Japan (status 1998)

Table 7
Limit values for passenger cars with ≤ 10 seating positions.
Driving schedule: 10·15-step test for toxic components as well as 3-step acceleration and smoke test.

Driving schedule	10·15-step hot test					Smoke test
Valid as from [2]	Vehicle reference weight kg	HC g/km	NOₓ g/km	CO g/km	Particulates g/km	Filter-paper blackening [1] %
10.97 10.98	Max. ≤ 1265 > 1265	Max./Mean 0.62/0.4 0.62/0.4	Max./Mean 0.55/0.4 0.55/0.4	Max./Mean 2.7/2.1 2.7/2.1	0.14/0.08 0.14/0.08	25 25

[1] Full-load measurement at 3 stipulated engine speeds, or with engine accelerated freely.
[2] Different validity dates for domestic-production vehicles and imported vehicles.

Table 8
Limit values for commercial vehicles and buses with ≥ 11 seating positions.
Gross weight rating (GWR) > 2,5 t. 13-step test for toxic components, and 3-step acceleration and smoke test.

Driving schedule	13-step test					Smoke test
Valid as from [2]	Vehicle reference weight kg	HC g/km	NOₓ g/km	CO g/km	Particulates g/km	Filter-paper blackening [1] %
10.94	DI [2] IDI [3]	2.9 2.9	6.0 5.0	7.4 7.4	0.7 0.7	40 40
10.98	DI [2] und IDI [3]	2.9	4.5	7.4	0.25	25

[1] Full-load smoke measurement at 3 stipulated engine speeds, or with engine accelerated freely.
[2] Direct-injection engines.
[3] Indirect injection (prechamber) engines.

Diesel fuel-injection systems: An overview

Fields of application

Diesel engines are characterized by their high levels of economic efficiency. This is of particular importance in commercial applications. Diesel engines are employed in a wide range of different versions (Fig. 1 and Table 1), for example as:
- The drive for mobile electric generators (up to approx. 10 kW/cylinder),
- High-speed engines for passenger cars and light commercial vehicles (up to approx. 50 kW/cylinder),
- Engines for construction, agricultural, and forestry machinery (up to approx. 50 kW/cylinder),
- Engines for heavy trucks, buses, and tractors (up to approx. 80 kW/cylinder),
- Stationary engines, for instance as used in emergency generating sets (up to approx. 160 kW/cylinder),
- Engines for locomotives and ships (up to approx. 1,000 kW/cylinder).

Technical requirements

More and more demands are being made on the diesel engine's injection system as a result of the severe regulations governing exhaust and noise emissions, and the demand for lower fuel-consumption. Basically speaking, depending on the particular diesel combustion process (direct or indirect injection), in order to ensure efficient air/fuel mixture formation, the injection system must inject the fuel into the combustion chamber at a pressure between 350 and 2,050 bar, and the injected fuel quantity must be metered with extreme accuracy. With the diesel engine, load and speed control must take place using the injected fuel quantity without intake-air throttling taking place.
The mechanical (flyweight) governing principle for diesel injection systems is in-

Fig. 1

Overview of the Bosch diesel fuel-injection systems

M, MW, A, P, ZWM, CW in-line injection pumps in order of increasing size; **PF** single-plunger injection pumps; **VE** axial-piston distributor injection pumps; **VR** radial-piston distributor injection pumps; **UPS** unit pump system; **UIS** unit injector system; **CR** Common Rail system.

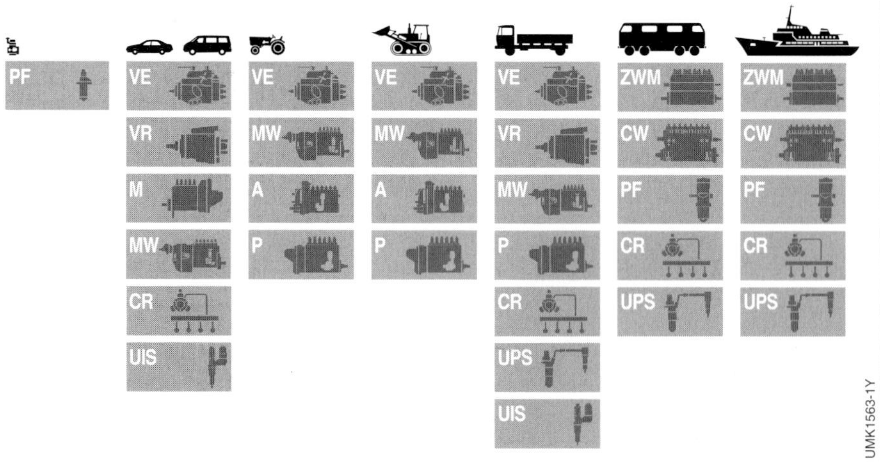

creasingly being superseded by the Electronic Diesel Control (EDC). In the passenger-car and commercial-vehicle sector, new diesel fuel-injection systems are all EDC-controlled.

According to the latest state-of-the-art, it is mainly the high-pressure injection systems listed below which are used for motor-vehicle diesel engines.

Table 1
Diesel fuel-injection systems: Properties and characteristic data

Fuel-injection system Type	Injection				Engine-related data			
	Injected fuel quantity per stroke mm^3	Max. nozzle pressure bar	Mechanical / Electronic / Electromechanical / Solenoid valve (m, e, em, MV)	Direct injection / Indirect injection (DI, IDI)	Pilot injection / Post injection (VE, NE)	No. of cylinders	Max. speed min^{-1}	Max. power per cylinder kW
In-line injection pumps								
M	60	550	m, e	IDI	–	4…6	5,000	20
A	120	750	m	DI / IDI	–	2…12	2,800	27
MW	150	1,100	m	DI	–	4…8	2,600	36
P 3000	250	950	m, e	DI	–	4…12	2,600	45
P 7100	250	1,200	m, e	DI	–	4…12	2,500	55
P 8000	250	1,300	m, e	DI	–	6…12	2,500	55
P 8500	250	1,300	m, e	DI	–	4…12	2,500	55
H 1	240	1,300	e	DI	–	6…8	2,400	55
H 1000	250	1,350	e	DI	–	5…8	2,200	70
Axial-piston distributor injection pumps								
VE	120	1,200/350	m	DI / IDI	–	4…6	4,500	25
VE…EDC [1]	70	1,200/350	e, em	DI / IDI	–	3…6	4,200	25
VE…MV	70	1,400/350	e, MV	DI / IDI	–	3…6	4,500	25
Radial-piston distributor injection pump								
VR…MV	135	1,700	e, MV	DI	–	4.6	4,500	50
Single-plunger injection pumps								
PF(R)…	150… 18,000	800… 1,500	m, em	DI / IDI	–	arbitrary	300… 2,000	75… 1,000
UIS 30 [2]	160	1,600	e, MV	DI	VE	8 [3a]	3,000	45
UIS 31 [2]	300	1,600	e, MV	DI	VE	8 [3a]	3,000	75
UIS 32 [2]	400	1,800	e, MV	DI	VE	8 [3a]	3,000	80
UIS-P1 [3]	62	2,050	e, MV	DI	VE	6 [3a]	5,000	25
UPS 12 [4]	150	1,600	e, MV	DI	VE	8 [3a]	2,600	35
UPS 20 [4]	400	1,800	e, MV	DI	VE	8 [3a]	2,600	80
UPS (PF[R])	3,000	1,400	e, MV	DI	–	6…20	1,500	500
Common Rail accumulator injection system								
CR [5]	100	1,350	e, MV	DI	VE [5a]/NE	3…8	5,000 [5b]	30
CR [6]	400	1,400	e, MV	DI	VE [6a]/NE	6…16	2,800	200

[1] EDC Electronic Diesel Control; [2] UIS unit injector system for comm. vehs. [3] UIS unit injector system for pass. cars; [3a] With two ECU's large numbers of cylinders are possible; [4] UPS unit pump system for comm. vehs. and buses; [5] CR 1st generation for pass. cars and light comm. vehs.; [5a] Up to 90° crankshaft BTDC, freely selectable; [5b] Up to 5,500 min^{-1} during overrun; [6] CR for comm. vehs., buses, and diesel-powered locomotives; [6a] Up to 30° crankshaft BTDC.

Injection-pump designs

In-line fuel-injection pumps

All in-line fuel-injection pumps have a plunger-and-barrel assembly for each cylinder. As the name implies, this comprises the pump barrel and the corresponding plunger. The pump camshaft integrated in the pump and driven by the engine, forces the pump plunger in the delivery direction. The plunger is returned by its spring.

The plunger-and-barrel assemblies are arranged in-line, and plunger lift cannot be varied. In order to permit changes in the delivery quantity, slots have been machined into the plunger, the diagonal edges of which are known as helixes. When the plunger is rotated by the movable control rack, the helixes permit the selection of the required effective stroke. Depending upon the fuel-injection conditions, delivery valves are installed between the pump's pressure chamber and the fuel-injection lines. These not only precisely terminate the injection process and prevent secondary injection (dribble) at the nozzle, but also ensure a family of uniform pump characteristic curves (pump map).

PE standard in-line fuel-injection pump

Start of fuel delivery is defined by an inlet port which is closed by the plunger's top edge. The delivery quantity is determined by the second inlet port being opened by the helix which is diagonally machined into the plunger.

The control rack's setting is determined by a mechanical (flyweight) governor or by an electric actuator (EDC).

Control-sleeve in-line fuel-injection pump

The control-sleeve in-line fuel-injection pump differs from a conventional in-line injection pump by having a "control sleeve" which slides up and down the pump plunger. By way of an actuator shaft, this can vary the plunger lift to port closing, and with it the start of delivery and the start of injection. The control sleeve's position is varied as a function of a variety of different influencing variables. Compared to the standard PE in-line injection pump therefore, the control-sleeve version features an additional degree of freedom.

Distributor fuel-injection pumps

Distributor pumps have a mechanical (flyweight) governor, or an electronic control with integrated timing device. The distributor pump has only <u>one</u> plunger-and-barrel asembly for all the engine's cylinders.

Axial-piston distributor pump

In the case of the axial-piston distributor pump, fuel is supplied by a vane-type pump. Pressure generation, and distribution to the individual engine cylinders, is the job of a central piston which runs on a cam plate. For one revolution of the driveshaft, the piston performs as many strokes as there are engine cylinders. The rotating-reciprocating movement is imparted to the plunger by the cams on the underside of the cam plate which ride on the rollers of the roller ring.

On the conventional VE axial-piston distributor pump with mechanical (flyweight) governor, or electronically controlled actuator, a control collar defines the effective stroke and with it the injected fuel quantity. The pump's start of delivery can be adjusted by the roller ring (timing device). On the conventional solenoid-valve-controlled axial-piston distributor pump, instead of a control collar an electronically controlled high-pressure solenoid valve controls the injected fuel quantity. The open and closed-loop control signals are processed in two ECU's. Speed is controlled by appropriate triggering of the actuator.

Radial-piston distributor pump

In the case of the radial-piston distributor pump, fuel is supplied by a vane-type pump. A radial-piston pump with cam ring and two to four radial pistons is responsible for generation of the high pressure and for fuel delivery. The injected fuel quantity is metered by a high-pressure solenoid valve. The timing device rotates the cam ring in order to adjust the start of delivery. As is the case with the solenoid-valve-controlled axial-piston pump, all open and closed-loop control signals are processed in two ECU's. Speed is controlled by appropriate triggering of the actuator.

Single-plunger fuel-injection pumps

PF single-plunger pumps
PF single-plunger injection pumps are used for small engines, diesel locomotives, marine engines, and construction machinery. They have no camshaft of their own, although they correspond to the PE in-line injection pumps regarding their method of operation. In the case of large engines, the mechanical-hydraulic governor or electronic controller is attached directly to the engine block. The fuel-quantity adjustment as defined by the governor (or controller) is transferred by a rack integrated in the engine.

The actuating cams for the individual PF single-plunger pumps are located on the engine camshaft. This means that injection timing cannot be implemented by rotating the camshaft. Here, by adjusting an intermediate element (for instance, a rocker between camshaft and roller tappet) an advance angle of several angular degrees can be obtained.

Single-plunger injection pumps are also suitable for operation with viscous heavy oils.

Unit-injector system (UIS)
With the unit-injector system, injection pump and injection nozzle form a unit. One of these units is installed in the engine's cylinder head for each engine cylinder, and driven directly by a tappet or indirectly from the engine's camshaft through a valve lifter.

Compared with in-line and distributor injection pumps, considerably higher injection pressures (up to 2050 bar) have become possible due to the omission of the high-pressure lines. Such high injection pressures coupled with the electronic map-based control of duration of injection (or injected fuel quantity), mean that a considerable reduction of the diesel engine's toxic emissions has become possible together with good shaping of the rate-of-discharge curve.

Electronic control concepts permit a variety of additional functions.

Unit-pump system (UPS)
The principle of the UPS unit-pump system is the same as that of the UIS unit injector. It is a modular high-pressure injection system. Similar to the UIS, the UPS system features one UPS single-plunger injection pump for each engine cylinder. Each UP pump is driven by the engine's camshaft. Connection to the nozzle-and-holder assembly is through a short high-pressure delivery line precisely matched to the pump-system components.

Electronic map-based control of the start of injection and injection duration (in other words, of injected fuel quantity) leads to a pronounced reduction in the diesel engine's toxic emissions. The use of a high-speed electronically triggered solenoid valve enables the characteristic of the individual injection process, the so-called rate-of-discharge curve, to be precisely defined.

Accumulator injection system

Common-Rail system (CR)
Pressure generation and the actual injection process have been decoupled from each other in the Common Rail accumulator injection system. The injection pressure is generated independent of engine speed and injected fuel quantity, and is stored, ready for each injection process, in the rail (fuel accumulator). The start of injection and the injected fuel quantity are calculated in the ECU and, via the injection unit, implemented at each cylinder through a triggered solenoid valve.

PE in-line fuel-injection pumps

Fuel-injection systems

Assignments

The fuel-injection system is responsible for supplying the diesel engine with fuel. To do so, the injection pump generates the pressure required for fuel injection. The fuel under pressure is forced through the high-pressure fuel-injection tubing to the injection nozzle which then injects it into the combustion chamber.

The fuel-injection system (Figs. 1 and 2) includes the following components and assemblies: The fuel tank, the fuel filter, the fuel-supply pump, the injection noz-zles, the high-pressure injection tubing, the governor, and the timing device (if re-quired). The combustion process in the diesel engine depends to a large degree upon the quantity of fuel which is injected and upon the method of introducing this fuel to the combustion chamber.

The most important criteria in this respect are the fuel-injection timing and the duration of injection, the fuel's distribution in the combustion chamber, the moment in time when combustion starts, the amount of fuel metered to the engine per degree cankshaft, and the total injected fuel quantity in accordance with the engine loading. The optimum interplay of all these parameters is decisive for the faultless functioning of the diesel engine.

Fig. 1

Fuel-injection system with mechanically (flyweight) governed in-line fuel-injection pump

1 Fuel tank, **2** Fuel-supply pump, **3** Fuel filter, **4** In-line fuel-injection pump, **5** Timing device, **6** Governor, **7** Nozzle holder with nozzle, **8** Fuel-return line, **9** Glow plug (GSK), **10** Battery, **11** Glow-plug and starter switch, **12** Glow control unit (GZS).

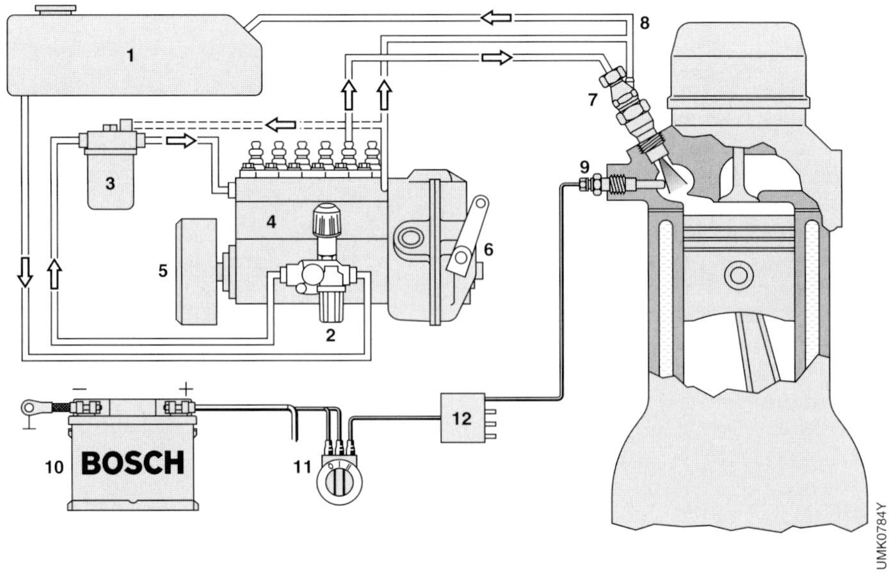

BOSCH

UMK0784Y

Fuel-injection techniques

Fuel metering

Depending upon the diesel combustion process in question, to ensure a good air-fuel mixture, the injection pump must inject the fuel at very high pressures, as well as metering it with maximum-possible precision. An accuracy of approx. 1°cks (crankshaft) for start of injection is necessary if optimal matching is to be reached between fuel economy, exhaust-gas emissions, and running noise.

With the standard in-line injection pump, a timing device is used to control start of injection and to compensate for pressure-wave propagation time in the high-pressure fuel-injection line. This adjusts the pump's start of delivery (port closing) in the "advance" direction as pump speed increases. In special cases, load-dependent

Fig. 2

Injection-pump assembly, mechanically governed

1 Fuel-injection pump, **2** Governor, **3** Fuel-supply pump, **4** Timing device.

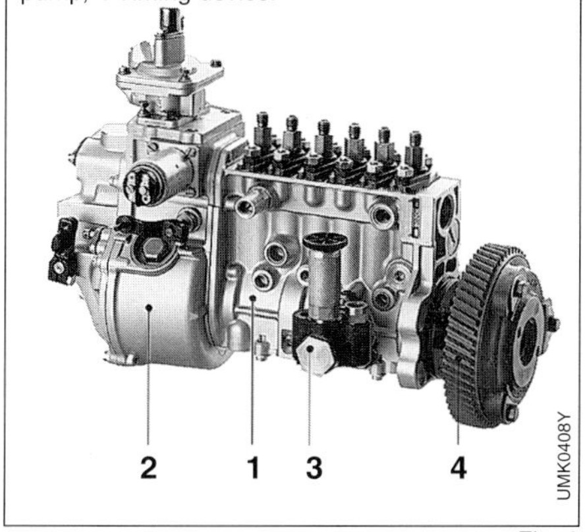

2   1   3   4

Fig. 3

control is provided. The diesel engine's load and speed control are determined by the injected fuel quantity (correct terms for fuel quantity without intake-air throttling are: Injected fuel volume [mm^3/stroke] and injected fuel mass [mg/stroke]) without intake-air throttling (Fig. 3).

Fuel-injection system with electronically controlled in-line fuel-injection pump

1 Fuel tank, **2** Fuel-supply pump, **3** Fuel filter, **4** In-line fuel-injection pump, **5** Electrical shutoff device (ELAB), **6** Fuel-temperature sensor, **7** Rack-travel sensor, **8** Actuator with linear magnet, **9** Rotational-speed sensor, **10** Injection nozzle, **11** Coolant-temperature sensor, **12** Accelerator-pedal sensor, **13** Switch for clutch, brake, retarder, **14** Operator unit, **15** Warning lamp and diagnosis connection, **16** Trip recorder or vehicle-speed sensor, **17** ECU, **18** Air-temperature sensor, **19** Boost-pressure sensor, **20** Exhaust-gas turbocharger, **21** Battery, **22** Glow-plug and starter switch.

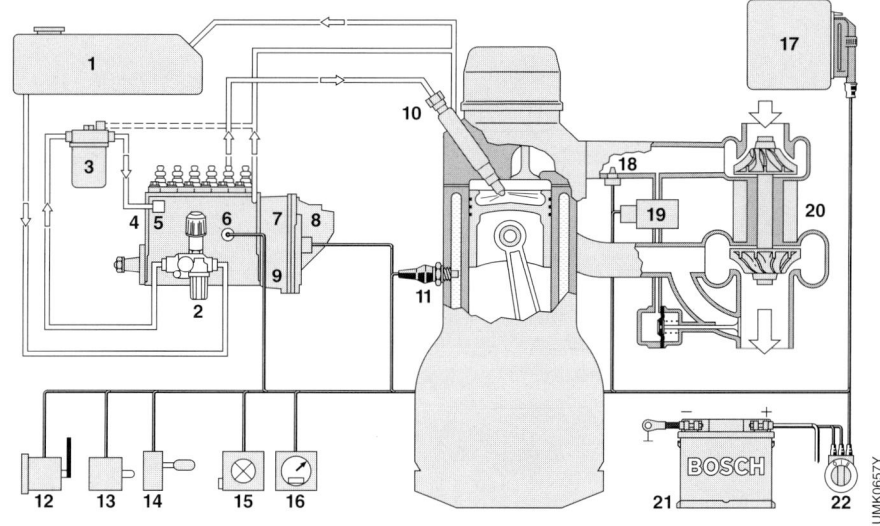

Plunger stroke-phase sequence

With the pump plunger at BDC, the pump-barrel inlet ports are open. Through them, the fuel which is under supply-pump pressure flows from the pump's fuel gallery and into the high-pressure chamber. As it moves up in the barrel, the pump plunger passes the inlet ports and closes them. This plunger stroke is termed the prestroke. During the further course of the plunger stroke, fuel pressure is increased so that the de-livery valve opens. If a constant-volume valve is used (see "Delivery valves" Chapter), the plunger travels through a further stroke known as the retraction stroke. During the effective (working) stroke, the fuel is forced through the high-pressure line to the nozzle. The effective stroke is terminated as soon as the plunger's helix opens the spill port (or in-let port). From this instant, no more fuel is delivered to the nozzle because during the remaining plunger stroke (residual stroke), the fuel is forced back through the plunger's vertical slot and into the fuel gallery (Fig. 4).

Following reversal of plunger travel at TDC, the fuel continues to flow through the vertical slot into the barrel until the plunger helix closes the spill port (or inlet port) again. During the plunger's con-tinuing return stroke, a vacuum is gene-rated in the pump barrel, and fuel flows into the high-pressure chamber as soon as the plunger opens the inlet port. The cycle can begin again.

Fig. 4

Plunger-stroke phases

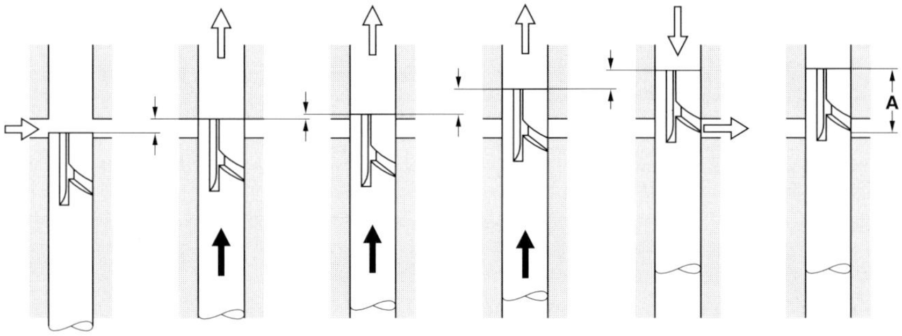

1 Bottom dead center (BDC)	2 Prestroke	3 Retraction stroke	4 Effective stroke	5 Residual stroke	6 Top dead center (TDC)
Fuel flows from the injec-tion pump's fuel gallery and into the high-pressure chamber of the plunger-and-barrel assembly.	Plunger stroke from BDC to closure of the inlet port by the top edge of the plunger (variable de-pending upon plunger-and-barrel assembly).	Plunger stroke from end of the prestroke until the deli-very valve opens (only if a constant-volume valve is used).	Plunger stroke from opening of the delivery valve to open-ing of the inlet port by the plunger helix (overflow).	Plunger stroke from opening of the inlet port to TDC.	Reversal of plunger travel. **A** Total stroke

UMK0421E

Fuel-delivery control

Using a toothed control rack. **a** Zero delivery, **b** Partial delivery, **c** Maximum delivery.
1 Pump barrel, **2** Inlet port, **3** Pump plunger, **4** Helix, **5** Control rack.

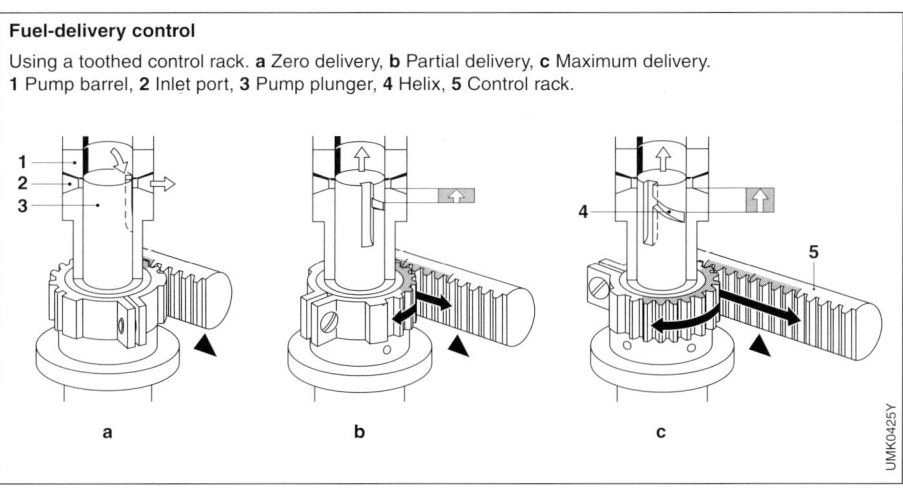

Fig. 5

Among other things, the power delivered by a diesel engine depends upon the injected fuel quantity. It is the injection pump's job to always meter the appropriate quantity of fuel to the engine in accordance with its loading.

The injected fuel quantity is varied by changing the plunger's effective stroke. To do so, the control rack turns the pump plunger in the barrel so that the helix, which runs diagonally around the plunger circumference, can open the inlet port sooner or later and in doing so change the end-of-delivery point and with it the injected fuel quantity. In the case of maximum delivery, port opening cannot take place until the maximum effective stroke has been reached, in other words at the maximum possible delivery quantity. With partial delivery, port opening occurs sooner depending upon the exact position of the pump plunger. At the zero-delivery position, the vertical slot in the plunger is directly opposite to the inlet port. This means that for the complete plunger stroke the high-pressure chamber is connected to the fuel gallery through the plunger's vertical slot. No fuel is delivered. This is the position to which the plungers are rotated when the engine is switched off (Fig. 5).

The PE..A in-line injection pump uses a toothed rack to turn the plungers and vary the injected fuel quantity (Fig. 6).

Fig. 6

PE..A...-Type in-line fuel-injection pump

1 Control-sleeve gear, **2** Control sleeve,
3 Spring-chamber cover, **4** Delivery-valve holder,
5 Valve holder, **6** Delivery valve,
7 Pump barrel, **8** Pump plunger,
9 Control rack, **10** Plunger control arm,
11 Plunger return spring, **12** Spring seat,
13 Adjusting screw, **14** Roller tappet,
15 Camshaft.

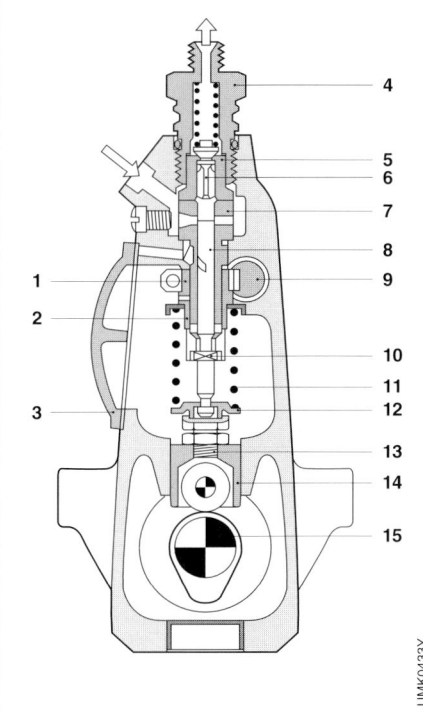

Cam shapes

Different combustion systems and combustion chambers necessitate individually tailored conditions of injection, in other words, the injection process must be specifically matched to the particular engine type. The plunger speed, and therefore the duration of injection, depend upon the plunger actuating cam's lift relative to the angle of cam rotation. This is why a wide variety of different cam contours are required for everyday operation. In order to improve the conditions of injection, such as rate-of-discharge curve and pressure loading, special cam shapes can be generated by calculation.

Use is made of symmetrical cams, asymmetrical cams, and cams (back-kick cam) which prevent the engine being started in the false direction of rotation (Fig. 7).

Plunger-and-barrel assemblies (pumping elements)

Plunger-and barrel assembly:
Basic version
The pump plunger together with the pump barrel form the plunger-and-barrel assembly. This utilises the overflow principle in conjunction with port-and-helix control.

The pump plunger has been so finely lapped into the plunger barrel that it provides an adequate seal even at high pressures and low speeds, and no additional sealing elements are required.

In addition to its vertical slot, the pump plunger has also been machined on the side and the resulting diagonal edge cut in the plunger's wall is termed the control helix (Figs. 8a and 8b).

A single helix suffices for injection pressures up to 600 bar, but above this figure the plunger needs two diametrically opposite helixes. This measure serves to prevent plunger "seizure" because the plunger is no longer forced against the barrel wall by the injection pressure. The barrel is provided with one or two inlet ports for fuel inlet and end of delivery.

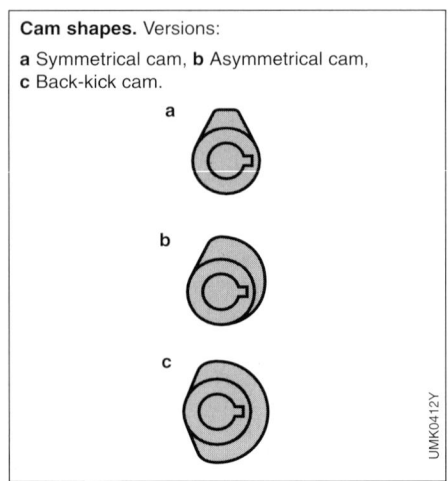

Cam shapes. Versions:

a Symmetrical cam, b Asymmetrical cam,
c Back-kick cam.

Fig. 7

Fig. 8

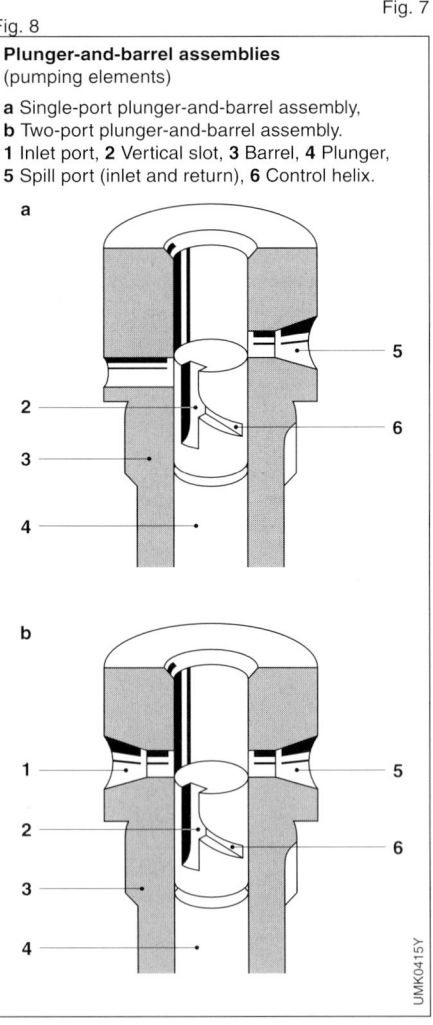

Plunger-and-barrel assemblies
(pumping elements)

a Single-port plunger-and-barrel assembly,
b Two-port plunger-and-barrel assembly.
1 Inlet port, 2 Vertical slot, 3 Barrel, 4 Plunger,
5 Spill port (inlet and return), 6 Control helix.

Due to the precise matching of the plunger to the barrel, it is essential that only complete plunger-and-barrel assemblies are exchanged and never the plunger or the barrel alone.

Plunger-and-barrel assembly with leakage-return duct

It the fuel-injection pump is connected to the engine's lube-oil circuit, under certain circumstances leak fuel can result in lube-oil dilution. This is avoided to a great extent by plunger-and-barrel assemblies with a leakage-return duct to the pump's fuel gallery. In such cases, the barrel is provided with a ring-shaped groove which is connected to the fuel gallery through a separate passage, or the leak oil is collected in a ring-shaped plunger groove and then returned to the fuel gallery through appropriate slots in the plunger (Fig. 9).

Versions

Special requirements calling for the reduction of noise or exhaust emissions for instance, dictate that some form of load-dependent start-of-delivery (port closing) is provided. Plunger versions which in addition to the lower helix also have an upper helix permit the start of delivery to be adjusted as a function of load (Fig. 10). In order to improve the starting performance of certain engine types,

Fig. 9

Plunger-and-barrel assembly with leakage-return duct

a 1 Leakage-return slot,
2 Ring-shaped plunger groove,
b 1 Leakage-return duct,
2 Ring-shaped barrel groove.

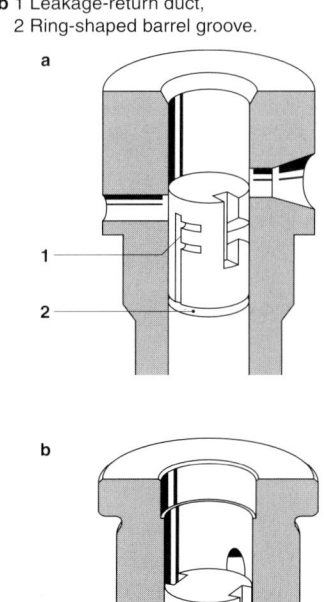

Fig. 10

Pump-plunger versions

a Lower helix,
b Upper and lower helix,
c Lower helix with starting groove (1).

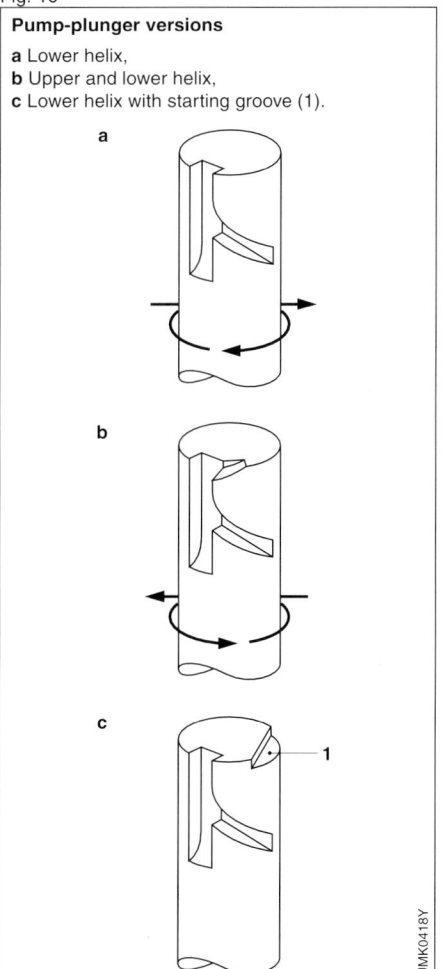

UMK0417Y

UMK0418Y

special plungers are used which have a special starting groove. This starting groove is machined into the plunger's top edge and is effective only when the plunger is in the start position. It results in a start of delivery which is retarded by 5...10° relative to the crankshaft setting.

Delivery valves

It is the job of the delivery valve to interrupt the high-pressure circuit between the high-pressure fuel-injection lines and the pump plunger, as well as to relieve the high-pressure lines and the nozzle space by reducing pressure to a given static level. This reduction in pressure causes the nozzle to close rapidly and precisely, as well as preventing undesirable fuel dribble. During the actual injection process, the pressure which is generated in the nozzle's high-pressure chamber causes the delivery valve's cone to lift from its seat in the valve holder, and the pressurized fuel is forced through the delivery-valve holder and the high-pressure lines to the injection nozzle. As soon as the plunger's helix terminates the fuel delivery, pressure drops in the nozzle's high-pressure chamber, and the delivery-valve spring forces the valve cone back onto its seat. This separates the chamber above the pump plunger from the high-pressure circuit until the plunger's next effective (working) stroke (Fig. 11).

Constant-volume valve without return-flow restriction

In the constant-volume valve part of the valve-element stem is shaped like a piston (retraction piston) and is precisely lapped into the valve-stem guide. When the plunger's helix terminates the fuel delivery, and the spring closes the delivery valve, the piston enters the valve-stem guide and closes off the high-pressure line from the nozzle's high-pressure chamber. This means that the volume available to the fuel in the high-pressure line is increased by the retraction piston's stroke volume. This retraction volume is matched to the length of the high-pressure line, which means that the line length is not to be changed.

Fig. 11

Delivery-valve holder with delivery valve

a Closed,
b During fuel delivery.
1 Delivery-valve holder, **2** Delivery-valve spring,
3 Delivery valve, **4** Valve seat,
5 Valve holder.

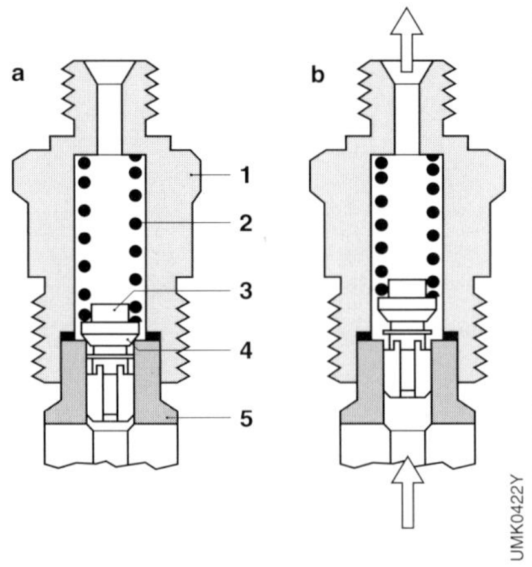

UMK0422Y

Fig. 12

Constant-volume delivery-valve element

a Normal,
b With compensation.
1 Valve seat, **2** Retraction piston,
3 Ring-shaped groove, **4** Valve stem,
5 Vertical slot, **6** Machined section.

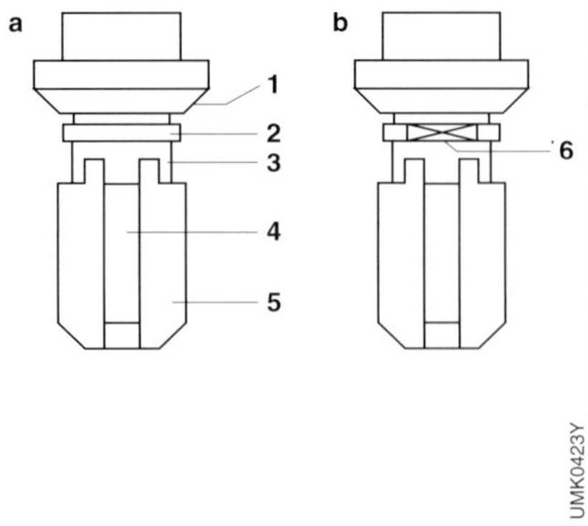

UMK0423Y

In order to achieve specific fuel-delivery characteristics, compensation valves can be used in special cases. These have an extra machined section on the retraction piston (Fig. 12).

Constant-volume valve with return-flow restriction

The return-flow restriction can be used in addition to the constant-volume valve. The reverse pressure waves which are generated when the injection nozzle closes can cause wear and cavitation in the delivery-valve's high-pressure chamber. This can be reduced or even prevented completely by the damping effect of the return-flow restriction which is located in the upper section of the delivery-valve holder, in other words between the constant-volume valve and the nozzle. There is a narrow restriction passage in the valve body which on the one hand provides the required throttling effect and on the other for the most part prevents pressure-wave reflection. The valve opens in the delivery direction and there is no restriction or throttling effect. A plate (Fig. 13) is used as the valve body

for pressures up to approx. 800 bar, and for higher pressures a guided cone.

Constant-pressure valve

The constant-pressure valve (Fig.14) is used with high-pressure fuel-injection pumps which develop pressures above approx. 800 bar, and small high-speed direct-injection (DI) engines. It comprises a forward-delivery valve in the delivery direction, and a pressure-holding valve in the return-flow direction. Between injections, the latter maintains the static line pressure as constant as possible under all operating conditions. The advantage of the constant-pressure valve lies in the avoidance of cavitation and in improved hydraulic stability.

It the constant-pressure valve is to be employed efficiently, this necessitates more precise adjustments and governor modifications.

Fig. 13

Delivery-valve holder with return-flow restriction

1 Delivery-valve holder,
2 Delivery-valve spring,
3 Valve plate,
4 Valve holder.

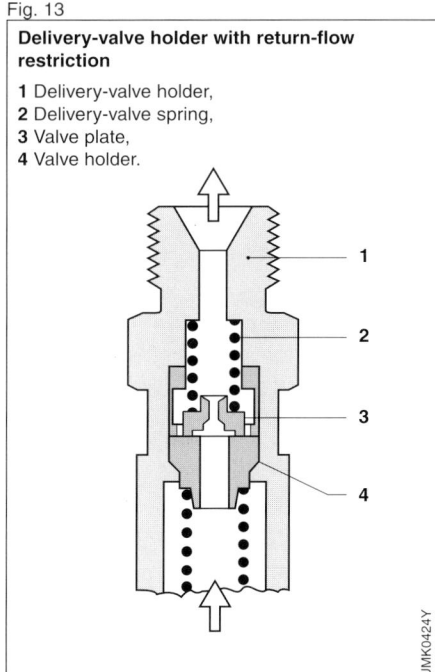

Fig. 14

Constant-pressure valve

1 Valve holder, 2 Valve element,
3 Valve spring, 4 Filler piece,
5 Compression spring, 6 Spring seat,
7 Ball, 8 Restriction passage.

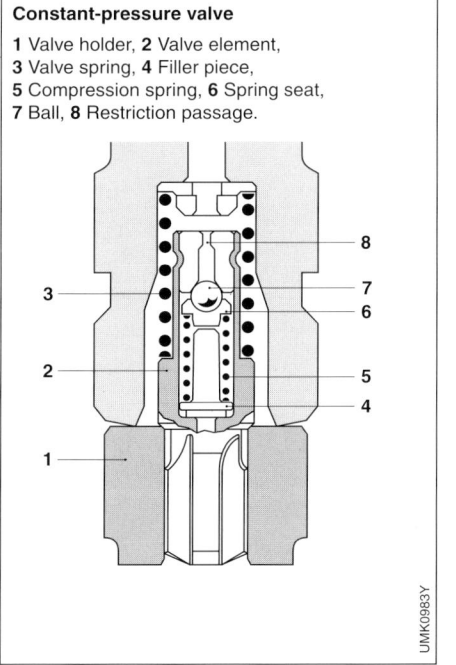

PE standard in-line fuel-injection pumps

Design and construction

The standard PE in-line injection pumps incorporate their own camshaft, and a plunger-and-barrel assembly (pumping element) for each engine cylinder (Fig. 1).

The complete fuel-injection system is comprised of:
- A fuel-injection pump,
- A mechanical (flyweight) or electronic governor for control of engine-speed and injected fuel quantity,
- A timing device (if required) for varying the start of delivery as a function of engine speed,
- A fuel-supply pump for delivering the fuel from the fuel tank, through the fuel filter and the fuel line, to the injection pump,
- A number of high-pressure fuel-injection lines, corresponding to the number of engine cylinders, connecting the injection pump and the injection nozzles,
- The injection nozzles.

The injection pump's camshaft is driven by the diesel engine. Injection-pump speed and crankshaft speed are identical for 2-stroke engines. For 4-stroke engines, pump speed is the same as engine camshaft speed, in other words half crankshaft speed.

The drive between injection pump and engine must be as torsionally rigid as possible if today's high injection pressures are to be generated.

There are a number of different sizes of in-line injection pumps for the various engine outputs.

The injected fuel quantity depends upon the swept volume of the injection-pump barrel, and maximum (pump-side) injection pressures are between 400 and 1,150 bar.

Fig. 1

PES in-line fuel-injection pump

1 Delivery-valve holder, **2** Filler piece, **3** Delivery-valve spring, **4** Pump barrel, **5** Delivery valve, **6** Inlet port and spill port, **7** Control helix, **8** Pump plunger, **9** Control sleeve, **10** Plunger control arm, **11** Plunger return spring, **12** Spring seat, **13** Roller tappet, **14** Cam, **15** Control rack.

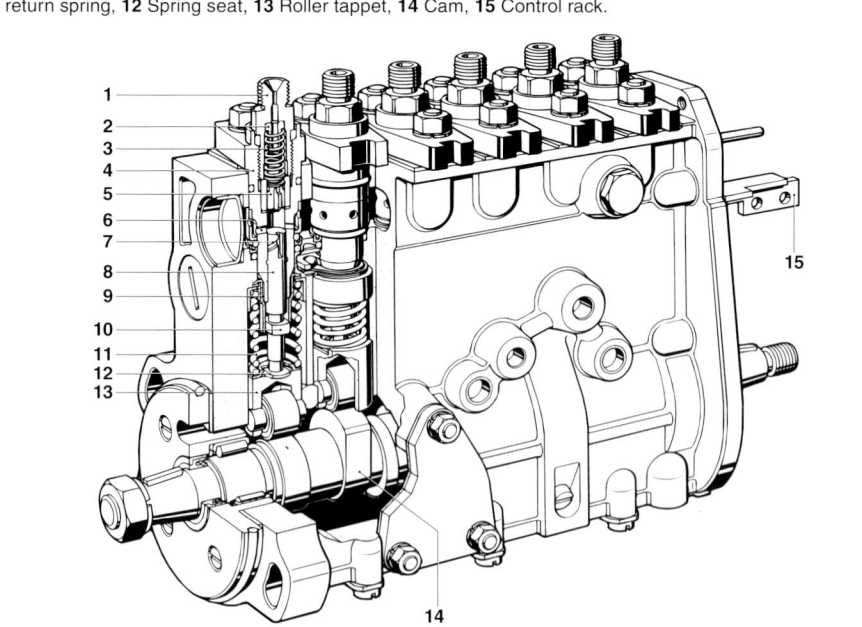

To lubricate the moving injection-pump components (e.g. camshaft, roller tappets etc.) there must be a certain amount of oil in the injection pump. The injection pump is connected to the diesel engine's lube-oil circuit, and oil circulates through the pump during operation.

Each pump type is allocated to a given type series, which in some cases overlap with respect to their power ranges. These will be described in the following chapters.

Two different construction principles are used for in-line injection pumps: The principle for the M and A pumps, and that for the MW and P pumps.

The power outputs of diesel engines equipped with in-line injection pumps range from 10 to 70 kW per cylinder. This broad power-output range is made possible by the availability of a wide variety of different pump versions. The pump sizes A, M, MW, and P are manufactured in large batches (Fig. 2).

The pump sizes ZW, P9, and P10 are available for even higher cylinder power outputs.

Method of operation

Interaction between the components

The camshaft of the PE in-line injection pump is integrated in the aluminum pump housing. It is connected to the diesel engine either through a timing device, through a coupling element, or directly. A roller tapper with spring seat is located above each camshaft cam. The spring seat provides a positive-drive connection between pump plunger and roller tappet. The pump plunger moves up and down in the pump barrel, and together these two components form the plunger-and-barrel assembly (pumping assembly).

Tab. 1
Overview

Features	PE in-line injection pumps				
	M	A	MW	P1...3000	P7100...8000
Injection pressure in bar (pump side)	550	750	1100	950	1300
Application	Passenger cars and vans	Light to medium commercial vehicles, tractors, industrial engines.			Heavy commercial vehicles, industrial engines.
Output per cylinder in kW/cylinder	20	27	36	60	160

Fig. 2

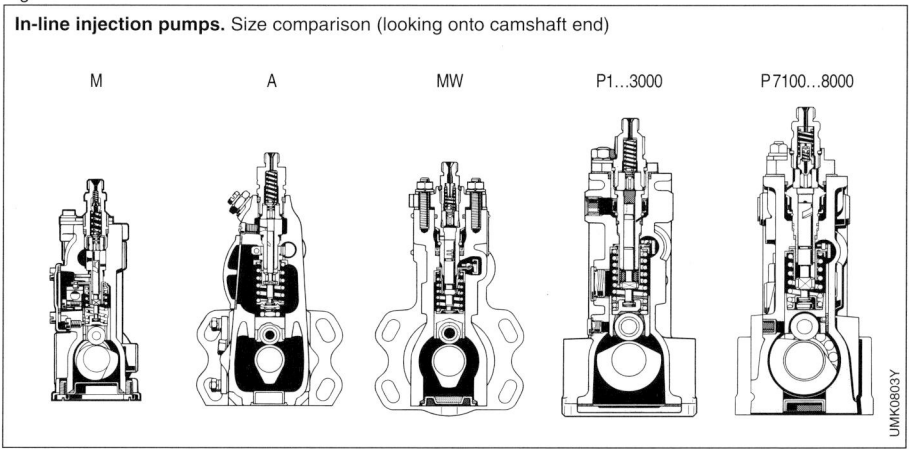

In-line injection pumps. Size comparison (looking onto camshaft end)

M   A   MW   P1...3000   P7100...8000

UMK0803Y

The injection-pump barrel has either one or two inlet ports which lead from the pump's fuel gallery into the pump barrel. The delivery-valve holder complete with delivery valve is located at the top of the plunger-and-barrel assembly. The control sleeve is the connection between the pump plunger and the control rack. The control rack, which is free to move lengthways in the pump housing as dictated by the governor, engages with the control-sleeve gear or with a linkage lever to turn the "control-sleeve/pump plunger unit" in accordance with the governor output as described in the Chapter "Engine-Speed Governing". This permits the precise control of injected fuel quantity.

Injection-pump drive

In the in-line injection pump, camshaft rotation is converted directly to the vertical lift of the roller-tappet which results in the reciprocating plunger movement (Fig. 3).

The pump plunger's total lift cannot be varied, although the effective stroke, and with it the delivery quantity, can be changed by using the control rack to rotate the pump plunger.

The plunger is forced up to TDC by the cam, and back down again to BDC by the plunger-return spring. This spring must be selected so that even at maximum pump speed the roller tappet cannot jump off of the cam. This must be avoided at all costs because the impact caused when the roller "hits" the cam again is bound to lead to cam or roller damage.

The angular offset between adjoining cams ensures that the injection sequence agrees exactly with the engine's firing sequence and firing interval.

Fig. 3

Plunger-and-barrel assembly (pumping element). Drive

a BDC position, **b** TDC position. **1** Cam, **2** Roller tappet, **3** Lower spring seat, **4** Plunger return spring, **5** Upper spring seat, **6** Control sleeve, **7** Pump plunger, **8** Pump barrel.

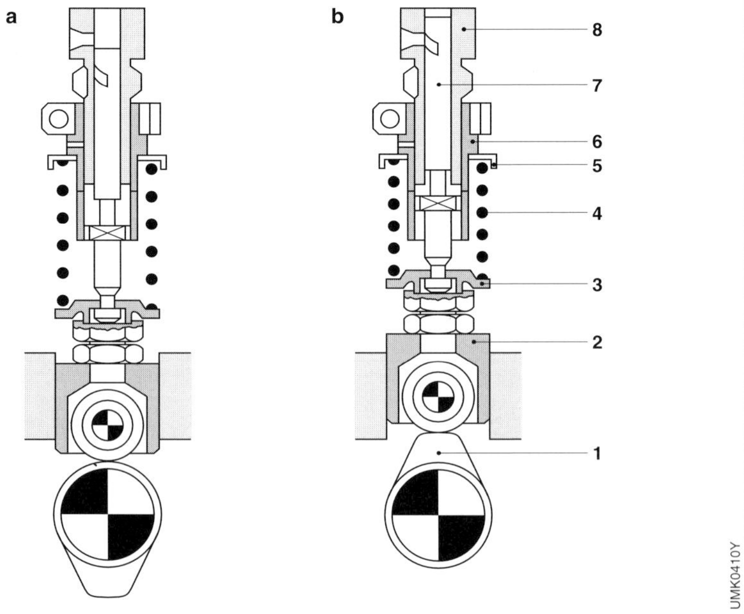

Additional components

Engine-speed control (governing)

The governor's main task is to limit the engine's maximum speed (maximum no-load speed). It must limit engine speed to the maximum permitted by the engine manufacturer, because otherwise the unloaded diesel engine will speed up out of control until it destroys itself. It must also be possible to maintain specific engine speeds inside a given engine-speed range or within the complete range. Depending upon governor design, this can apply for instance to idle speed and maximum speed.

The governor also has a number of other functions: Changing full-load delivery as a function of engine speed (torque control), or as a function of atmospheric pressure of charge-air pressure, or provision of the injected fuel quantity needed for starting. To do so, the governor shifts the control rack so that the pump plunger is rotated to the appropriate setting for the required delivery quantity (Fig. 4). For

Fig. 4

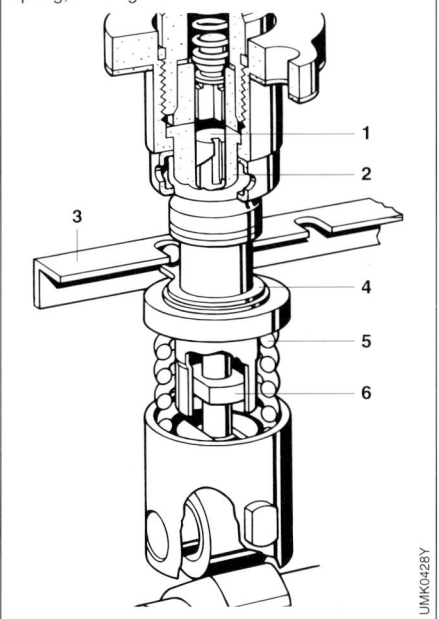

Fuel-delivery control

Using a control-sleeve lever. **1** Plunger, **2** Barrel, **3** Control rack, **4** Control sleeve, **5** Plunger return spring, **6** Plunger control arm.

UMK0428Y

governing on in-line injection pumps, mechanical (flyweight) governors, or Electronic Diesel Control (EDC), are used. Pneumatic governors are no longer used because they cannot comply with the severe requirements made on a modern diesel engine.

The above governors are described in the Technical Instruction manual "Governors for diesel in-line fuel-injection pumps".

Mechanical engine-speed control

There are a variety of different mechanical governor types in use:
– Maximum-speed governor, for limiting the maximum speed (high-idle speed).
– Minimum-maximum-speed governor (mainly for automotive applications), governs only at the upper and lower limits of the engine speed range, but not in between. The driver changes the injected fuel quantity by means of the accelerator pedal.
– Variable-speed governor, governs throughout the complete speed range in addition to the maximum (high-idle) and idle speeds.

Developments in diesel fuel-injection techniques are today determined by the steadily increasing demands made on exhaust-gas "quality", fuel economy, driveability and comfort, and engine power. Consequently, the demands made on the fuel-injection system, and particularly on the governor, are also becoming more severe.

Electronic engine-speed control

The Electronic Diesel Control (EDC) complies in full with the increasing demands made upon the engine-speed governing system. As well as permitting electrical measurement and electronic data processing, EDC incorporates control loops with electrical actuators which, compared to the mechanical governor, provide for more functions as well as improving existing functions.

The EDC comprises:
– A variety of sensors,
– The electronic control unit (ECU), and
– The actuator fitted to the injection pump.

Injection timing

The most important criteria for diesel engine optimization are:

– Low exhaust emissions,
– Low combustion noise, and
– Low specific fuel consumption.

The moment at which the injection pump starts to deliver fuel is known as the "start of delivery" (or port closing). This moment in time is selected according to injection lag and ignition lag. These are variable parameters which are a function of the particular operating point. Injection lag is defined as the lag between start of delivery and start of injection. Ignition lag as the lag between start of injection and start of combustion. Start of injection is defined as the crankshaft angle in the TDC area at which the nozzle injects fuel into the combustion chamber.

Start of combustion is defined as the moment of A/F mixture ignition which can be influenced by start of injection.

With the PE fuel-injection pump, the speed-dependent adjustment of the start of delivery (port closing) is best performed using a timing device.

Assignments

Because it directly changes the start-of-delivery point, the timing device should be termed a start-of-delivery adjuster. The timing device (eccentric-type) transfers the drive torque to the injection pump while at the same time performing its timing function. The drive torque required by the injection pump depends upon pump size, number of barrels, injected fuel quantitiy, injection pressure, plunger diameter, and cam contour. The fact that drive torque has a direct effect upon the timing characteristic, must be taken into account during design work, as well as the work capacity.

Fig. 5

Cylinder pressure

A Start of injection, **B** Start of combustion, **C** Ignition lag.
1 Induction stroke, **2** Compression stroke, **3** Power stroke, **4** Exhaust stroke.

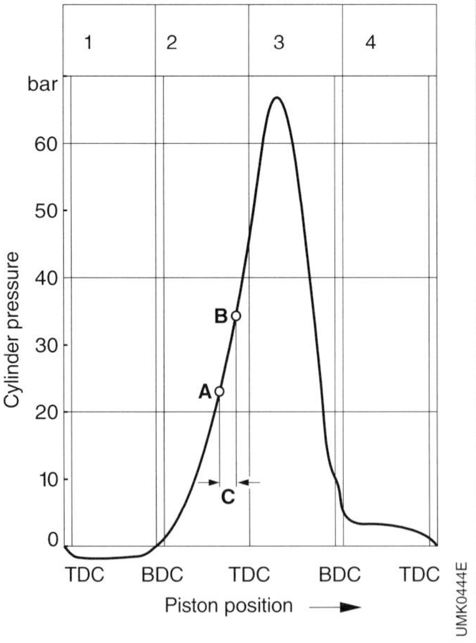

Fig. 6

Timing device
Design and construction.

1 Drive element, **2** Hub, **3** Housing, **4** Adjusting eccentric element, **5** Compensating eccentric element, **6** Hub bolt, **7** Flyweights, **8** Backing disk.

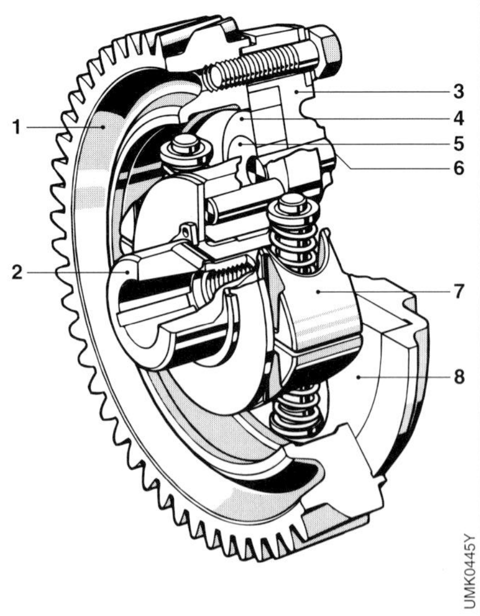

Fig. 7

Timing device
Function.

a In initial position,
b Setting at low speeds,
c Setting at medium speeds,
d Final setting at high speeds,
α = Advance angle.

a

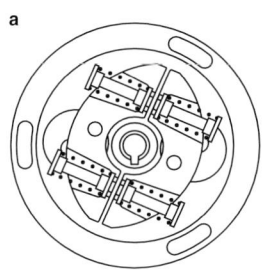

b

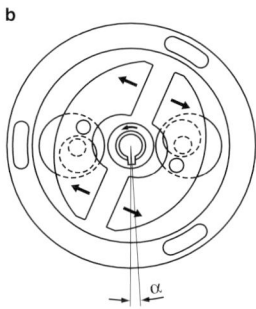

c

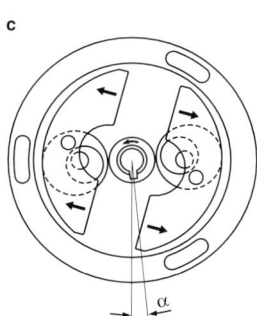

d

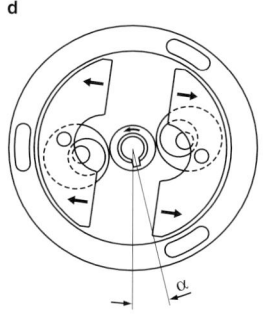

UMK0446Y

Standard in-line fuel-injection pumps

Design and construction

The timing device for the in-line injection pump is mounted directly on the end of the pump's camshaft. Basically, one differentiates between the open-type and the closed-type of timing device.

The "closed" timing device has its own lube-oil reservoir which makes it independent of the engine's lube-oil circuit.

The "open" design on the other hand is connected directly to the engine's lube-oil circuit. Its housing is screwed to a toothed gear, and the compensating and adjusting eccentrics are mounted in the housing so that they are free to pivot. The compensating and adjusting eccentrics are guided by a pin which is rigidly connected to the housing. Apart from costing less, the "open" type has the advantage of needing less room and of being more efficiently lubricated.

Operating principle

The timing device is driven by a toothed gear which is accomodated in the engine timing case. The connection between the input and drive output (hub) is through interlocking pairs of eccentric elements (Fig. 6). The largest of these, the adjusting eccentric elements (4), are located in holes in the backing disk (8) which in turn is bolted to the drive element (1). The compensating eccentric elements (5) fit in the adjusting eccentric elements (4) and are guided by them and the hub bolt (6). On the other hand the hub bolt is directly connected with the hub (2). The flyweights (7) engage with the adjusting eccentric element (4) and are held in their starting positions by progressive springs (Fig. 7).

Sizes

The size of the timing device as defined by its outside diameter and depth, determines the installable flyweight mass, the distance between centers of gravity, and the possible flyweight travel. These three factors also define the timing device's work capacity and the application range.

Pump sizes

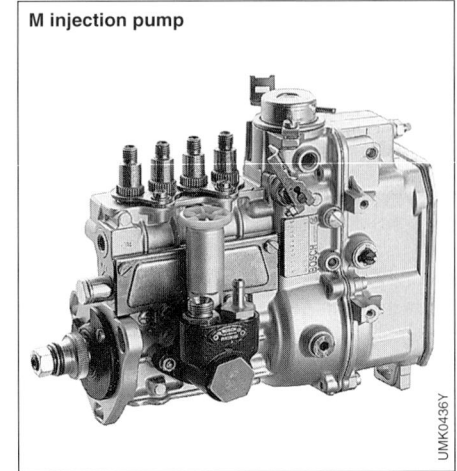

M injection pump

Fig. 8

Size M pump

The size M injection pump (Figs. 8 and 9) is the smallest pump in the PE range of in-line pumps. It has a light-metal housing and is fastened to the engine with a flange.

Access to the interior of the pump is possible after removal of a base plate and a side cover, and the M pump is therefore referred to as an "open-type" pump. Peak injection pressure is limited to 400 bar.

After removing the pump's side cover, the delivery quantities of the plunger-and-barrel assemblies can be adjusted and equalized.

Individual adjustment is by shifting the clamping pieces on the control rod.

During operation, the pump-plunger settings and with them the delivery quantities are adjusted by the control rod within the range defined by the pump construction. The M pump control rod is a flattened round-steel bar on which are mounted the grooved clamping pieces. A lever is firmly attached to each pump-barrel control sleeve, and the pin rivetted to its end engages with the groove of the control-rod clamping piece. This design is known as lever control.

The pump plungers are in direct contact with the roller tappets, and the prestroke adjustment is by selection of tappet rollers with the appropriate diameter.

Lubrication of the M pump is through a common lube-oil supply with the engine. The M pump is available with either 4, 5, or 6 pump barrels (4, 5, or 6-cylinder pump), and is only for use with diesel fuel.

Fig. 9

M injection pump (section drawing).

1 Delivery valve, **2** Pump barrel, **3** Control-sleeve lever, **4** Control rod, **5** Clamping piece, **6** Roller tappet, **7** Camshaft, **8** Camshaft lobe.

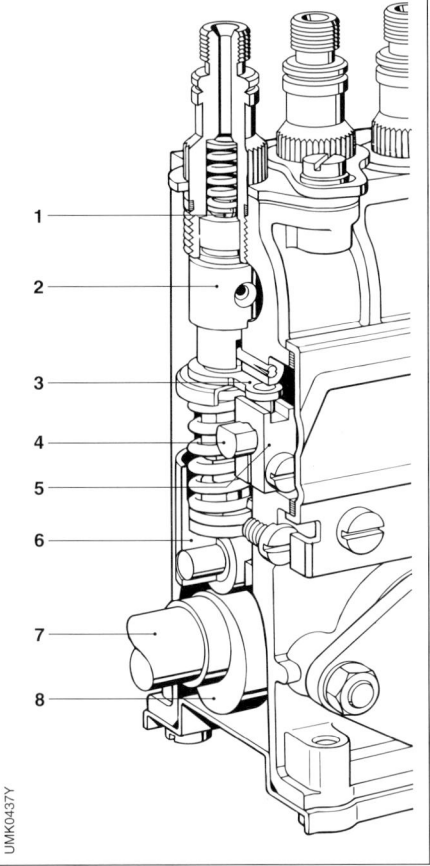

Size A pump

The size A in-line injection pumps (Figs. 10 and 11) with their larger delivery ranges follow directly after the size M pumps.

This pump also has a light-metal housing and can be attached to the engine by flange or cradle mounting.

The A pump is also of the "open" design, and the pump barrels are inserted directly into the aluminum housing from above, whereby the delivery-valve assembly is pressed against the pump housing by the delivery-valve holder. The sealing pressures, which are far in excess of the hydraulic delivery pressures, must be absorbed by the pump housing. For this reason, the A pump's peak injection pressure is limited to 600 bar.

In contrast to the M pump, the A pump is provided with an adjusting screw (with locknut) in each roller tappet for setting the prestroke.

For adjusting the delivery quantity by means of the control rack, the A pump also differs from the M pump in being equipped with pinion control, instead of lever control. A gear segment clamped on the plunger's control sleeve engages with the control rack, and to adjust the plunger-and-barrel assemblies for equal delivery quantities, the locking screws must be released and the control sleeve turned relative to the gear segment and thus relative to the control rack.

All adjustment work on this type of pump must be carried out with the pump at standstill and the housing open. Similar to the M pump, the A pump has a side-mounted spring-chamber cover which must be removed in order to gain access to the pump's interior.

For lubrication, the pump is connected to the engine's lube-oil circuit.

The A pump is available in versions with up to 12 cylinders, and in contrast to the M pump is suitable for multifuel operation.

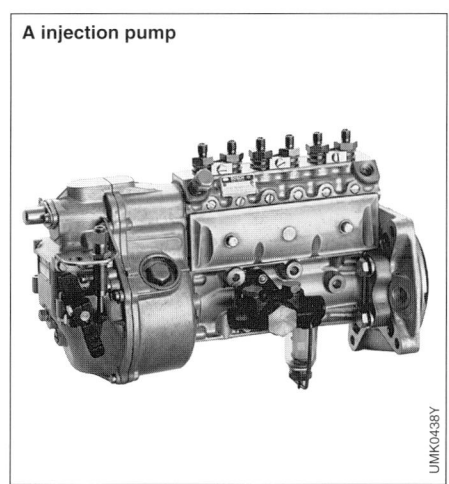

A injection pump

Fig. 10

Fig. 11

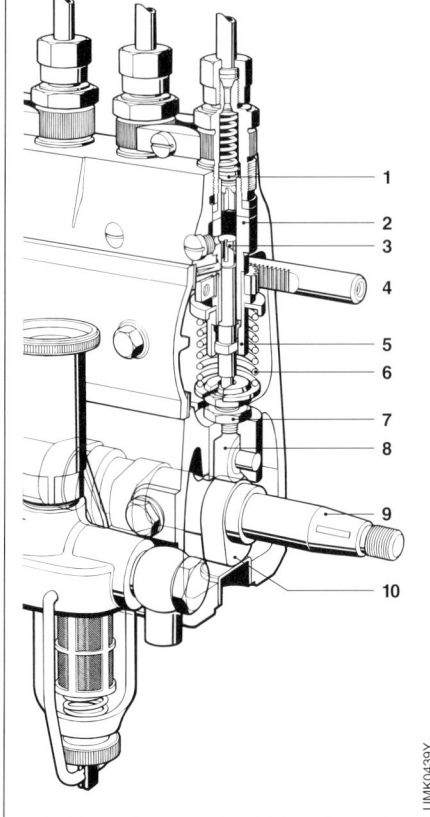

A injection pump (section drawing).

1 Delivery valve, **2** Pump barrel,
3 Pump plunger, **4** Control rack, **5** Control sleeve,
6 Plunger return spring, **7** Adjusting screw,
8 Roller tappet, **9** Camshaft, **10** Camshaft lobe.

Size MW pump

The MW in-line injection pump was developed to satisfy the need for higher peak injection pressures (Figs. 12 and 13).

The MW pump is a "closed-type" in-line injection pump, and its peak injection pressure is limited to 900 bar. It also has a light-metal housing and is fastened to the engine using either cradle-, flatbed-, or flange mounting.

The design of the MW pump differs considerably to that of the A and M pumps. The major difference being the use of a plunger-and-barrel assembly comprising the pump barrel, delivery valve, and delivery-valve holder. This is assembled outside the pump and inserted into the pump housing from above. On the MW pump, the delivery-valve holder is screwed directly into the pump barrel which has been extended upwards. The prestroke is adjusted by shims of varying thicknesses which are inserted between the housing and the barrel-and-valve assembly. The adjustment of uniform delivery between the individual plunger barrels is carried out from outside the pump by turning the barrel-and-valve assemblies. The assembly fastening flanges are provided with slots for this purpose.

The pump plunger's position remains unchanged when the barrel-and-valve assembly is turned.

The MW pump is available in versions with up to max. 8 barrels (8 cylinders) and is suitable for a variety of different mounting methods.

It operates with diesel fuel, and lubrication is through the engine's lube-oil circuit.

MW injection pump

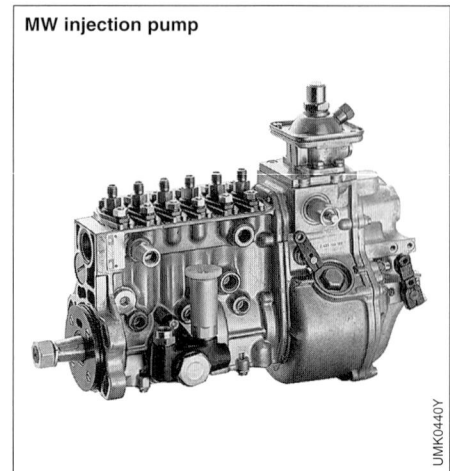

Fig. 12

Fig. 13

MW injection pump (section drawing).
1 Fastening flange for the plunger-and-barrel assembly, **2** Delivery valve, **3** Pump barrel, **4** Pump plunger, **5** Control rack, **6** Control sleeve, **7** Roller tappet, **8** Camshaft, **9** Camshaft lobe.

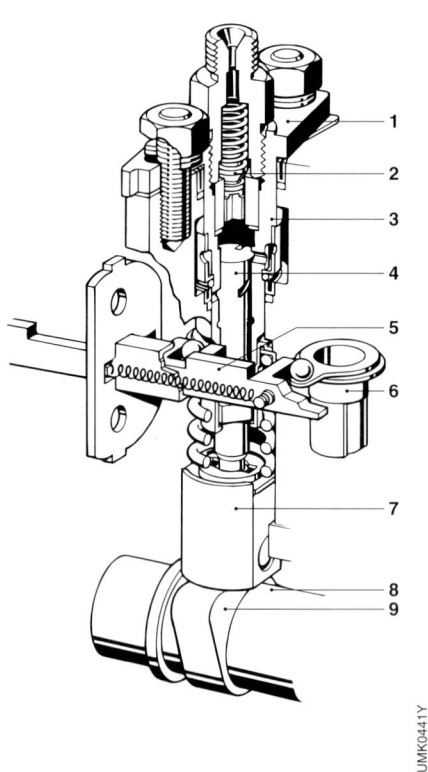

Size P pump

The size P in-line injection pump was also developed to provide higher peak injection pressures (Figs. 14 and 15). Similar to the MW pump, it is of the "closed" type and is fastened to the engine using a base or flange mounting. In the case of P pumps designed for a peak injection pressure of 850 bar, the pump barrel is inserted in a flange bushing which is already provided with the threads for the delivery-valve holder. With this version of pump-barrel installation, the sealing forces put no load on the pump housing. Prestroke adjustment is performed the same as with the MW pump.

Low-pressure in-line injection pumps use conventional fuel-gallery flushing. Here, the fuel flows through the fuel galleries of the individual barrels one after another, and in the direction of the pump's longitudinal axis. Fuel enters the galleries at the fuel inlet and leaves at the fuel return. Taking the P8000 version of the P pump, which is designed for pump-side injection pressures of up to 1150 bar as an example, this flushing method would lead to an excessive fuel-temperature gradient (of up to 40 °C) inside the pump between the first and last barrel. Since the fuel's energy density decreases along with its increasing temperature and its resulting increase in volume, this would lead to the injection of differing "quantities of energy" into the engine's combustion chambers. These injection pumps therefore use "cross-flushing", a method whereby the fuel galleries of the individual pump barrels are separated from each other by means of throttling points. This means that they can be flushed parallel to each other (at right angles to the pump's longitudinal axis) under practically identical temperature conditions.

This injection pump is also connected to the engine's lube-oil circuit for lubrication. The P pump is also available in versions for up to 12 barrels (cylinders), and is suitable for diesel fuel as well as for multifuel operation.

P injection pump

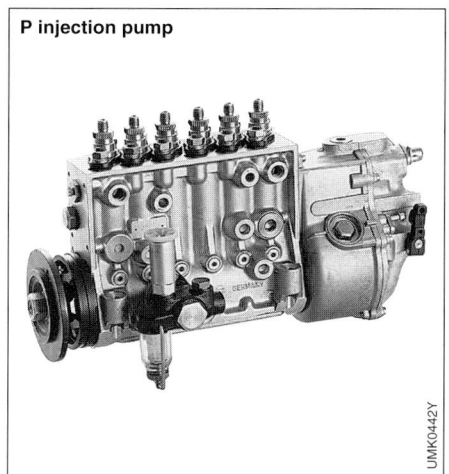

Fig. 14

Fig. 15

P injection pump (section drawing).

1 Delivery valve, **2** Pump barrel, **3** Control rod, **4** Control sleeve, **5** Roller tapper, **6** Camshaft, **7** Camshaft lobe.

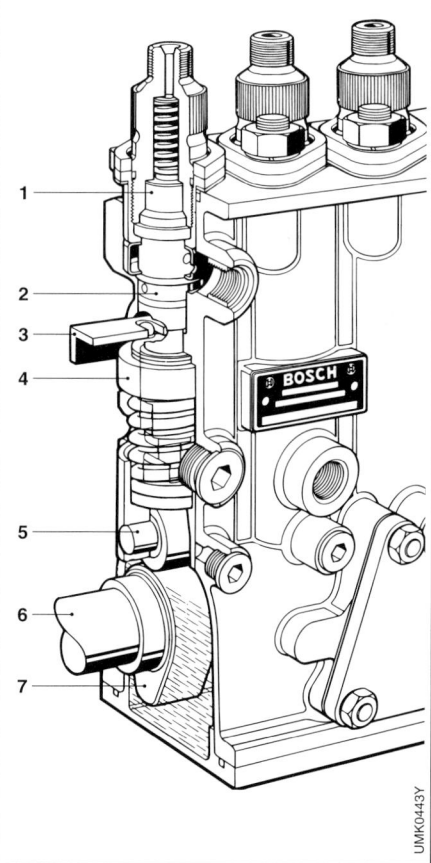

PE in-line fuel-injection pumps for alternative fuels

Application, design, and construction

Specially designed diesel engines can operate on alternative fuels, whereby one differentiates between:
- Multifuel engines which in addition to diesel fuel, can also run on gasoline (petrol) or kerosene,
- Alcohol engines which operate with methanol or ethanol (ethyl alcohol),
- Engines which run on biomass fuels.

The change from one type of fuel to another necessitates adaptive measures in the fuel-metering in order to avoid excessive variations in power output. The most important fuel characteristics are: Viscosity, boiling point, lubricity, density, and self-ignition point. Constructional measures are necessary at the fuel-injection equipment and at the engine in order that these characteristics can be optimally matched to each other.

If diesel fuels with a considerably reduced sulphur content are used with Bosch in-line fuel-injection pumps no negative effects are to be expected. Alternative fuels though have a lower boiling point which means that the injection pump's fuel gallery must be flushed more intensively and at higher pressures. A special fuel-supply pump is available for this purpose.

Subassemblies

Injection pumps for multi-fuel engines
In the case of low-density fuels (gasoline/petrol), the full-load delivery is increased by means of a switchable control-rod stop. On the other hand, in order to prevent fuel losses when low-viscosity fuels are used, the pump's plunger-and-barrel assemblies are provided with an oil block in the form of two ring-shaped grooves in the pump barrel, the upper groove being connected with the pump's fuel gallery through a passage. During the effective (working) stroke, the fuel that leaks between the plunger and the barrel wall flows through this passage and back to the fuel gallery.

There is an inlet passage in the bottom groove through which engine lube-oil is forced into the groove under pressure via a fine filter. At normal engine operating speeds, this pressure exceeds the fuel pressure in the fuel gallery so that the plunger-and-barrel assembly is sealed-off effectively. A non-return valve prevents fuel entering the lube-oil circuit when the oil pressure drops below a certain level at idle.

Injection pumps for alcohol engines
The in-line injection pumps are also suitable for operation with methanol and ethanol (ethyl alcohol) provided they have been modified beforehand. Such modifications include:

- Fitting special seals and gaskets,
- Providing special protection for surfaces in contact with alcohol,
- Fitting rustproof springs,
- Operating with special lubricants.

In order that the engine receives an amount of energy equivalent to that provided by diesel fuel, the delivery quantity for methanol is 2.3 times higher and that for ethanol 1.7 times higher. And compared to diesel fuel, considerably more wear must be expected at the delivery-valve seats and nozzle-needle seats.

Injection pumps for biomass fuels
Serious complications in the diesel fuel-injection installation are not to be expected when an engine is run with rape oil (RME). Although, depending upon the specifications of the fuel in use special measures may be required.

PE in-line control-sleeve fuel-injection pumps

Increasing attention is being paid to reducing the toxic content of commercial-vehicle exhaust gases, and engine designers are concentrating on preventing the generation of toxic substances at the source, or at least on reducing it.

With commercial-vehicle diesel engines, high injection pressures and optimum start-of-injection timing are making major contributions towards achieving this target. A direct result of these endeav-

ours has been the development of a new generation of high-pressure injection pumps, the in-line control-sleeve injection pumps (Fig. 1).

Design and construction

Due to its "control sleeve" which can be moved up and down on the pump plunger, the control-sleeve injection pump differs from the conventional in-line injection pump both in operating principle and in design and construction. The control sleeve used with this pump supersedes the timing device as described in the Chapter "Injection timing" on Page 16. Technically speaking the design and con-

Fig. 1

In-line control-sleeve injection pump

1 Pump barrel, **2** Control sleeve, **3** Control rod, **4** Pump plunger, **5** Camshaft,
6 Port-closing actuator solenoid, **7** Control-sleeve setting shaft, **8** Rod-travel actuator solenoid,
9 Inductive rod-travel sensor, **10** Plug-in connection, **11** Disk for port-closing block and for oil return pump.

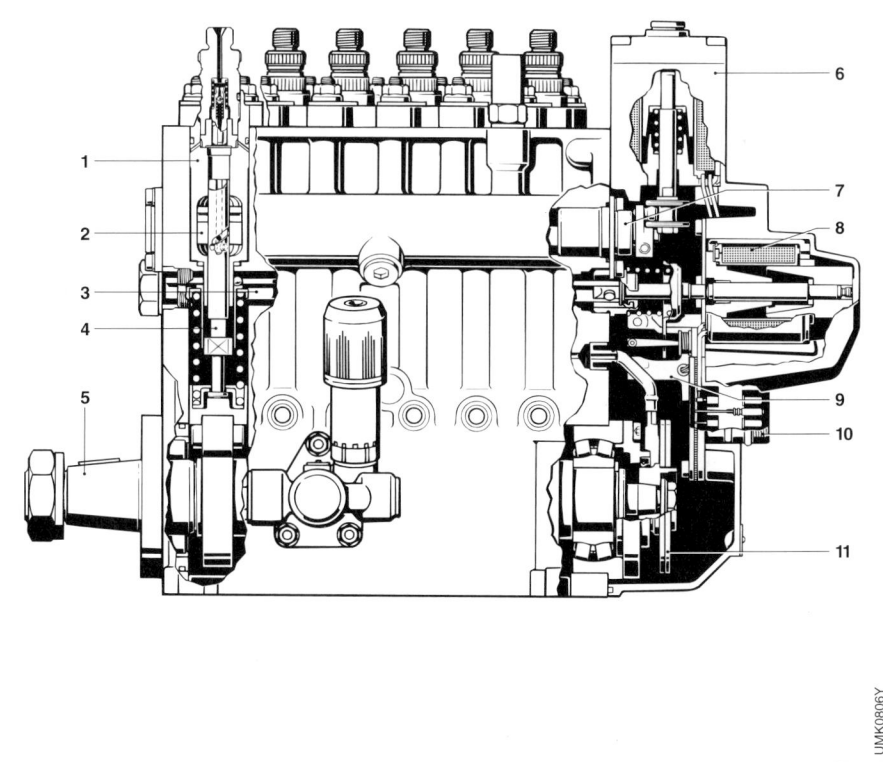

UMK0806Y

struction of the rest of the pump has remained unchanged. The in-line control-sleeve injection pump operates with injection pressures of approx. 1150 bar. Thanks to the technology used, it is possible to freely program the start-of-delivery timing. The essential feature of this pump principle is that the start-of-delivery timing is practically independent of injected fuel quantity, and is implemented by adjustments carried out simultaneously at all control-sleeve elements.

This means that the former "rigid" timing device attached to the end of the camshaft and designed to handle the injection pump's high driving torques is no longer necessary (Fig. 2).

Fig. 2

Control-sleeve adjustment mechanism
1 Pump plunger, **2** Control sleeve,
3 Control-sleeve shaft, **4** Control rod.

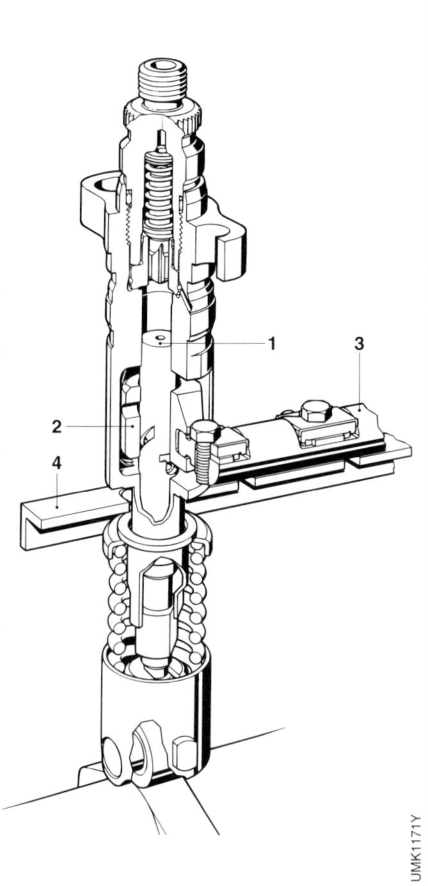

Operating principle

In order to change the start of delivery (port closing) or start of injection, the pre-stroke is adjusted by means of a control sleeve which can be moved up and down on the pump plunger. This means that compared to the conventional in-line injection pump, electronics has been introduced as a second method to govern the injection pump.

Each pump plunger is equipped with a control sleeve which incorporates the spill port. All sleeves are adjusted simultaneously by the control-sleeve levers which engage with them through "windows" in the barrel. The levers are fastened to the control-sleeve shaft. Depending upon the control sleeve's vertical position, start of delivery (port closing) begins sooner or later relative to the camshaft lobe. The injected fuel quantity is metered using the port-and-helix control familiar from the conventional in-line injection pump.

The in-line control-sleeve injection pump is an element in the electrical servo-system with which start-of-injection and injected fuel quantity can be programmed as a function of a variety of different influencing variables. This form of control permits pollutant emissions to be reduced to a minimum (important for instance, with respect to the severe US limits), optimization of fuel economy in all operating states, precise fuel metering, and further improvement of the starting phase and in particular of the warm-up phase.

Electronic control

Operating-data acquisition
A number of sensors are attached at various points on the control-sleeve in-line injection pump itself, on the vehicle's engine, and on the vehicle. These measure the temperature of the air, the fuel, and the engine, as well as the air pressure, the intake-air quantity, the engine speed, and the accelerator-pedal position (driver input). The sensors convert these environmental and operating

UMK1171Y

parameters into electrical signals which are inputted into the ECU.

Operating-data processing

A computer-based electronic control unit (ECU) uses the incoming input data to generate the desired value for the pump setting.

The ECU then forwards this information to the pump in the form of electrical signals through the various solenoid actuators in the pump's actuator system.

The setpoint injected fuel quantity outputted by the ECU is set by means of the rod-position control loop. The ECU specifies a setpoint rod-travel and receives the feedback of the actual rod position via the rod-travel sensor. To complete the closed control loop, the ECU repeatedly calculates the current which must be inputted to the injection pump's actuator system in order for the setpoint value and the actual value to conform with each other. Safety considerations dictate that a return spring is fitted to bring the control rod back to the "zero-delivery" position when no current is applied to the actuator system.

Start of delivery (port closing) is also adjusted using a closed control loop. One of the nozzle holders is equipped with a needle-motion sensor which signals the start-of-injection to the computer-based ECU which then calculates the actual value for the moment of injection, taking into account the camshaft position. This is then compared with a setpoint value, and by means of current control the computer-controlled port-closing actuator system is adjusted so that actual and setpoint values are identical. Since the port-closing actuator system is "structurally stable", a special position check-back signalling unit is not required. Structurally stable means that the lines of force of the solenoid actuator and of the return spring always have a clear intersection point, so that the travel of the solenoid actuator is proportional to the applied current. This is equivalent to closing the control loop.

Fuel delivery

Port closing (start of delivery)

As soon as the pump plunger has travelled a certain distance towards TDC, the bottom edge of the control sleeve closes the pump plunger's spill port. Pressure can now build up in the high-pressure chamber above the plunger, and fuel delivery commences.

Port opening (end of delivery)

Following the plunger's remaining stroke to TDC, the pump-plunger helix and the spill port in the control sleeve terminate the fuel delivery. Port opening and with it the injected fuel quantity can be varied by rotating the plunger by means of the control rod.

Port-closing (start of delivery) adjustment

The control sleeve must be moved in the direction of TDC (or BDC) in order to adjust the port closing and with it the start of injection. Whereas moving the control sleeve to a position nearer to TDC results in increased prestroke and therefore in delayed start of delivery (injection), moving it nearer to BDC reduces the prestroke and advances the start of delivery (injection).

Depending upon the shape of the camshaft lobe, not only the delivery velocity is changed but also the delivery rate (theoretical quantity of fuel delivered per degree camshaft) and the injection pressure.

Fuel supply and delivery

Fuel tank

The fuel tank must be of noncorroding material, and must remain free of leaks at double the operating pressure and in any case at 0.3 bar. Suitable openings or safety valves must be provided, or similar measures taken, in order to permit excess pressure to escape of its own accord. Fuel must not leak past the filler cap or through pressure-compensation devices. This applies when the vehicle is subjected to minor mechanical shocks, as well as when cornering, and when standing or driving on an incline.

The fuel tank and the engine must be so far apart from each other that in case of an accident there is no danger of fire. In addition, special regulations concerning the height of the fuel tank and its protective shielding apply to vehicles with open cabs, as well as to tractors and buses.

Fuel filter

The service life of a diesel fuel-injection system is determined to a large extent by the high quality of the fuel filter and by regular servicing and maintenance. The pressure-generating elements in the plunger-and-barrel assembly, and the nozzles themselves, have been manufactured to an accuracy of a few thousandths of a millimeter and are matched precisely to each other. This means that their correct functioning is endangered if the fuel reaching them contains contaminants of this magnitude. In other words, if the fuel is not adequately filtered, the fuel-injection components are damaged and subjected to premature wear. Further negative results are:

– Inefficient combustion,
– High fuel consumption,
– Poor starting,
– Rough idle, and
– Reduced engine power.

Fuel filter (Two-stage box-type filter).

Fig. 1

Fig. 2

Multi-stage filter
(With spiral V-form filter element).

1 Filter cover with mounting, **2** Course filter,
3 Fine filter.

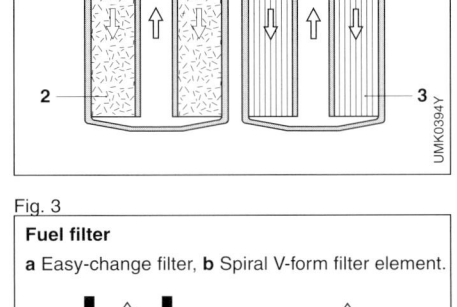

Fig. 3

Fuel filter
a Easy-change filter, **b** Spiral V-form filter element.

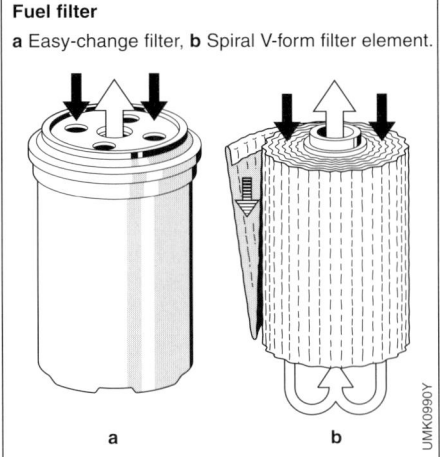

a   b

This is why the trouble-free operation of the fuel-injection system, and consequently of the diesel engine, necessitates the fuel being perfectly filtered. For this purpose, special filters are required which have been designed in accordance with the in-line injection pump's specific requirements (Figs. 1, 2, 3). The filter element is a spiral V-shaped paper element with a pore size of 8 µm. In addition to single-stage filters, multi-stage filters (with higher filtering efficiency) and parallel filters (larger active filtering surface) are also available for special applications. The filters are mounted using a variety of different filter covers (with straight flange or angle flange), and there are a number of different connection possibilities.

Since the contamination removed by the filter remains in the filter, it is imperative that the stipulated filter-exchange intervals are complied with. Even under the most critical operational and servicing conditions, it is a simple matter to replace an easy-change filter, a step that excludes the possibility of damage to the system due to inadequate filtering. An electrical heater can be installed to ensure trouble-free winter operation.

Gravity-feed fuel tank

Usually, gravity-feed fuel tanks (operation without fuel-supply pump) are used with agricultural tractors and small diesel engines. With this method, the fuel flows through the filter to the injection pump solely due to the effects of gravitational force. If there is very little difference in height between fuel tank and fuel filter or injection pump, it is recommended that larger-diameter fuel pipes are used so that an adequate supply of fuel can be guaranteed. In the case of gravity feed, a shutoff cock should be installed in the line between fuel tank and fuel filter, so that the fuel supply can be switched off during maintenance and/or repair work, thus making it unnecessary to drain the fuel tank.

Fuel-supply pumps

A fuel-supply pump must be fitted in vehicles with inadequate gradient between fuel tank and injection pump, or when the fuel lines are too long. Normally, the fuel-supply pump is flanged directly to the injection pump. Depending upon the operating conditions and upon engine-specific factors, different fuel-line configurations are necessary, two of which are shown in Figs. 4 and 5.

It the fuel tank is in the immediate vicinity of the engine, the engine's radiated heat can lead to the formation of vapor bubbles inside the fuel-line system. To prevent this, the injection pump's fuel gallery is continually flushed with fuel so that the pump is cooled. With this line configuration, excess fuel is returned to the fuel tank through the overflow valve and the return line.

Fig. 4

Diesel fuel-injection system
With overflow valve fitted to the injection pump.

1 Fuel tank,
2 Fuel-supply pump,
3 Fuel filter,
4 Fuel-injection pump,
5 Injection nozzle,
6 Overflow valve.

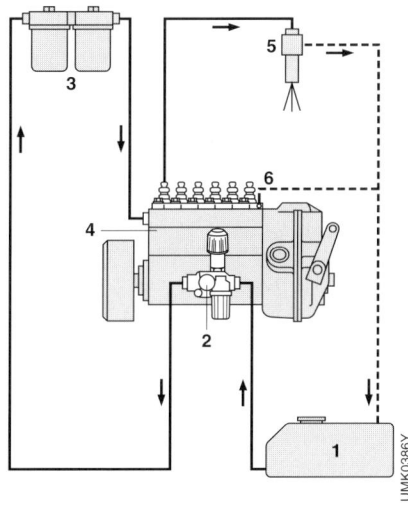

If, in addition, the under-hood temperatures are very high, a fuel-line configuration as per Fig. 5 is possible. Here, the fuel filter is equipped with an overflow restriction through which during operation some of the fuel returns to the tank and takes gas and vapor bubbles with it. Gas bubbles which form in the injection pump's fuel gallery are returned to the fuel tank by the excess fuel leaving the injection pump through the overflow valve and the fuel-return line.

Fuel is drawn from the fuel tank by the fuel-supply pump and transferred under pressure through the fuel filter and into the injection pump's fuel gallery. In the majority of cases, mechanical piston pumps are used which are mounted on the injection pump, or more rarely on the engine.

A cam on the injection pump or engine camshaft drives the supply-pump piston through a spring-loaded roller tappet. In addition to the fuel-supply pumps dealt with in this publication, there are also electrical supply pumps on the market as well as models for multifuel operation.

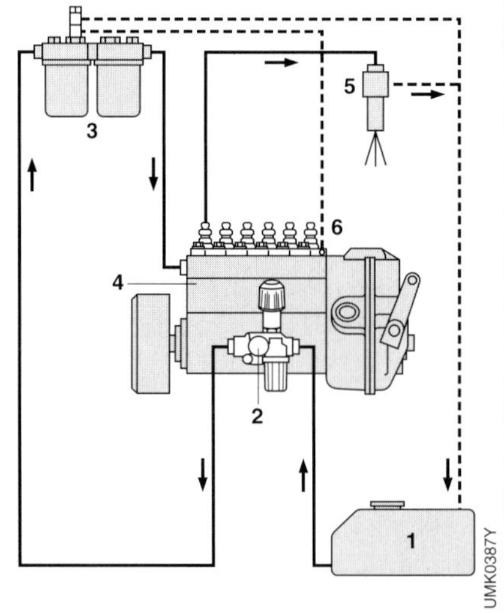

The choice of the correct fuel-supply pump is determined by the following criteria:
- The type of fuel-injection pump,
- The supply pump's required delivery rate,
- The fuel-line configuration,
- The space available in the engine compartment.

Single-acting or double-acting fuel-supply pumps can be used depending upon the amount of fuel required by the engine.

Single-acting fuel-supply pumps
The single-acting fuel-supply pump uses the throughflow principle and is available for use with pump sizes M, A, MW, and P (Fig. 6).

Via the sliding tappet, the rotating cam forces the fuel-pump piston upwards against the force of the spring. The vacuum generated by piston movement in the pump's fuel gallery causes the suction valve integrated in the piston to

Fig. 5

Diesel fuel-injection system
Additional overflow restriction on the fuel filter.

1 Fuel tank,
2 Fuel-supply pump,
3 Fuel filter with overflow restriction,
4 Fuel-injection pump,
5 Injection nozzle,
6 Overflow valve.

Fig. 6

Fuel-supply pump (single-acting).

1 O-ring, **2** Spring seat,
3 Aluminum pump housing, **4** Suction valve,
5 Tappet body, **6** Sliding tappet,
7 O-ring, **8** O-ring, **9** Pump piston,
10 Spacer ring, **11** Pressure connection,
12 Delivery valve, **13** Spring,
14 Spring seat, **15** Suction connection.

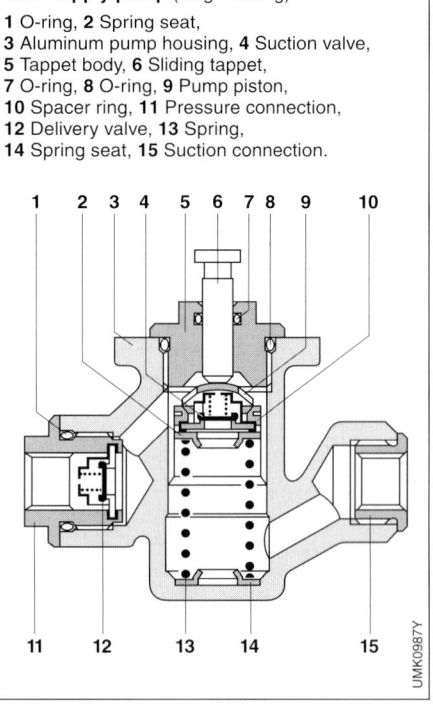

open, and the fuel can enter the fuel gallery between suction valve and pressure valve.

When the camshaft rotates far enough for the spring to force the plunger back down again, the suction valve closes and the pressure valve opens. Fuel is forced through the high-pressure line to the injection pump (Fig. 8).

Double-acting fuel-supply pumps

Double-acting fuel-supply pumps have a higher delivery rate than their single-acting counterparts and are used with injection pumps having large numbers of barrels (cylinders) and correspondingly higher delivery quantities. They are suitable for use with P- and ZW-pumps (Fig. 7).

In contrast to the single-acting pump, the double-acting pump delivers fuel to the injection pump during both pump-plunger strokes. In other words, twice per camshaft revolution (Fig. 9).

Hand pumps
The hand (primer) pump has the following functions:

– To fill the injection system's intake side before the system is taken into operation for the first time,
– For refilling and bleeding the system after repair or maintainance work,
– For refilling and bleeding the system after the vehicle's fuel tank has run dry.

The hand pump is usually integrated in the fuel supply pump, although there are versions which are installed in the line between the fuel tank and the supply pump.

Preliminary filters
The function of the preliminary filter is to protect the supply pump from course contaminants. Under rough operating conditions, for instance when refuelling takes place from barrels, we recommend that a strainer is also installed, either in the fuel tank or in the line to the fuel-supply pump.

The preliminary filter can either be integrated in the fuel-supply pump, or installed at the supply-pump inlet or between the fuel tank and the supply pump.

Fig. 7

Fuel-supply pump (double-acting).

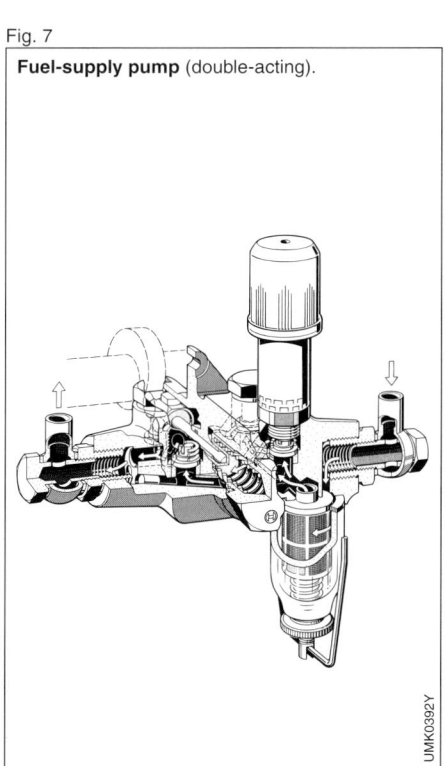

UMK0392Y

Fuel-supply pump. Operating principle (single-acting).

a Cam stroke, **b** Spring stroke.
1 Drive eccentric, **2** Camshaft, **3** Pressure chamber, **4** Suction chamber.

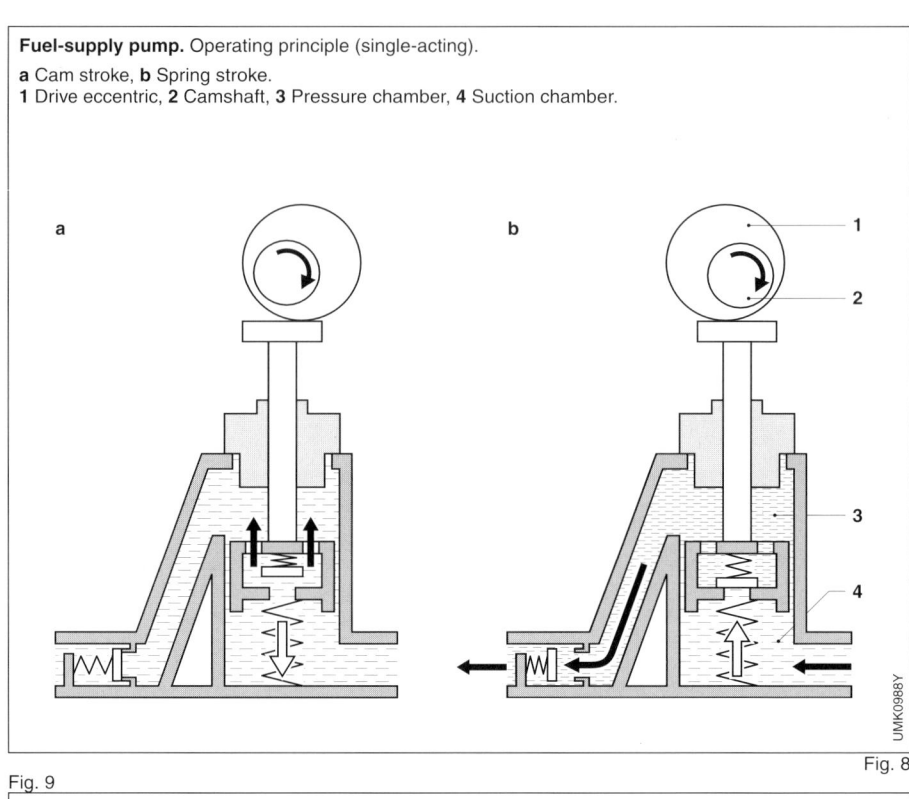

Fig. 8

Fig. 9

Fuel-supply pump. Operating principle (double-acting).

a Cam stroke, **b** Spring stroke.
1 Camshaft, **2** Drive eccentric, **3** Pressure chamber, **4** Suction chamber.

UMK0989Y

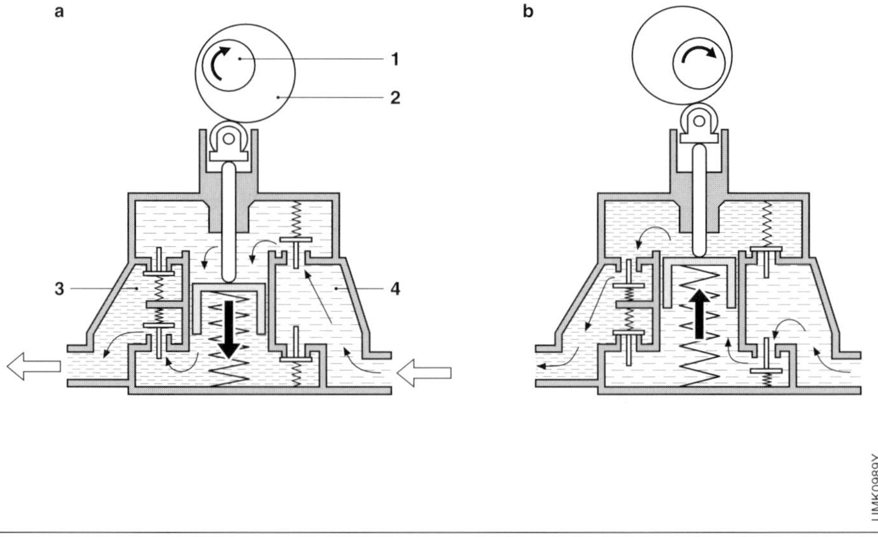

Fuel lines/high-pressure delivery lines

Steel tubing must be used for the high-pressure delivery lines in the fuel-injection system's high-pressure stage. On the other hand, flame-inhibiting, steel-braid-armored flexible fuel lines can be used for the low-pressure stage. These must be routed to ensure that they cannot be damaged mechanically, and fuel which has dripped or evaporated must not be able to accumulate nor must it be able to ignite. In the high-pressure stage, the high-pressure delivery lines represent the connection between the fuel-injection pump and the nozzles. Apart from being routed without sharp bends, they must be as short as possible, and their bend radii must be at least 50 mm.

On vehicle engines, the high-pressure lines are normally fastened with clamping pieces at regular intervals. This means that external vibrations are not trans-ferred to the pressure lines, or only to a limited degree. Seamless steel tubing is used for the high-pressure lines.

These can have different dimensions depending upon pump size, and with respect to their length, internal diameter, and wall thickness they must be matched to the injection process. This results in specified line dimensions which must be precisely complied with.

The pipe sealing cone is attached to the end of the high-pressure line (Fig. 10). Special delivery lines are required for high-pressure fuel injection (with nozzle pressures of up to 1400 bar). The compression pulsating fatigue strength of these lines depends upon the material used and upon the maximum peak-to-valley height of the internal wall roughness.

It is also possible to install specially treated high-pressure delivery lines. In order to increase their internal strength, these lines are bent to the required shape and subjected to very high pressure (up to 3800 bar) which is then suddenly released. This leads to material compression on the inner walls with a resulting increase in strength.

Fig. 10

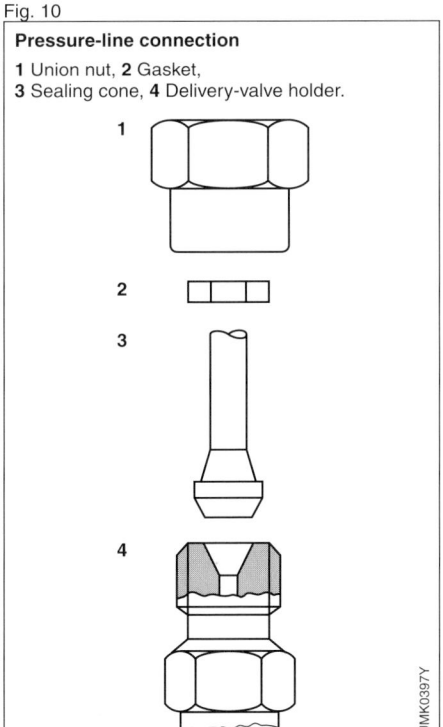

Pressure-line connection
1 Union nut, **2** Gasket,
3 Sealing cone, **4** Delivery-valve holder.

1
2
3
4

UMK0397Y

Operation of the in-line fuel-injection pump

For trouble-free operation, the injection pump must be correctly adjusted, it must be bled completely, it must be connected to the engine lube-oil system, and it must be timed to the engine.

These steps are an absolute necessity if the diesel engine's optimal fuel-consumption/power-output ratio is to be achieved, and if compliance with increasingly severe exhaust-gas legislation is to be ensured. These facts make an injection-pump test bench indispensable. The specified adjustment values for the injection pump are ascertained during engine trials.

Pump adjustment

On the test bench

The adjustment of the plunger-and-barrel assemblies for the same prestroke and for identical delivery quantity, as well as the adjustment of the governor and timing device is carried out on the Bosch injection-pump test bench. The test benches are equipped with all the necessary measuring devices and the drive speed is infinitely variable. The repair and test instructions appropriate to the test bench in question, together with the test specifications, contain all the necessary data for service and repair work.

Fig. 1

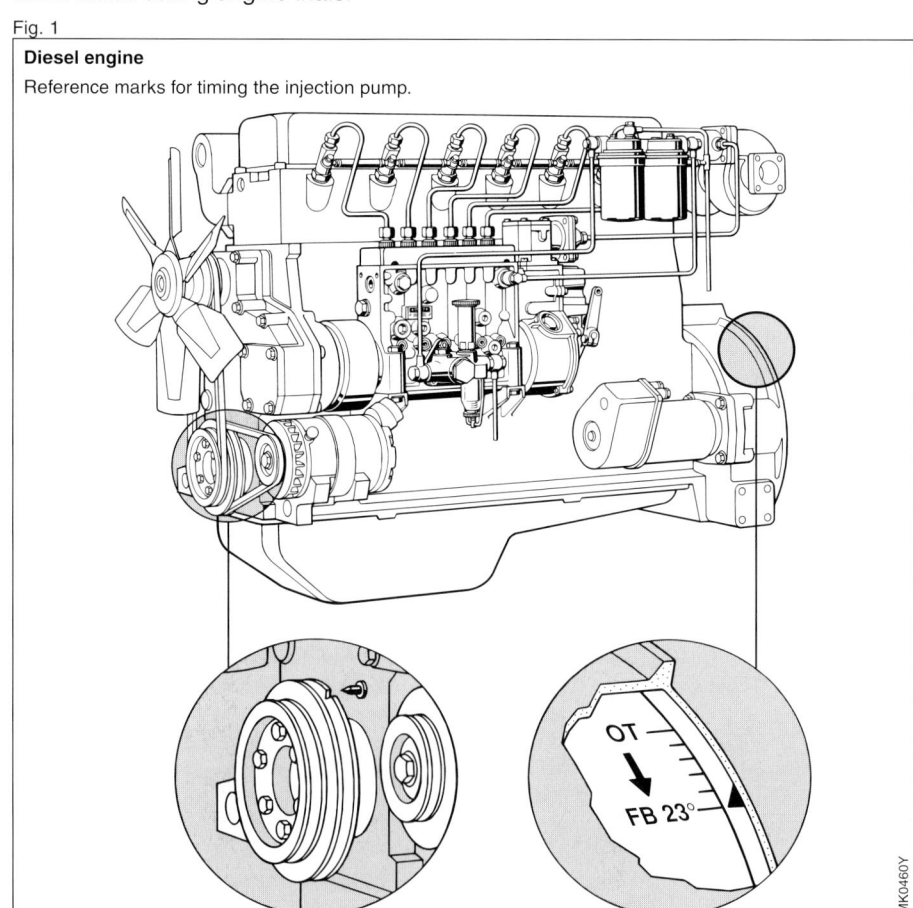

Diesel engine

Reference marks for timing the injection pump.

UMK0460Y

On the engine

The injection pump is timed to the engine with the help of the reference marks for start of injection (port closing). These markings are on the engine as well as on the injection pump (Fig. 1).

Usually, the engine's compression stroke is used as the basis for the timing adjustments, although for reasons specific to the engine other possibilities may be considered. It is therefore important that the manufacturer's instructions are observed. In most cases, the port-closing reference mark is on the engine's flywheel, on its V-belt pulley or on its vibration damper. There are a number of possibilities for adjusting the injection pump to the correct port-closing (start-of-injection) setting:

– The fuel-injection pump is delivered with its camshaft locked in a given position. After being bolted to the engine with the engine's crankshaft in the appropriate position, the pump camshaft is released. This well-proven method is inexpensive and is gaining more and more in popularity.

– The fuel-injection pump is equipped with a port-closing indicator on the governor end which must be aligned with the reference marks when the injection pump is mounted.

– There is a port-closing mark on the timing device or on the clutch, which must be brought into alignment with a mark on the pump housing. This method though is not as accurate as the two methods described above.

– After the injection pump has been mounted on the engine, the high-pressure overflow method is applied at one of the pump outlets in order to find the port-closing point (the instant in time when the pump plunger closes the inlet port). This "wet" method is also being increasingly superseded by the first two methods described above.

Bleeding the injection system

Air bubbles in the fuel can impair injection-pump operation, or even make it impossible. Therfore, installations which are to be taken into service for the first time, or which have been shut down temporarily, must be bled thoroughly.

If the fuel-supply pump is equipped with a hand (primer) pump, this is used to fill the suction line, delivery line, fuel filter, and fuel-injection pump with fuel. Whereby the vent screws on the filter cover and the fuel-injection pump are to remain open until the fuel flows out completely free of bubbles.

The installation is to be bled every time the filter has been changed or other work carried out on the system.

During actual operation, the installation automatically bleeds itself via the overflow valve (Fig. 2) on the Bosch fuel filter (permanent venting). A restriction is used instead if the pump is not equipped with an overflow valve.

Fig. 2

Fuel-injection system
With flushing of the pump's fuel gallery.

1 Vent screw,
2 Overflow valve, overflow restriction.

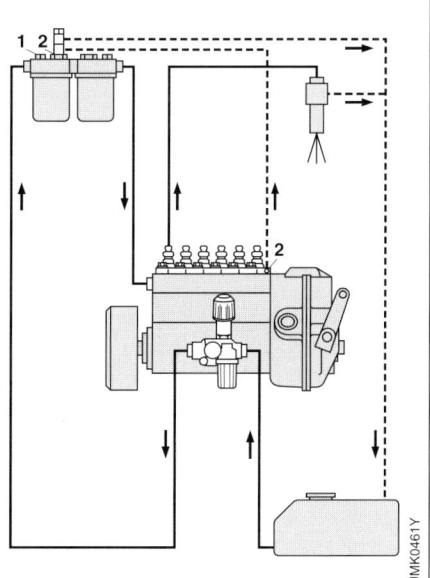

UMK0461Y

Lubrication

The injection pump and the governor are best connected to the engine's lube-oil circuit, because with this form of lubrication the injection pump remains maintenance-free.

Filtered engine oil is supplied to the injection pump and the governor through a pressure line and an inlet passage via the roller-tappet gap, or by means of a special oil supply valve. In the case of base and cradle-mounted injection pumps, the lube-oil return to the engine is via a return line (Fig. 3). In the case of front-flange mounting, lube-oil return can take place through the camshaft bearing or through special passages.

Before being taken into operation for the first time, the injection pump and the governor must be filled with the same oil as the engine. In the case of injection pumps without direct connection to the engine's lube-oil circuit, the oil is poured in through the cover after removal of the bleeder cap or the bleeder filter. The pump oil level is checked along with the engine-oil change (at the oil-change intervals prescribed by the engine manufacturer). To check the pump oil level the oil-level screw on the governor is removed. Excess oil (increased by leakage oil) is drained off, and if oil is missing fresh oil is added. When the injection pump is removed, or when the engine receives a major overhaul, the lube-oil must be replaced. For oil-level checks, injection pumps and governors with separate oil supplies are provided with their own dipstick.

Shutting down the pump for an extended period

If the engine, and therefore the fuel-injection pump, are to remain out of service for an extended period of time, no diesel fuel is to remain in the pump because with time this becomes gummy and the pump plungers and delivery valves would tend to stick and might even corrode. For this reason, before shut-down about 10% of a reliable rust-inhibiting oil is to be added to the fuel in the fuel tank, the same proportion being added to the oil in the injection pump's camshaft chamber. The engine is then run for about 15 minutes during which time the last remains of the normal diesel fuel are flushed out of the injection pump which at the same time the pump is efficiently safeguarded against gumming and corrosion. New injection pumps which have already been provided with effective corrosion protection at the factory are marked with a "p".

Fig. 3

Injection-pump (Lubrication).

a Return via drive-side bearing plate, **b** Return via return line.

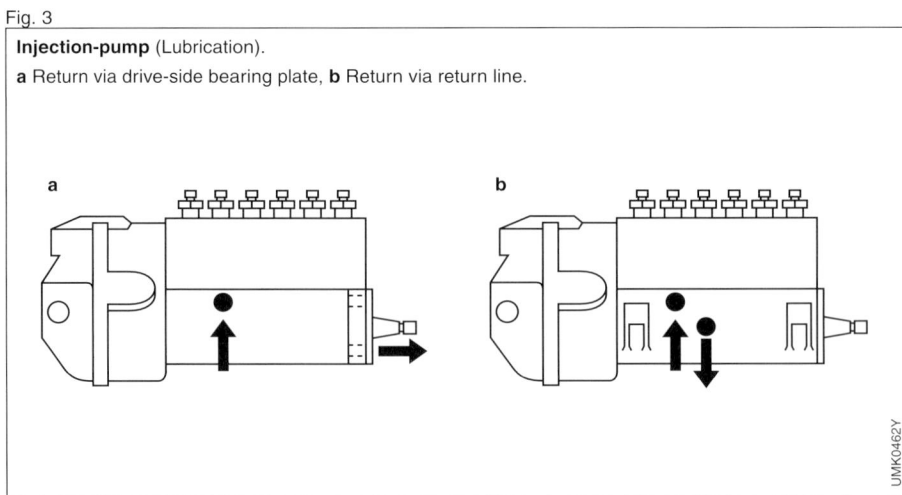

The diesel story

It is common knowledge that the "Diesel" is an economical long-lasting vehicle. But few people are aware of the fact that it was the fuel-injection pump from Robert Bosch which finally put the diesel engine on the road.

Rudolf Diesel presented his invention to the public in 1895:
An engine with compression ignition.
This new engine had a number of distinct advantages compared to the already well-established spark-ignition engine: On the one hand it used considerably less fuel, which in any case was relatively cheap compared to the fuel for the SI engine, and on the other it could be designed for far higher power outputs.
Diesel's invention quickly established itself, and soon was the automatic choice for marine engines and stationary engines. The diesel engine had one great disadvantage though – it could not be run at higher speeds. But as the diesel engine became more popular, and its advantages increasingly well known, more and more demand arose for a small, high-speed version.
The fuel supply to the engine was the major obstacle in the search for more speed. With the method used at that time, in which fuel was blown into the engine with compressed air, any worthwhile increase in engine speed was impossible. Furthermore, the "air pump" was a complicated piece of equipment which could not be substantially reduced in size or weight.
It was at the end of 1922 that Robert Bosch decided to develop a fuel-injection system for diesel engines. The technical foundations had been laid:

Bosch already had considerable experience with the internal-combustion engine, production engineering was highly developed and, above all, the experience gained with the manufacture of lubricating pumps could be used to advantage.
In early 1923, about a dozen different injection-pump designs had been drawn up, and the first trials were performed on an engine in the middle of 1923. In Summer 1925, the final design was approved and in 1927 the first series-production fuel-injection pumps left the factory.
This fuel-injection pump developed by Bosch finally put the diesel engine on the road and paved the way for a breakthrough of unforeseen proportions. The diesel engine became increasingly popular and came to the forefront in more and more applications, particularly in the automotive sector.
Since then, there has been no let-up in the development of the diesel engine and its fuel-injection system. At the start of the sixties, the distributor injection pump with automatic timing device developed by Bosch gave fresh impetus to the diesel engine. A decade later, following extensive research work, the Electronic Diesel Control (EDC) from Bosch was ready for series production.
Precision metering of the minutest quantities of fuel, in the right place at the right time, made possible by decades of innovative work at Bosch, solved the diesel's problems of long ago. And the Diesel is still unsurpassed when it comes to fuel economy and fuel efficiency.
Rudolf Diesel's vision has become reality.

Nozzles and nozzle holders

The injection nozzles and their respective nozzle holders are vitally important components situated between the in-line injection pump and the diesel engine. Their assignments are as follows:
– Metering the injection of fuel,
– Management of the fuel,
– Defining the rate-of-discharge curve,
– Sealing-off against the combustion chamber.

Considering the wide variety of combustion processes and the different forms of combustion chamber, it is necessary that the shape, "penetration force", and atomization of the fuel spray injected by the nozzle are adapted to the prevailing conditions. This also applies to the injection time, and the injected fuel quantity per degree camshaft.

Since the design of the nozzle-holder combination makes maximum use of standardized components and assemblies, this means that the required flexibility can be achieved with a minimum of components. The following nozzles and nozzle holders are used with in-line injection pumps:
– Pintle nozzles (DN..) for indirect-injection (IDI) engines, and
– Hole-type nozzles (DLL../DSLA..) for direct-injection (DI) engines.
– Standard nozzle holders (single-spring nozzle holders), with and without needle-motion sensor, and
– Two-spring nozzle holders, with and without needle-motion sensor.

Pintle nozzles

Application

Pintle nozzles are used with in-line injection pumps on indirect-injection engines (pre-chamber and whirl-chamber engines).

In this type of diesel engine, the air/fuel mixture is for the most part formed by the air's vortex work. The injected fuel spray serves to support this mixture-formation process.

The following types of pintle nozzle are available:
– Standard pintle nozzles (Fig. 1),
– Throttling pintle nozzles, and
– Flat-cut pintle nozzles (Fig. 2).

Design and construction

All pintle nozzles are of practically identical design, the only difference being in the pintle's geometry:

Standard pintle nozzles

On the standard pintle nozzle, the nozzle needle is provided with a pintle which extends into the injection orifice of the nozzle body in which it is free to move with a minimum of play. The injection spray can be matched to the engine's requirements by appropriate choice of dimensions and pintle designs.

Fig. 1

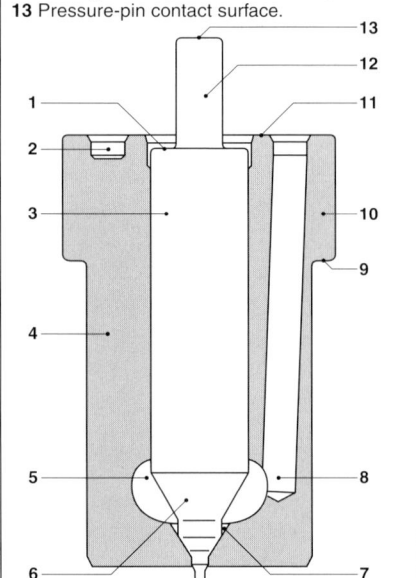

Standard pintle nozzle

1 Lift stop surface, **2** Ring groove, **3** Needle guide, **4** Nozzle-body shaft, **5** Pressure chamber, **6** Pressure shoulder, **7** Seat lead-in, **8** Inlet port, **9** Nozzle-body shoulder, **10** Nozzle-body collar, **11** Sealing surface, **12** Pressure shaft, **13** Pressure-pin contact surface.

Throttling pintle nozzle

The throttling pintle nozzle is a pintle nozzle with special pintle dimensions. The special pintle design serves to define the shape of the rate-of-discharge curve. When the nozzle needle lifts, it first of all opens a small annular gap so that only a small amount of fuel is injected (throttling effect).

As needle lift increases (due to pressure rise), the spray orifice is opened increasingly until the major portion of the injection (main injection) takes place towards the end of needle lift. Since the pressure in the combustion chamber rises less sharply, this shaping of the rate-of-injection curve leads to "softer" combustion. This results in quieter combustion in the part-load range. In other words, it is possible to shape the required rate-of-discharge curve by means of the pintle shape, the characteristic of the nozzle needle's spring, and the throttling gap.

Flat-cut pintle nozzles

This nozzle's pintle has a ground surface which opens a flow cross-section in addition to the annular gap when the pintle opens (only slight needle lift). The resulting increased flow volume prevents deposits forming in this flow channel. This is the reason why flat-cut pintle nozzles coke-up far less, and any coking which does take place is more uniform. The annular gap between spray orifice and throttling pintle is very small (less than 10 μm). Very often, the flat-cut pintle surface is parallel to the nozzle-needle axis. Referring to Fig. 3, with an additional inclined cut on the pintle, the gradient of the injected-fuel-quantity curve's flat portion (Curve 2) can be increased so that the transition to full nozzle opening is less abrupt. Specially shaped pintles, such as the "radius" or "profile surface" types, can be applied to match the flow curve to engine-specific requirements. Part-load noise and vehicle driveability are both improved as a result.

Fig. 2

Flat-cut pintle nozzle

a Side view, **b** Front view.
1 Needle seat, **2** Nozzle-body floor,
3 Throttling pintle, **4** Flat cut, **5** Injection orifice,
6 Profiled pintle, **7** Total overlap,
8 Cylindrical overlap, **9** Nozzle-body seat.

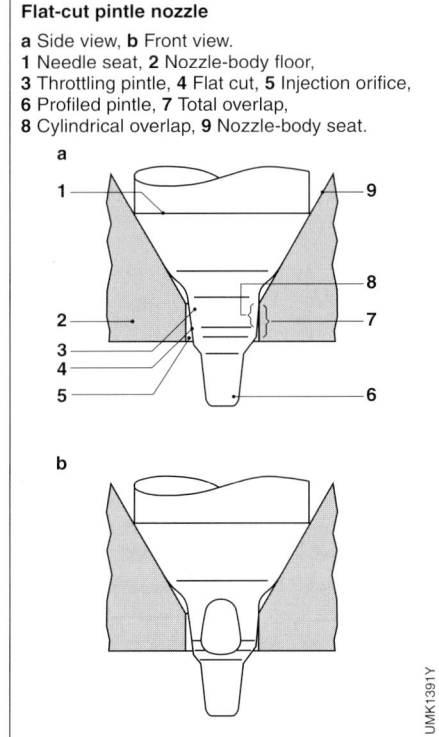

Fig. 3

Flow quantity as a function of needle lift and nozzle version

1 Throttling pintle nozzle,
2 Throttling pintle nozzle with inclined cut on pintle (flat-cut pintle nozzle)

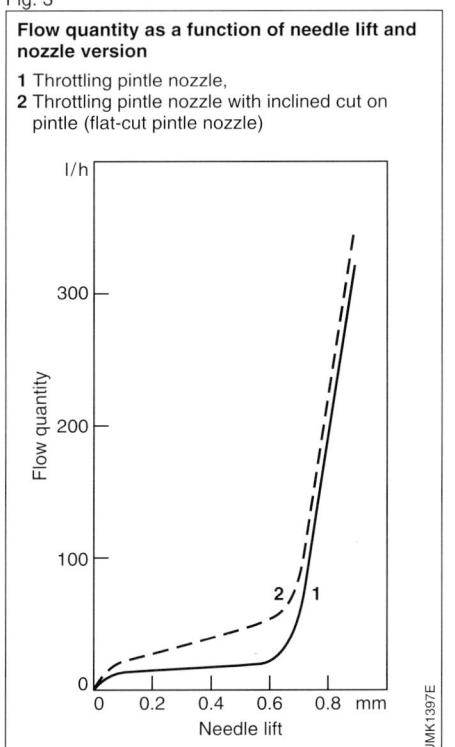

Hole-type nozzles

Application

Hole-type nozzles are used with in-line injection pumps on direct-injection engines.

One differentiates between:
- Sac-hole, and
- Seat-hole nozzles.

The hole-type nozzles also vary according to their size:
- Type P with 4 mm needle diameter, and
- Type S with 5 and 6 mm needle diameters.

Design and construction

The spray holes are located on the envelope of a spray cone (Fig. 4). The number of spray holes and their diameter depend upon:
- The injected fuel quantity,
- The combustion-chamber shape, and
- The air swirl in the combustion chamber.

The input edges of the spray holes can be rounded by hydro-erosive (HE) machining.

Fig. 4

Spray cone

γ Spray-cone offset angle, δ Spray cone.

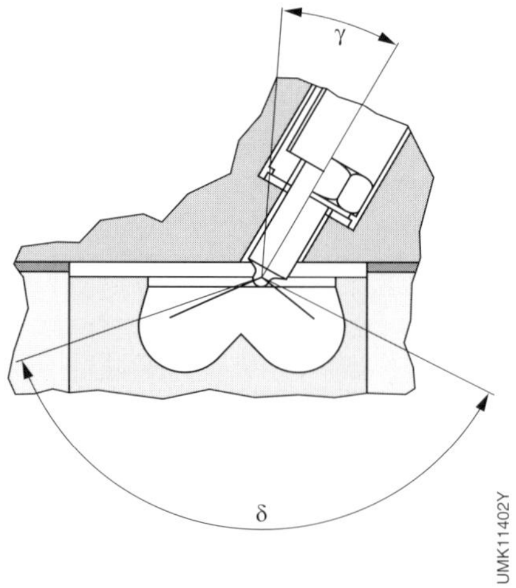

At those points where high flow rates occur (spray-hole entrance), the abrasive particles in the hydro-erosive (HE) medium cause material loss.

This so-called HE-rounding process can be applied to both sac-hole and seat-hole nozzles, whereby the target is:
- Prevent in advance the edge wear caused by abrasive particles in the fuel, and/or
- Reduce the flow tolerance.

For low hydrocarbon emissions, it is highly important that the volume filled with fuel (residual volume) below the edge of the nozzle-needle seat is kept to a minimum. Seat-hole nozzles are therefore used.

Designs

Sac-hole nozzle

The spray holes of the sac-hole nozzle (Fig. 5) are arranged in the sac hole.

In the case of a round nozzle tip (Fig. 6a), depending upon design the spray holes are drilled mechanically or by means of electrochemical machining (e.c.m.).

Sac-hole nozzles with conical tip (Figs. 6b and 6c) are always drilled using e.c.m. methods.

Sac-hole nozzles are available
- With cylindrical, and
- Conical sac holes

in a variety of different dimensions.

Sac-hole nozzle with cylindrical sac hole and round tip (Fig. 6a):

This nozzle's sac hole has a cylindrical and a semispherical portion, and permits a high level of design freedom with respect to
- Number of spray holes,
- Spray-hole length, and
- Injection angle.

The nozzle tip is semispherical, and together with the shape of the sac hole, ensures that the spray holes are of identical length.

Sac-hole nozzle with cylindrical sac hole and conical tip (Fig. 6b):

This type of nozzle is used exclusively with spray-hole lengths of 0.6 mm. The tip's conical shape enables the wall thickness to be increased between the throat radius and the nozzle-body seat with an attending improvement of nozzle-tip strength.

Sac-hole nozzle with conical sac hole and conical tip (Fig. 6c):

Due to the conical shape of this nozzle's sac hole, its volume is less than that of a nozzle with cylindrical sac hole. The volume is between that for a seat-hole nozzle and a sac-hole nozzle with cylindrical sac hole. In order to achieve uniform tip-wall thickness, the tip's conical design corresponds to that of the sac hole.

Fig. 5

Sac-hole nozzle

1 Pressure shaft, **2** Needle-lift stop face,
3 Inlet passage, **4** Pressure shoulder,
5 Needle shaft, **6** Nozzle tip,
7 Nozzle-body shaft, **8** Nozzle-body shoulder,
9 Pressure chamber, **10** Needle guide,
11 Nozzle-body collar, **12** Locating hole,
13 Sealing surface,
14 Pressure-pin contact surface.

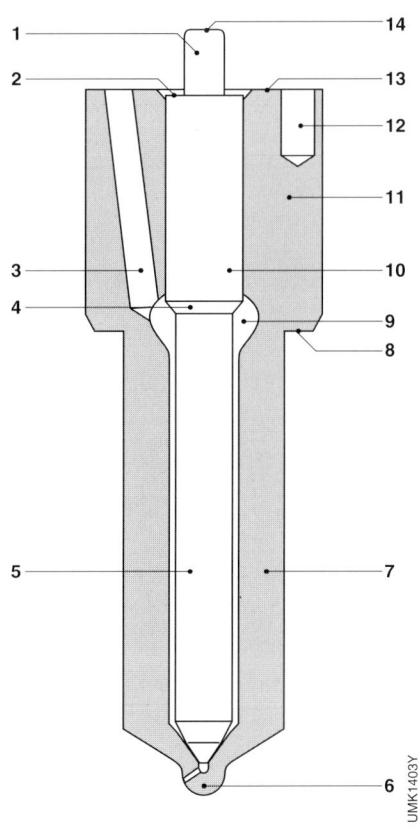

Fig. 6

Sac-hole shapes

a Cylindrical sac hole with round tip,
b Cylindrical sac hole with conical tip,
c Conical sac hole with conical tip.
1 Shoulder, **2** Seat entrance, **3** Needle seat,
4 Needle tip, **5** Injection orifice,
6 Injection-orifice entrance, **7** Sac hole,
8 Throat radius, **9** Nozzle-tip cone,
10 Nozzle-body seat, **11** Damping cone.

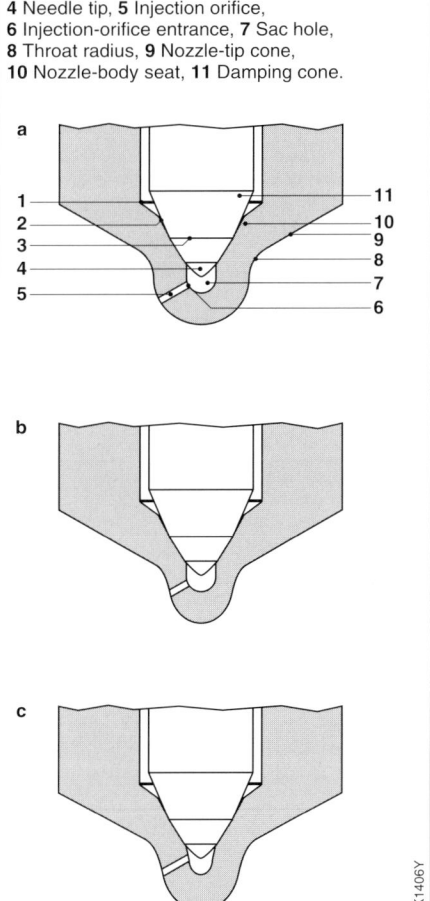

Seat-hole nozzle

In order to minimise the residual volume – and therefore the HC emissions – the start of the spray hole is located in the seat taper, and with the nozzle closed it is covered almost completely by the nozzle needle. This means that there is no direct connection between the sac hole and the combustion chamber (Figs. 7 and 8). The sac-hole volume here is much lower than that of the sac-hole nozzle. Compared to sac-hole nozzles, seat-hole nozzles have a much lower loading limit and are therefore only manufactured as Size P with a spray-hole length of 1 mm.

For reasons of strength, the nozzle tip is conically shaped. The spray holes are always formed using e.c.m. methods.

Fig. 7

Seat-hole nozzle

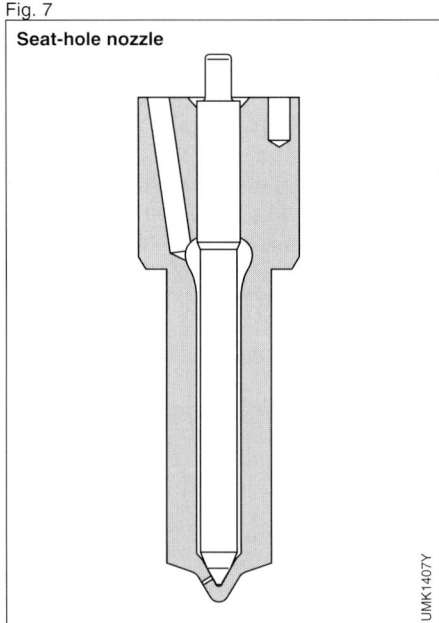

Fig. 8

Seat-hole nozzle: Tip shape

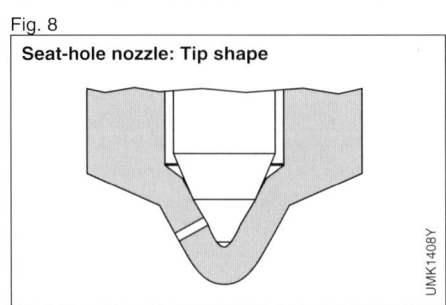

Standard nozzle holders

Assignments and designs

Nozzle holders with hole-type nozzles in combination with an in-line injection pump are used on DI engines.

With regard to the nozzle holders, one differentiates between
– Standard nozzle holders (single-spring nozzle holders) with and without needle-motion sensor, and
– Two-spring nozzle holders, with and without needle-motion sensor.

Application

The nozzle holders described here have the following characteristics:
– Cylindrical external shape with diameters between 17 and 21 mm,
– Bottom-mounted springs (leads to low moving masses),
– Pin-located nozzles for direct-injection engines, and
– Standardised components (springs, pressure pin, nozzle-retaining nut) make combinations a practical matter.

Design

The nozzle-and-holder assembly comprises the injection nozzle and the nozzle holder.

The nozzle holder contains the following components (Fig. 9):
– Nozzle-holder body,
– Intermediate element,
– Nozzle-retaining nut,
– Pressure pin,
– Spring,
– Shim, and
– Locating pins.

The nozzle is centered in the nozzle body and fastened using the nozzle-retaining nut. When nozzle body and retaining nut are screwed together, the intermediate element is forced up against the sealing surfaces of nozzle body and retaining nut. The intermediate element serves as the needle-lift stop and with its locating pins centers the nozzle in the nozzle-holder body.

The nozzle-holder body contains the
– Pressure pin,
– Spring, and
– Shim.

The spring is centered in position by the pressure pin, whereby the pressure pin is guided by the nozzle-needle's pressure shaft.
The nozzle is connected to the injection pump's high-pressure line via the nozzle-holder feed passage, the intermediate element, and the nozzle-body feed passage. If required, an edge-type filter can be installed in the nozzle holder.

Method of operation
The nozzle-holder spring applies pressure to the nozzle needle through the pressure pin. The spring's initial tension defines the nozzle's opening pressure which can be adjusted using a shim.
On its way to the nozzle seat, the fuel passes through the nozzle-holder inlet passage, the intermediate element, and the nozzle nody. When injection takes place, the nozzle needle is lifted by the injection pressure and fuel is injected through the injection orifices into the combustion chamber. Injection terminates as soon as the injection pressure drops far enough for the nozzle spring to force the nozzle needle back onto its seat.

Two-spring nozzle holders

Application
The two-spring nozzle holder is a further development of the standard nozzle holder, and serves to reduce combustion noise particularly in the idle and part-load ranges.

Design
The two-spring nozzle holder features two springs located one behind the other. At first, only one of these springs has an influence on the nozzle needle and as such defines the initial opening pressure. The second spring is in contact with a stop sleeve which limits the needle's initial stroke.

Fig. 9

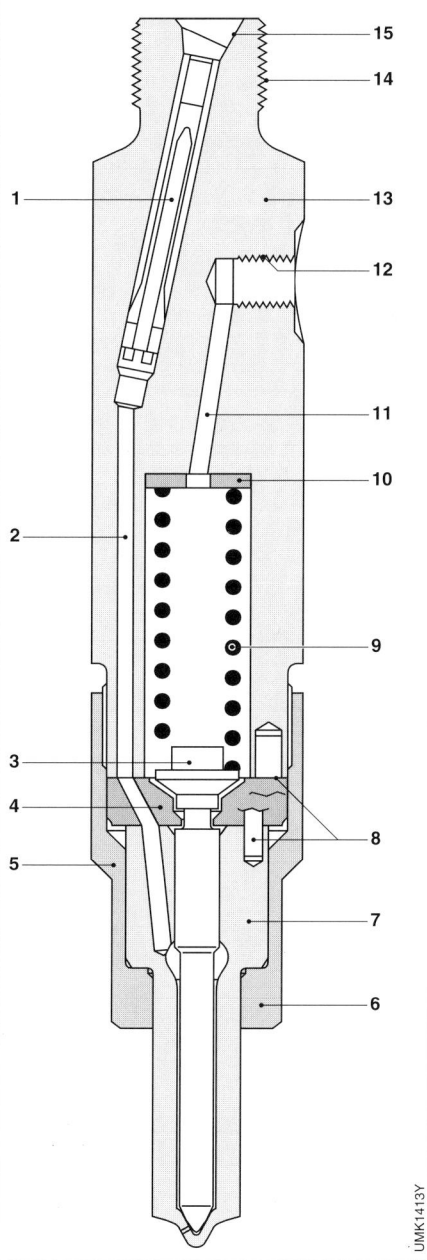

Standard nozzle holder

1 Edge-type filter, **2** Inlet passage,
3 Pressure pin, **4** Intermediate element,
5 Nozzle-retaining nut, **6** Wall thickness,
7 Nozzle, **8** Locating pins, **9** Spring,
10 Shim, **11** Leak-fuel passage,
12 Leak-fuel connection thread,
13 Nozzle-holder body, **14** Connection thread,
15 Sealing cone.

UMK1413Y

Fig. 10

Two-spring nozzle holder for direct-injection (DI) engines

1 Nozzle-holder body, 2 Shim, 3 Spring 1,
4 Pressure pin, 5 Guide element, 6 Spring 2,
7 Pressure pin, 8 Spring seat, 9 Shim,
     10 Intermediate element,
     11 Stop sleeve,
     12 Nozzle needle,
     13 Nozzle-retaining nut,
     14 Nozzle body.
     h_1 Initial stroke,
     h_2 Main stroke.

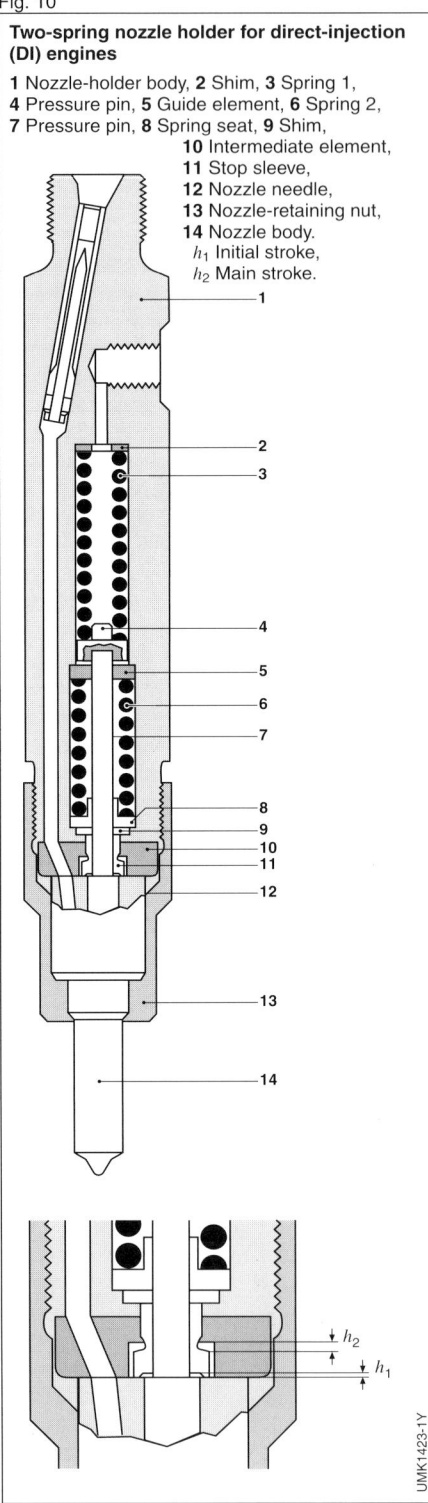

When strokes take place in excess of the initial stroke, the stop sleeve lifts and both springs have an effect upon the nozzle needle (Fig. 10).

Method of operation
During the actual injection process, the nozzle needle first of all opens an initial amount so that only a small volume of fuel is injected into the combustion chamber.
Along with increasing injection pressure in the nozzle holder though, the nozzle needle opens completely and the main quantity is injected (Fig. 11). This 2-stage rate-of-discharge curve leads to "softer" combustion and to a reduction in noise.

Nozzle holders with needle-motion sensor

Application
The start-of-injection point is an important parameter for optimum diesel-engine operation. For instance, its evaluation permits load and speed-dependent injection timing, and/or control of the exhaust-gas recirculation (EGR) rate.

Fig. 11

Comparison of needle-lift curves

a Standard nozzle holder (single-spring nozzle holder), **b** Two-spring nozzle holder.

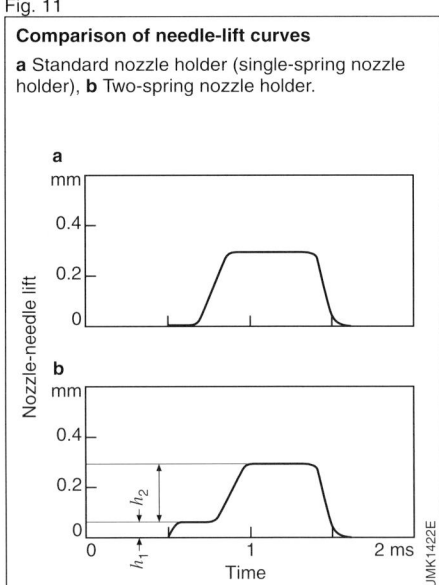

This necessitates a nozzle holder with needle-motion sensor (Fig. 13) which outputs a signal as soon as the nozzle needle opens.

Design

When the nozzle needle moves, the extended pressure pin enters the current coil.

The degree to which it enters the coil (overlap length "X" in Fig. 14) determines the strength of the magnetic flux.

Method of operation

The magnetic flux in the coil changes as a result of nozzle-needle movement and induces a signal voltage which is proportional to the needle's speed of movement but not to the distance it has travelled. This signal is processed directly in an evaluation circuit (Fig. 12).

When a given threshold voltage is exceeded, this serves as the signal to the evaluation circuit for the start of injection.

Fig. 13

Two-spring nozzle holder with needle-motion sensor for direct-injection (DI) engines

1 Nozzle-holder body, 2 Needle-motion sensor,
3 Spring 1, 4 Guide element, 5 Spring 2,
6 Pressure pin, 7 Nozzle-retaining nut.

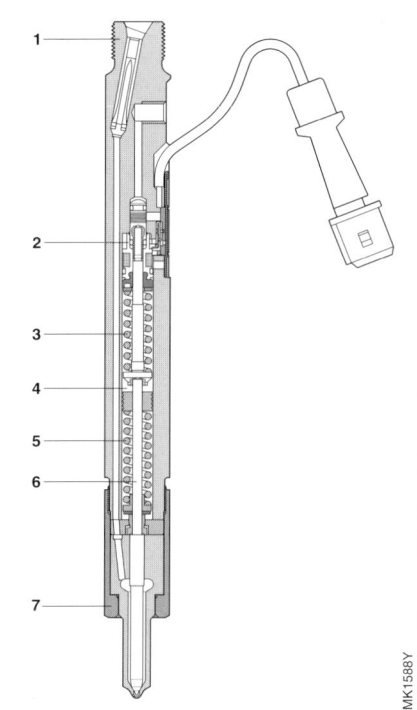

UMK1588Y

Fig. 12

Comparison between a needle-lift curve and the corresponding signal-voltage curve of the needle-motion sensor

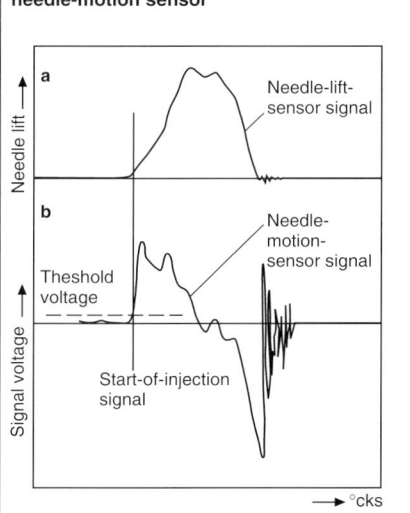

UMK1427E

Fig. 14

Needle-motion sensor in a two-spring nozzle holder for direct-injection (DI) engines

1 Adjusting pin, 2 Terminal,
3 Current coil, 4 Pressure pin,
5 Spring seat.
X Overlap length.

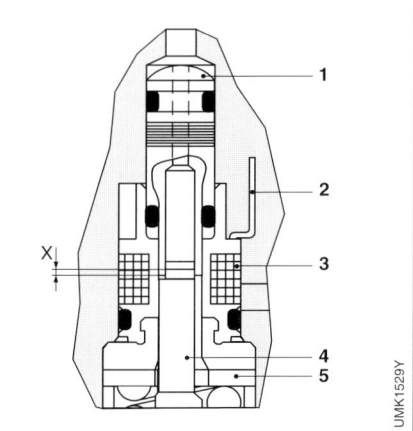

UMK1529Y

Mechanical governors for in-line fuel-injection pumps

Governing

Fuel injection

During its intake stroke, the diesel engine draws in only air. During the compression stroke, this air is heated to such a high temperature that the diesel fuel injected into the cylinder towards the end of the compression stroke ignites of its own accord. Fuel is metered to the engine by the fuel-injection pump, and is injected at high pressure through the injection nozzle into the combustion chamber.

Fuel injection must take place:
– In a precisely metered quantity according to the engine load,
– At the correct moment in time,
– For a precisely defined period of time, and
– In a manner suited to the particular combustion process concerned.

The fuel-injection pump and the governor connected to the control rack are responsible for these conditions being complied with. The amount of fuel injected per pump-plunger stroke is approximately proportional to the engine's torque.

Design, construction, and principle of operation of the Bosch PE in-line pump are described elsewhere.

If a mechanical (flyweight) governor is used in the vehicle, the control rack is connected with the accelerator pedal via the governor. With an electronic governor (EDC), the accelerator pedal is equipped with a sensor connected to the ECU. When the accelerator pedal is depressed, its movement is converted into the corresponding rack travel, with the momentary engine speed also being taken into account.

Why does the diesel engine need a governor?

With a diesel engine, there exists no single control-rack position which would permit the diesel engine to maintain its speed accurately without a governor. At idle, for example, without a governor the engine speed would either drop until the engine stalls, or it would continue to increase until the engine races, culminating in self-destruction.

The latter possibility is due to the diesel engine operating with an excess of air, meaning that there is no effective throttling of the cylinder charge as engine speed increases.

For instance, if a cold engine were started and left to run at idle while the initial fuel-delivery quantity continued to be injected, the engine's inherent friction would soon start to drop. The same applies to the drive resistance from engine-driven assemblies such as the alternator, air compressor, fuel-injection pump, etc.

This means that if the control-rack position were to remain unchanged, and the control rack were not retracted to reduce the fuel-delivery quantity (as a governor would do), the engine's speed would increase more and more (due to the above drop in friction) until it possibly reaches the point of self-destruction.

In other words, it is imperative that the diesel engine be equipped with a governor.

Nowadays, either mechanical (flyweight) governors or Electronic Diesel Control (EDC) are used for the in-line fuel-injection pumps.

Pneumatic governors, controlled by intake-manifold pressure, were formerly fitted to smaller injection pumps.

Fuel-injection system with mechanically governed PE in-line fuel-injection pump

1 Fuel tank, **2** Fuel-supply pump, **3** Fuel filter, **4** In-line fuel-injection pump, **5** Timing device, **6** Governor,
7 Nozzle holder with injection nozzle, **8** Fuel return line, **9** Sheathed-element glow plug (GSK),
10 Battery, **11** Glow-plug and starter switch, **12** Glow control unit.

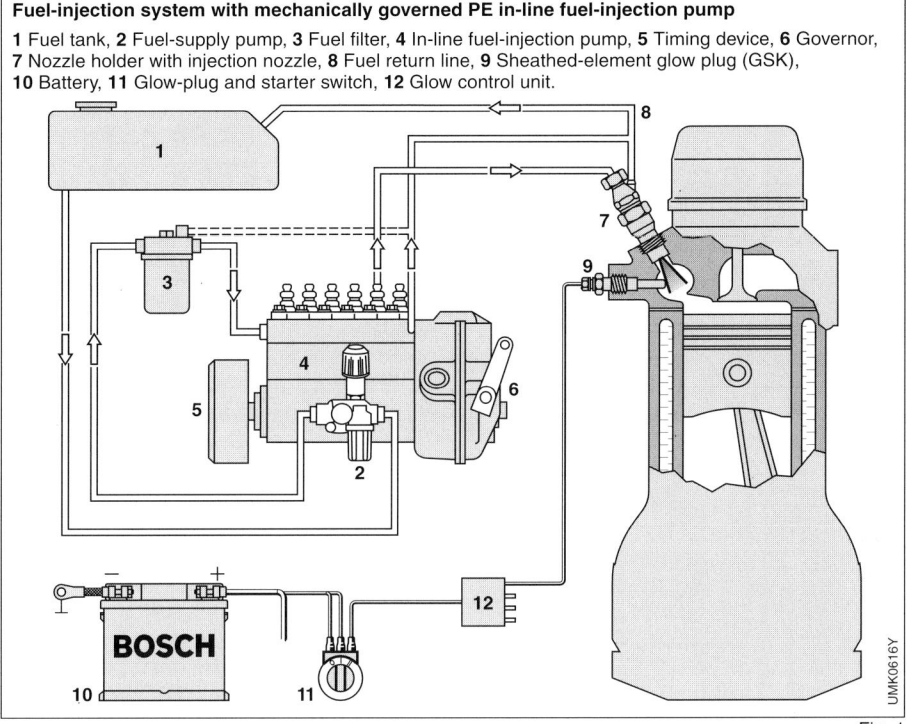

Fig. 1

Fig. 2

Closed control loop for mechanical diesel governing

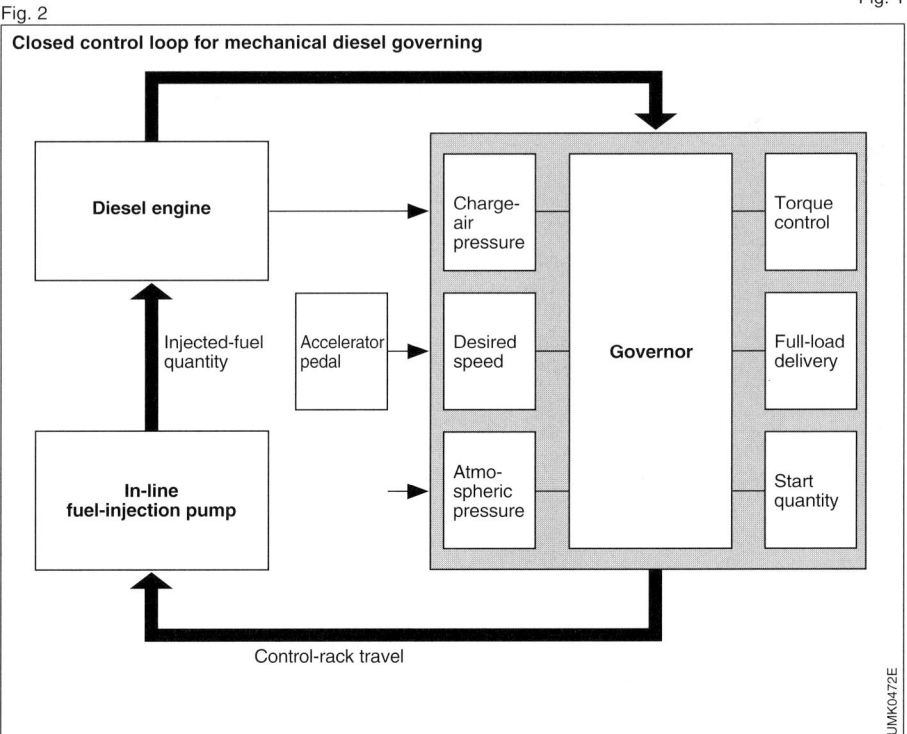

These have been discontinued as a result of the increased demands made on control (governing) precision and on governor functions. They are therefore not dealt with in this book.

Governor operation

No matter what the load is on the engine, the fuel-injection pump must always provide the engine with the correct amount of fuel. All in-line fuel-injection pumps have a plunger-and-barrel assembly (pumping element) for each engine cylinder. This comprises the pump barrel and the plunger. The plunger is forced in the fuel-delivery direction by an engine-driven camshaft and back again by its return spring. Since the plunger lift cannot be varied, the delivery quantity can only be adjusted by changing the plunger's effective stroke. To this end, the plungers are provided with an inclined helix so that the desired effective stroke is selected by rotating the plunger. Rotation is by means of the control rack which engages with the plunger, the control rack itself being shifted longitudinally by the governor. The plunger rotation positions the plunger helix to control the end of delivery (otherwise known as spill or port opening) and with it delivery quantity. Delivery commences when the top edge of the plunger closes the inlet port in the barrel wall.

In the case of maximum delivery, spill does not take place until maximum effective stroke, in other words with maximum-possible fuel delivery quantity. During partial delivery, spill takes place earlier depending upon the plunger's rotational setting. In the end position as required for zero delivery, that is, when the engine is to be switched off, the plunger's longitudinal slot is positioned directly opposite the inlet port. This means that the pressure chamber above the plunger is connected with the fuel gallery throughout the complete plunger stroke, so that no fuel is delivered (Fig. 3).

There are a number of different helix configurations. In the case of plungers with only a bottom helix, fuel delivery always

commences at the same plunger-lift point, whereas end of delivery takes place sooner or later depending upon the plunger's rotational setting. When the plunger has a top helix, start of delivery can be varied. There are also plungers available with both top and bottom helix.

Fig. 3

Fuel-delivery control

On in-line pumps by rotating the pump plunger
a Zero delivery, **b** Partial delivery,
c Maximum delivery,
1 Pump barrel, **2** Inlet port, **3** Pump plunger,
4 Helix, **5** Control rack.

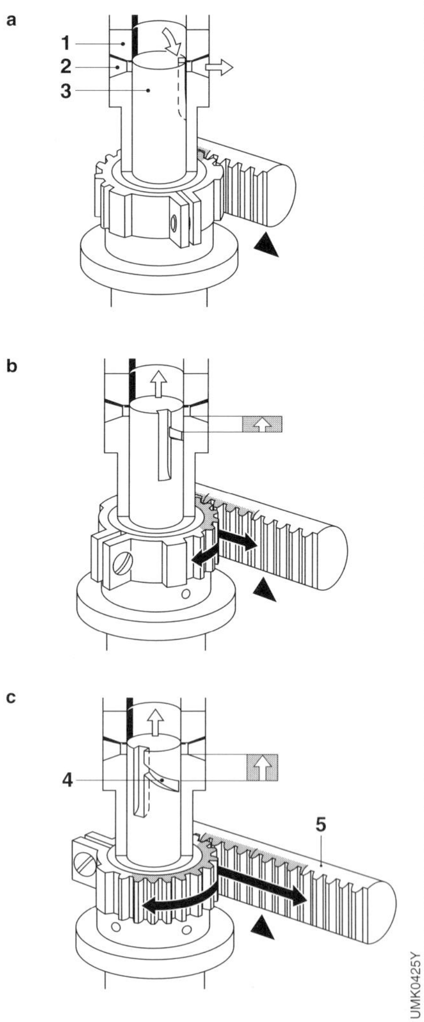

"Governor stories"

"It's a mistake to believe that the diesel engine is a crude machine that can only withstand crude handling!" [1])

Precision engineering and workmanship, as well as a high degree of intuitive feeling, are an absolute prerequisite if efficient functioning of the diesel engine is to be guaranteed and maintained.

At first, the form of diesel-engine control to which the above statement applies was carried out by the engine manufacturers themselves. But, in order to save the power take-off which was required at the engine, they came up with the demand for an in-line pump with attached speed control (governor).

It was at the end of the "Twenties" that Bosch took up this latest challenge. Following amazing feats of mechanical engineering, a flyweight-controlled minimum-maximum-speed governor went into series production in 1931. A slightly modified version of this governor came on the market a little later in the form of a variable-speed governor which was needed in particular for tractor and boat engines.

This form of flyweight governor appeared to be unsuitable for small high-revving automotive diesel engines though, and the concept of a pneumatic governor was born: "The control rack is connected with a leather diaphragm. The vacuum in the manifold, which is dependent upon engine speed, changes the diaphragm setting and with it the delivery quantity which thus becomes a function of throttle-valve setting (see Figure)." [2])

In the post-war years from 1946 to 1948 (variable-fulcrum governor) and as from 1955 (governor with external tensioning springs), a variety of considerably improved flyweight (mechanical) governor versions came onto the market (for instance, with integrated vibration dampers).

Supplementary devices for the adaptation of the full-load quantity to the desired engine torque curve (torque control) were fitted, as were devices for automatically setting the extra fuel needed for starting.

Today though, electronics plays a decisive role in the sector of diesel governing.

The electronic speed-control systems (Electronic Diesel Control or EDC) make the optimal operation of the diesel engine almost a matter of course.

[1]) Auer, Georg.
"Der Widerspenstigen Zähmung".
Diesel-Report. Robert Bosch GmbH.
Stuttgart, 1977. 8.
[2]) Schildberger, Friedrich.
Bosch und der Dieselmotor.
Stuttgart, 1950.

Pneumatic governor.

Original drawing taken from the 1950 publication "Bosch und der Dieselmotor".

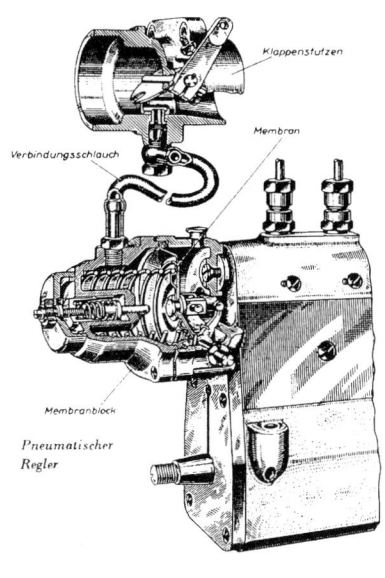

Speed droop

Every engine has a torque curve in accordance with its maximum loading capacity. And every engine speed is associated with a given maximum torque. If engine load is removed at a given engine speed, and the control rack is not adjusted accordingly, engine speed may only increase within the control range by the amount specified by the engine manufacturer (e.g. n_v = full-load speed to n_l = low idle speed). The increase in engine speed is proportional to the load change, i.e. the greater the engine-load reduction, the greater the engine-speed increase. The resulting phenomenon is known as speed droop", and one refers to governors with a "speed-droop" characteristic. The governor's speed droop is generally referred to the maximum full-load speed (= rated speed) and is calculated as follows:

$$\delta = \frac{n_{lo} - n_{vo}}{n_{vo}}$$

or in %:

$$\delta = \frac{n_{lo} - n_{vo}}{n_{vo}} \cdot 100\,\%$$

where δ = speed droop
n_{lo} = high-idle (maximum) speed
n_{vo} = maximum full-load speed

Example: (pump speeds)
n_{lo} = 1000 min⁻¹, n_{vo} = 920 min⁻¹

$$\delta = \frac{1000 - 920}{920} \cdot 100\,\% = 8.7\,\%$$

Generally speaking, a reasonably large speed droop increases the stability of the complete control loop (governor, engine, and driven machine or vehicle). On the other hand, the speed droop is limited by the operating conditions to about 0...5 % for engine-generator sets, and to about 6...15 % for vehicles.

In the subsequent figures, the following applies:
n_{vu} = minimum full-load speed
n_v = any full-load speed
n_{vo} = maximum full-load speed
n_{lu} = low idle speed
n_l = any idle speed
n_{lo} = high idle (maximum) speed

98

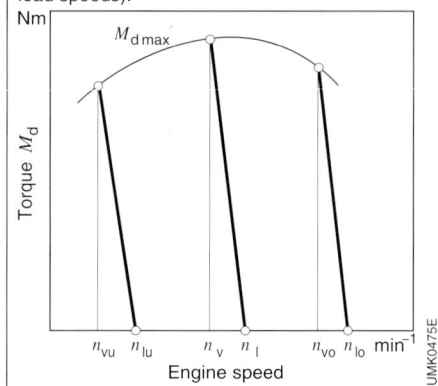

Full-load speeds

With the corresponding controlled idle speeds (no-load speeds).

Fig. 4

Fig. 5

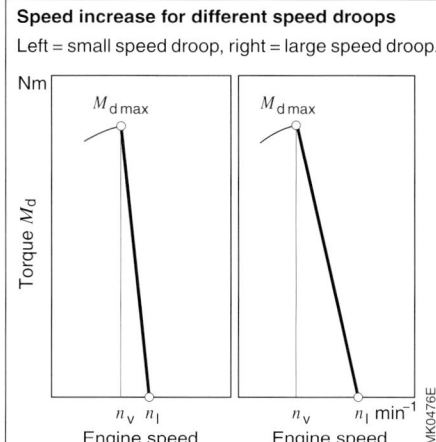

Speed increase for different speed droops

Left = small speed droop, right = large speed droop.

Fig. 6

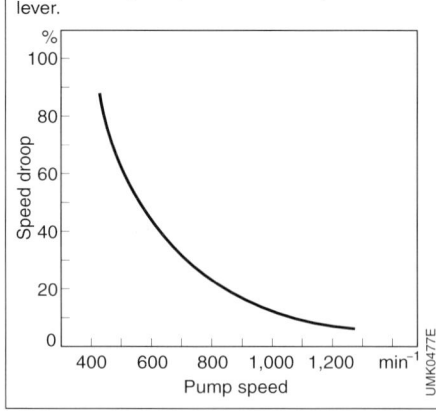

Speed droop of an RQV governor

At various engine speeds selected by the control lever.

Figure 7 provides a practical illustration of the effects of speed droop. With the desired engine speed set to a fixed value, actual engine speed varies within the speed-droop range as the engine load is changed (resulting, for instance, from a change in gradient).

Governor functions

The basic job of every governor is to limit the engine's maximum speed. In other words, the governor must ensure that engine speed never exceeds the maximum specified by the manufacturer. Depending upon its type, the governor can have further functions such as the maintenance of certain fixed speeds, such as idle, or maintaining the speed range between low idle and high idle (maximum) speed.

The governor can also have other responsibilities, the options provided by the electronic governor (EDC) being considerably more extensive than those of the mechanical (flyweight) governor.

The following chapters will focus on the mechanical governor. The electronic governor (EDC) is described in the Chapter "Electronic Diesel Control (EDC)".

The various demands made on the governors led to the development of the following governor types:

– Maximum-speed governor
  These governors are designed to limit the engine's maximum speed only.
– Minimum-maximum-speed governor
  In addition to maximum speed, these governors also control low idle speed.
– Variable-speed governor
  As well as the maximum and low idle speeds, these governors also control the speed range in between.
– Combination governors
  A combination of minimum-maximum-speed governor and variable-speed governor.
– Governors for stationary power units
  Designed for use with engine-generator sets as per DIN 6280.

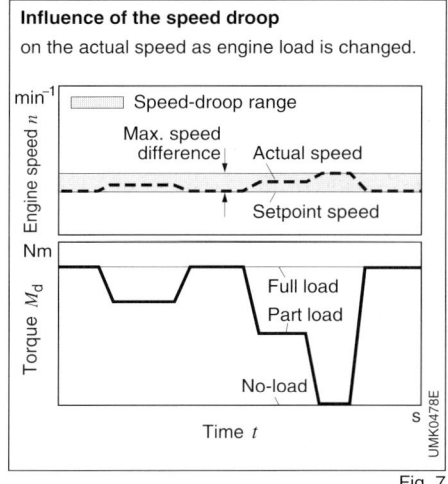

Influence of the speed droop on the actual speed as engine load is changed.

Fig. 7

Fig. 8

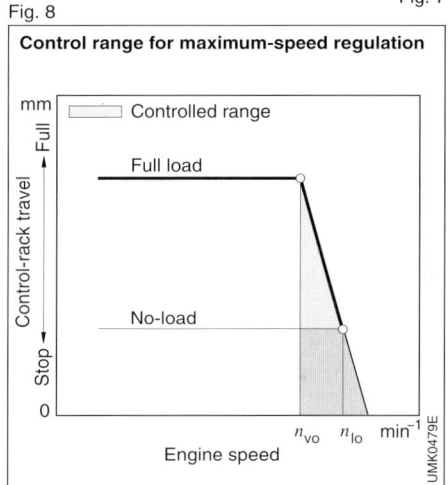

Control range for maximum-speed regulation

Apart from its basic function, the governor also has a number of other control functions. These include automatically starting and stopping the extra fuel required for starting (start quantity), and changing the full-load delivery quantity as a function of engine speed (torque control), charge-air pressure, or atmospheric pressure. Supplementary equipment is required for some of these functions. This will be dealt with later on in this book.

Maximum-speed regulation

Depending upon the speed droop, when the engine load is removed, maximum full-load speed n_{vo} is not to exceed n_{lo}

(high idle (maximum) speed). The governor complies with this stipulation by shifting the control rack in the Stop (shut-off) direction (Fig. 8). Governing (regulation) in the range between n_{vo} and n_{lo} is termed maximum-speed regulation.

The higher the speed droop, the higher the speed increase between $n_{vo} \ldots n_{lo}$.

Intermediate-speed regulation

When the specific application requires (e.g. in vehicles with auxiliary power take-off), the governor can maintain the engine speed within a stipulated range between idle speed and high idle (maximum) speed (Fig. 9).

Thus engine speed n only fluctuates within its operating range between n_v (full load) and n_l (no load), depending upon load.

Low-idle-speed control

Speed control (governing) can also take place in the engine's lowest speed range (Fig. 10). After cold starts, when the control rack shifts from starting position to B position, the engine's frictional resistance is still relatively high. This means that the fuel-delivery quantity to sustain engine operation is slightly above that which would normally correspond to the low-idle-speed adjustment point L, and the engine speed is slightly lower. During warm-up, the friction decrease causes the engine speed to increase, and the control rack shifts back to the L position. This is the low-idle-speed setting for the engine when it is at full operating temperature.

Torque control

Torque control is applied to permit full exploitation of the combustion air trapped in the cylinder. As such, it is not actually a control process, but rather one of the regulatory functions taken over by the governor. It has been designed for the full-load delivery quantity, i.e. the maximum amount of fuel injected in the engine's loadable range which can be combusted smoke-free. In general, the fuel requirement of a non-pressure-charged diesel engine drops along with increasing engine speed (reduced re-

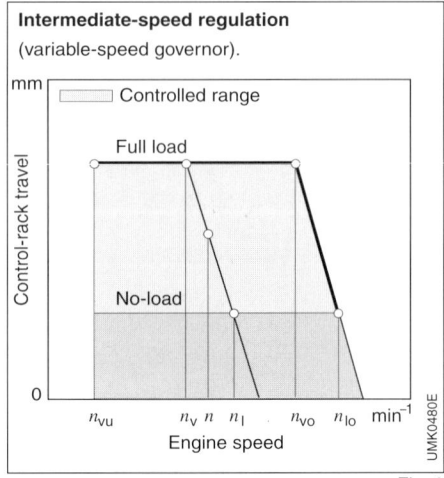

Intermediate-speed regulation
(variable-speed governor).

Fig. 9

Fig. 10

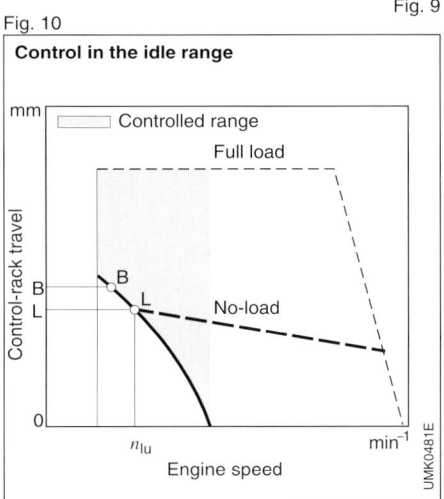

Control in the idle range

Fig. 11

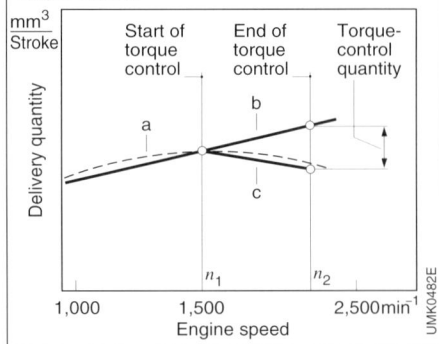

Fuel-requirement and fuel-delivery characteristics. With torque control

a Engine fuel requirement, **b** Full-load delivery without torque control, **c** Full-load delivery with torque control.

lative rate of air flow, thermal limits, changed mixture formation). On the other hand, at a constant control-rack position, the fuel quantity injected by the Bosch injection pump increases within a certain range as the speed increases. This is due to the throttling effect at the spill port of the pump's plunger-and-barrel assembly.

However, injection of excessive fuel results in smoke emissions and engine overheating. This means the injected fuel quantity must be adapted to the engine's fuel requirement (Fig. 11). On governors equipped with torque control, the control rack is shifted in the torque-control range by a fixed amount (the so-called torque-control travel) in the Stop (shutoff) direction. Thus, as the speed increases (from n_1 to n_2), the fuel-delivery quantity decreases (positive torque control or torque control in the direction of control). When engine speed drops (from n_2 to n_1), it increases (Fig. 12).

Design and location of the torque-control devices vary according to governor type. For details, refer to the particular governor's description.

The torque curve of a diesel engine with and without torque control is shown in Fig. 13. Maximum torque is developed throughout the complete speed range without the smoke limit being exceeded.

On engines fitted with an exhaust-gas turbocharger having a high boost ratio, the full-load fuel requirement in the lower speed ranges increases so much that the standard increase in fuel delivery from the injection pump no longer suffices. In such cases, torque control must be regulated as a function of engine speed or charge-air pressure. Depending upon the prevailing conditions, this is accomplished using either the governor or the manifold-pressure compensator (LDA), or both.

This form of torque control is termed negative control. Here, the injected fuel quantity is further increased when the speed increases (Fig. 14). In contrast, positive torque control indicates that the injected fuel quantity is reduced when speed increases.

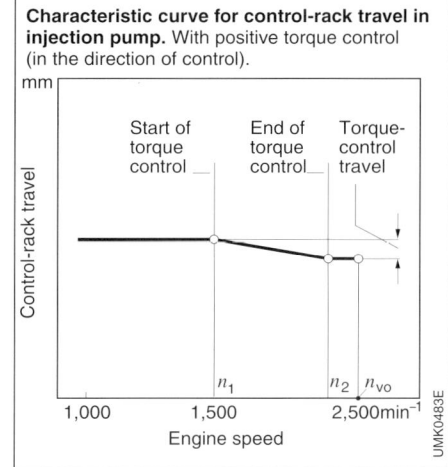

Characteristic curve for control-rack travel in injection pump. With positive torque control (in the direction of control).

Fig. 13

Fig. 12

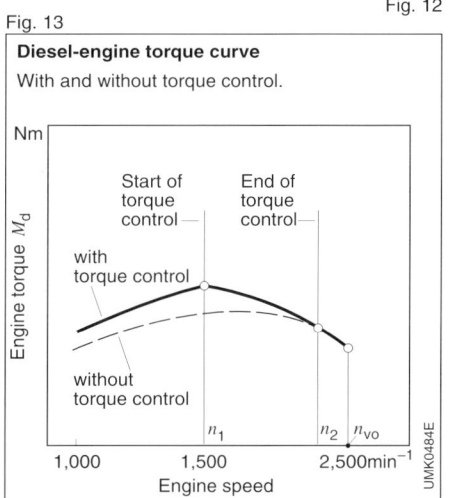

Diesel-engine torque curve
With and without torque control.

Fig. 14

Fuel-delivery characteristic

a Engine fuel requirement, **b** Full-load delivery without torque control, **c** Full-load delivery with torque control. c_1 Negative torque control, c_2 Positive torque control.

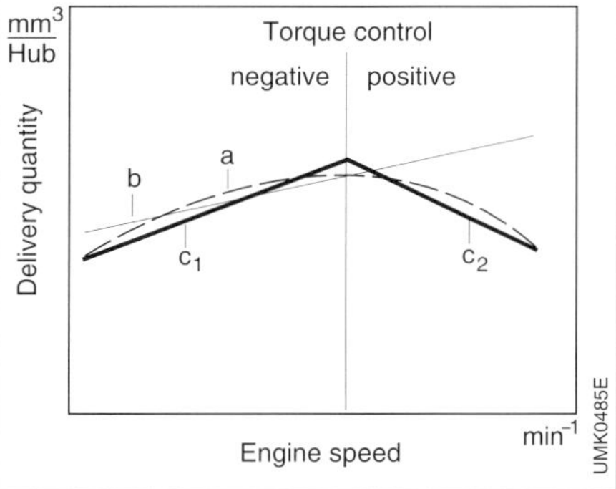

Governors –
An overview

Governor designation

The governor designation is given on its nameplate. It defines the governor's essential characteristics (e.g. governor type, idle/maximum speed etc.).
Using an example, the Table on the next page gives details of each component in the governor designation.

Maximum-speed governors

These governors are designed for use with diesel engines which power machinery at a specific rated speed. In such applications, the governor is responsible for maintaining the maximum speed. Idle-speed control and the control of a given start quantity are unnecessary.
As soon as the rated speed n_{vo} is exceeded due to a reduction in the engine load, the governor shifts the control rack towards Stop (shutoff) and reduces the fuel-delivery quantity. The speed increase and the reduction in rack travel follow the line A – B. The high idle (maximum) speed n_{lo} is reached when the load has been removed from the engine completely. The difference between n_{lo} and n_{vo} is determined by the governor's speed droop (Fig. 1).

Fig. 1

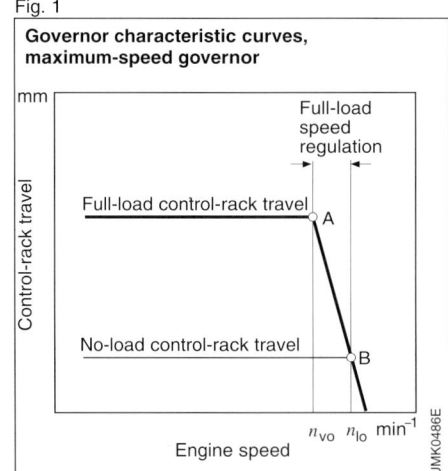

Governor characteristic curves, maximum-speed governor

mm

Control-rack travel

Full-load speed regulation

Full-load control-rack travel A

No-load control-rack travel B

n_{vo} n_{lo} min^{-1}

Engine speed

UMK0486E

Minimum-maximum-speed governors

With truck diesel engines, control (governing) is often unnecessary between low idle and high idle (maximum) speed. In this speed range, the driver selects the torque by using the accelerator pedal to position the injection pump's control rack directly. The governor ensures that the engine doesn't stall in the low-idle range and also governs the high idle (maximum) speed. Referring to the governor curves in Fig. 2, we see that the cold engine is started with the start quantity (A).
The driver presses the accelerator pedal to the floor, and when this is released, the control rack returns to the idle position (B). During warm-up, after fluctuating around the idle-speed control curve, low-idle speed finally levels off at L.
In general, upon completion of warm-up, maximum quantity is no longer required for restarting. In fact, some engines can be started even when the control lever is at idle. An add-on unit referred to as TAS (temperature-dependent starting stop) limits the start quantity for a warm engine even with fully depressed accelerator pedal. Once the engine is running, the control rack shifts to the full-load delivery setting when the accelerator pedal is pushed to the floor. Engine speed increases as a result and at n_1 torque control for the fuel-delivery quantity comes into effect and full-load delivery is slightly reduced. If engine speed continues to increase, torque control terminates at n_2.
With the accelerator pedal pressed to the floor, the full-load delivery quantity continues to be injected until the engine reaches its maximum full-load speed n_{vo}. At n_{vo}, full-load speed regulation comes into effect in accordance with the governor's speed droop, whereby engine speed increases slightly, rack travel is reduced and as a result the fuel-delivery quantity decreases.
The high idle (maximum) speed n_{lo} is reached when there is no load at all on the engine. During overrun (driving downhill), speed can increase further so that rack travel is reduced to zero.

Governor characteristic curves, minimum-maximum-speed governor

With torque control.

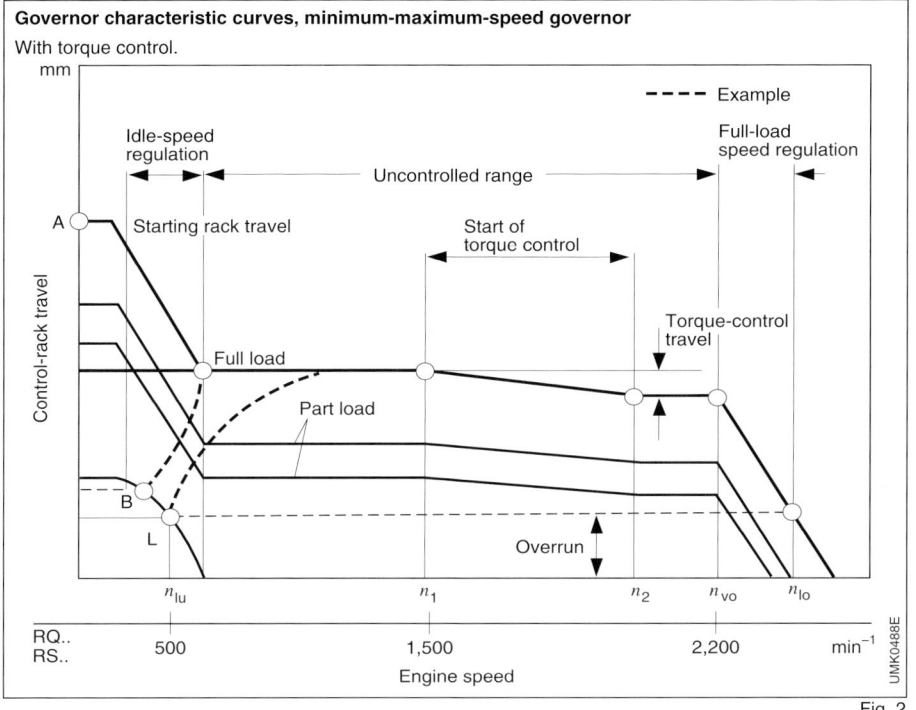

Fig. 2

Example for governor designation

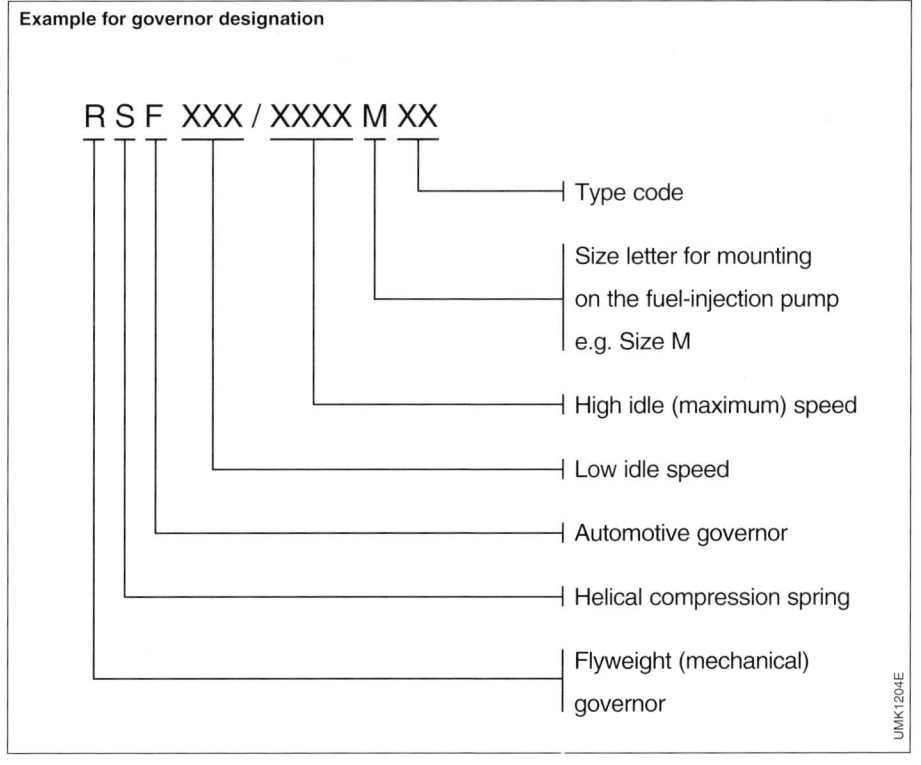

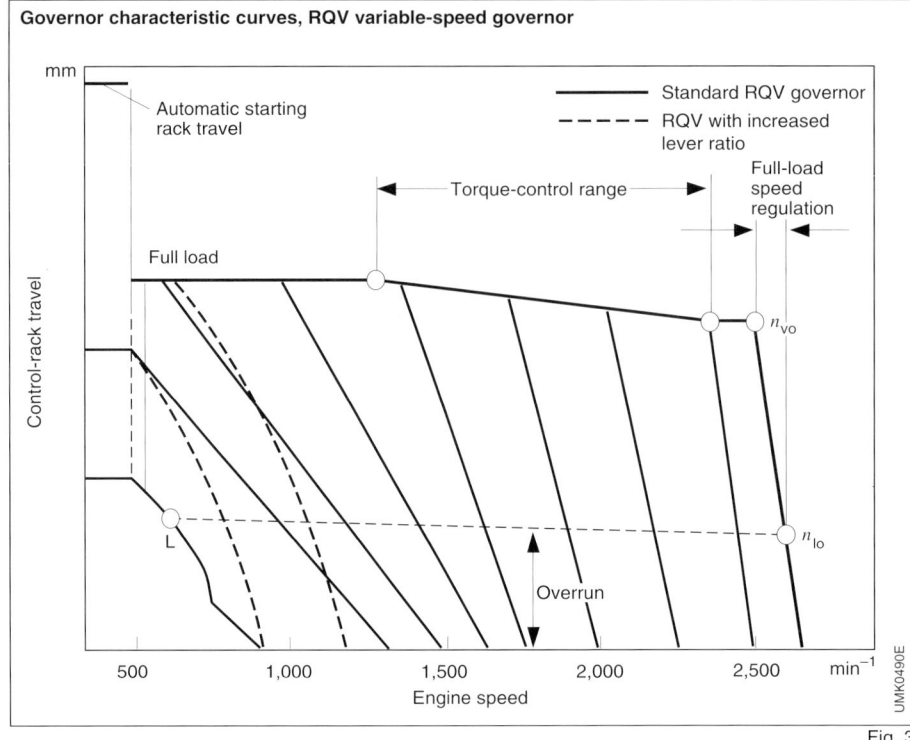

Governor characteristic curves, RQV variable-speed governor

Fig. 3

Variable-speed governors

There is a class of vehicles whose engines must maintain a given speed (e.g. agricultural tractors, road-sweeping machines, ships etc.), or which must provide power for an auxiliary drive (e.g. pumps, fire-fighting ladders etc.). Such engines are equipped with variable-speed governors.

In addition to the low-idle and high-idle (maximum) speeds, these governors also regulate intermediate speeds, independent of engine load. The speed is set at the governor control lever. Referring to the governor curves (Fig. 3), we see that the engine is started with the start quantity. Full-load regulation follows the full-load curve, and torque control takes place between n_1 and n_2 until speed-regulation breakaway starts at maximum full-load speed following the line from n_{vo} to n_{lo}.

The remaining curves show the breakaway characteristic for the intermediate speeds, whereby the increase in speed droop for decreasing engine speed is

evident. The dashed curves apply to vehicles whose auxiliary drives operate in the lower speed ranges. When the load increases, engine speed drops less than with a conventional governor (unbroken curves). This is due to a higher lever ratio.

Combination governors

When the normal speed droop in the upper or lower adjustment range of the RQV or RQUV governor is excessive for the particular application, and control is not required in the intermediate speed range, the governor's speed-sensing mechanism is designed for "stepped" regulation. Here, torque control is not possible in the non-regulated (maximum-speed governor) area of this governor's curve. Referring to the governor's curves (Fig. 4), it can be seen that they belong to a governor with the uncontrolled section of the curves in the lower speed range, and the controlled section in the upper speed range. Another governor type operates in the lower speed range as a

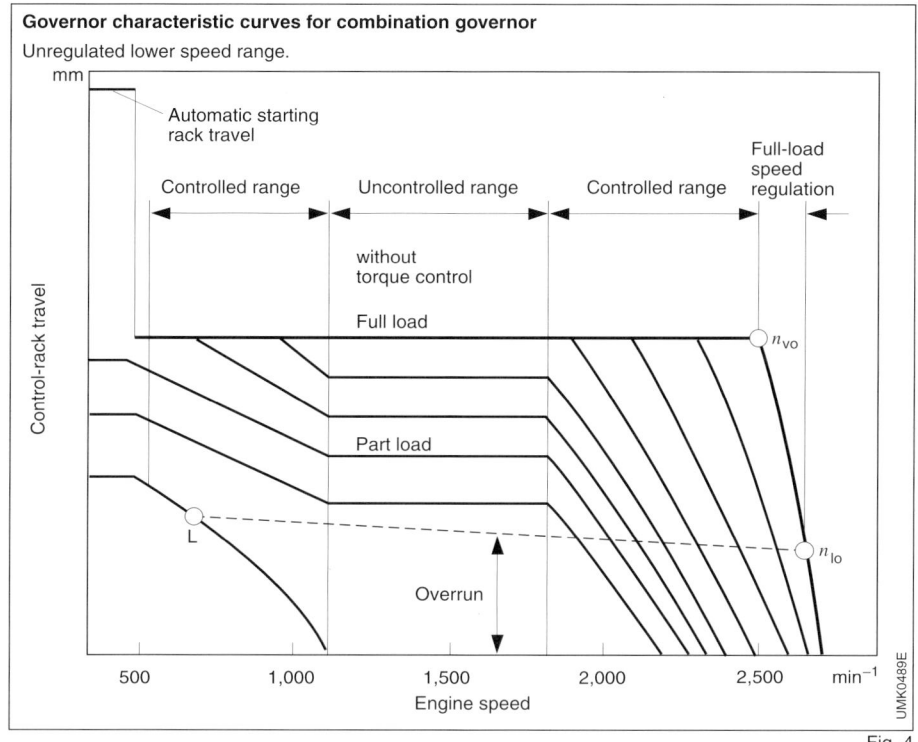

Governor characteristic curves for combination governor

Unregulated lower speed range.

Fig. 4

variable-speed governor (curves extending downwards) followed by an uncontrolled section till full-load speed regulation comes into effect (horizontal section of the curves). In both cases, the horizontal sections of the curves represent the control-rack travel for the various control-lever part-load settings.

The curves extending downward from the full-load line represent the speed regulation breakaway which has been set to operate from the respective intermediate speeds. The only design difference between combination governor and variable-speed governor lies in the different spring sets used.

Governors for engine-generator sets

Governor specifications for engine-generator sets are given in DIN 6280 (s. tab. on next pages). Type Classes 1, 2, and 3 can be operated with Bosch mechanical governors. Electronic governors are generally employed for Type Class 4,

which also includes generating sets with speed droop = 0 %.

Governor curves for engine-generator sets are shown in Fig. 5.

Provided that parallel operation is not required, the generators can be run at a fixed speed, whereby a simple maximum-speed governor suffices.

Fig. 5

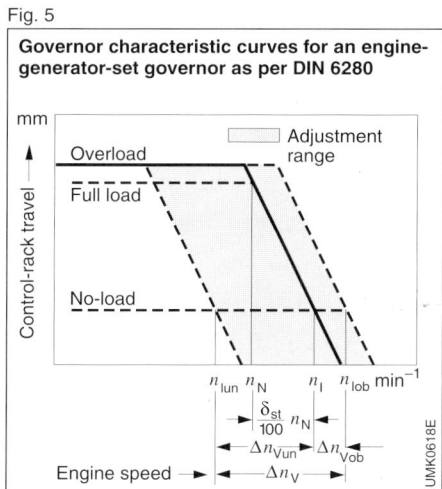

Governor characteristic curves for an engine-generator-set governor as per DIN 6280

Table 1. **Types of govenor**

Desig-nation	Type	Speed-sensing mechanism	Pump size	Torque control
RQ	Minimum-maximum speed governor, or maximum-speed governor only	Flyweights	A, MW, P	Positive
RQ	Generating-set governor	Flyweights	A, MW, P	None
RQU	Minimum-maximum-speed governor, or maximum-speed governor only	Flyweights [1]	ZW, P9, P10	Positive
RS	Minimum-maximum-speed governor	Flyweights	A, MW, P	Positive
RSF	Minimum-maximum-speed governor	Flyweights	M	Negative/Positive
RQV	Variable-speed governor or combination governor	Flyweights	A, MW, P	Positive
RQUV	Variable-speed governor	Flyweights [1]	ZW, P9, P10	Positive
RQV..K	Variable-speed governor	Flyweights	A, MW, P	Negative/Positive
RSV	Variable-speed governor	Flyweights	A, M, MW, P	Positive
RSUV	Variable-speed governor	Flyweights [1]	P	Positive
RE	Arbitrary governor characteristic curves	Solenoid	M, MW, P	Negative/Positive

[1] With gearing for low-speed engines.

Table 2. **Operating limits for type classes**
(applies only to engine-generator sets). (Excerpt from DIN 6280, Part 3)

No.	Designation	Symbol	Unit	Type class 1	2	3	4
4.2.4	Static speed deviation (speed droop)	δ_{st}	%	8	5	3	*
4.2.5	Speed-fluctuation range	v_n	%	–	1.5	0.5	*
4.2.1	Lower engine-speed control range	$\delta \cdot n_{Vun}$	%	$-(2.5+\delta_{st})$	$-(2.5+\delta_{st})$	$-(2.5+\delta_{st})$	*
4.2.2	Uper engine-speed control range	$\delta \cdot n_{Vob}$	%	+2.5	+2.5	+2.5	*
4.1.6	Frequency settling time*	t_{fzu}, t_{fab}	s	–	5	3	*

* After consultation with Bosch.

Table 3. **Definitions of rotational-speed terms**
(applies only to engine-generator sets) (Excerpt from DIN 6280, Part 4)

No.	Designation	Symbol	Definition
4.1	Rated speed	n_N	The engine speed assigned to the generator set's nominal frequency. The generator set's rated power refers to this rated speed.
4.3	Zero-power speed	n_l	Steady-state speed of the unloaded engine. The appropriate values for the rated-power speed and the part-load speed are with reference to an unchanged speed setting.
4.7	Minimum selectable zero-power speed	n_{lun}	Minimum steady-state speed of the un-loaded engine which can be set at the speed setpoint selector or at the governor.
4.8	Maximum selectable zero-power speed	n_{lob}	Maximum steady-state speed of the un-loaded engine which can be set at the speed setpoint selector or at the governor.
4.9	Engine-speed control range	Δn_V	Range between the selected maximum and minimum zero-power speeds. The engine-speed control range results from the addition of the values for the maximum and minimum engine-speed control ranges as per Sections 4.9.1 and 4.9.2.
4.9.1	Minimum selectable speed range	Δn_{Vun} δn_{Vun}	Range between selected minimum zero-power speed and the zero-power speed which results from unloading at the zero-power point with the setpoint unchanged. $\Delta n_{Vun} = n_l - n_{lun}$ The difference between these two speeds is given as a percentage of the rated speed. $\delta n_{Vun} = \dfrac{(n_l - n_{lun})}{n_N} \cdot 100$
4.9.2	Maximum selectable speed range	Δn_{Vob} δn_{Vob}	Range between selected maximum zero-power speed and the zero-power speed which results from unloading at the zero-power point with the setpoint unchanged. $\Delta n_{Vob} = n_{lob} - n_l$ The difference between these two speeds is given as a percentage of the rated speed. $\delta n_{Vob} = \dfrac{(n_{lob} - n_l)}{n_N} \cdot 100$
5.1	Steady-state speed droop	δ_{st}	The speed difference between zero-power speed n_l and rated speed n_N, expressed as a percentage of the rated speed. $\delta_{st} = \dfrac{(n_l - n_N)}{n_N} \cdot 100$

Mechanical governing

The Bosch mechanical (flyweight) governors are mounted on the fuel-injection pump whose control rack is connected with the governor linkage. The governor control lever is the connection to the accelerator pedal.

Speed-sensing mechanisms

Mechanical governors use two different designs of speed-sensing mechanism:
– RQ, RQV: The governor springs are incorporated in the flyweights, whereby each flyweight acts directly on a spring set which is designed specifically for a given rated speed and for the corresponding speed droop.
– RSV, RS, RSF: The centrifugal force is applied through a system of levers to the governor spring which is located outside the two flyweights. The force from both flyweights presses the sliding bolt against the tensioning lever which is being pulled in the opposite direction by the governor spring. With the RSV variable-speed governor, the driver tensions the governor spring via the control lever to select the desired speed. With the RS and RSF minimum-maximum-speed governors, the governor spring is fixed at maximum speed and cannot be influenced by the driver. The governor springs for both forms of speed-sensing mechanism have been selected so that spring force and centrifugal force are mutually balanced at the desired speed. When this is exceeded, the centrifugal force from the flyweights increases. A linkage system then adjusts the control rack to provide a corresponding reduction in the delivery quantity.

RQ Minimum-maximum-speed governors

Design

The injection pump's camshaft drives the governor hub through a vibration damper. The two flyweights with their bell cranks are held at one end in the governor hub. Each flyweight has its own built-in spring set. When the flyweights

move outwards due to centrifugal force, the bell cranks transform this (radial) movement into an axial movement at the sliding bolt. This axial movement is transferred to the so-called slider. The slider, which can only shift along a straight line due to its being held by the guide pin, provides the connection between the flyweight speed-sensing mechanism and the control rack by means of the variable-fulcrum lever. There is a sliding block guide in the variable-fulcum lever, and the bottom end of the lever is held in the sliding block. The movable guide block is guided radially by the linkage lever which itself is connected with the control lever on the same shaft. The control lever is shifted either by hand or through linkage from the accelerator pedal (Figs. 2 and 3). When the position of the control lever is changed, the guide block is shifted and the fulcrum lever tilts around the pivot point at the slider. When the governor comes into effect, the guide block becomes the pivot point for the fulcrum lever. The action of the sliding-block guide and the guide block allows a variable lever ratio for the fulcrum lever.

This arrangement ensures that enough force is always available to adjust the control rack, even in the low-idle range, in which the centrifugal forces exerted by the flyweights are still insufficient. The spring sets (governor springs) in the flyweights (Fig. 1) generally comprise

Fig. 1

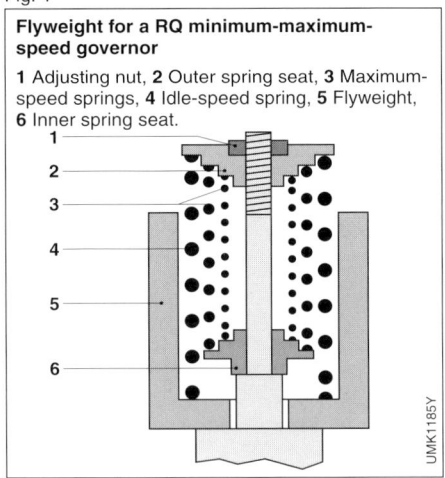

Flyweight for a RQ minimum-maximum-speed governor

1 Adjusting nut, **2** Outer spring seat, **3** Maximum-speed springs, **4** Idle-speed spring, **5** Flyweight, **6** Inner spring seat.

UMK1185Y

RQ Minimum-maximum-speed governor

1 Control rack, **2** Link fork, **3** Play-compensating spring, **4** Adjusting nut, **5** Governor springs, **6** Flyweight, **7** Bell crank, **8** Sliding bolt, **9** Slider, **10** Guide pin, **11** Control lever, **12** Fulcrum lever, **13** Guide block, **14** Linkage lever.

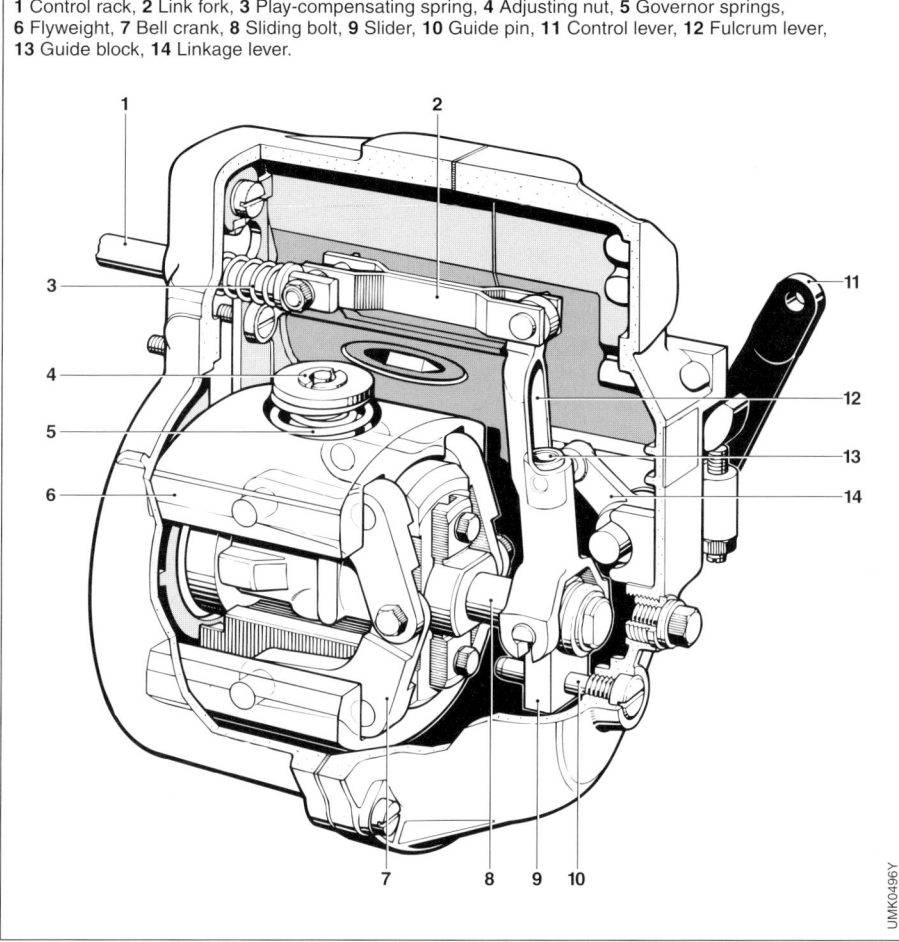

Fig. 2

Fig. 3

RQ Minimum-maximum-speed governor

Stop (shutoff) position.
 1 Shutoff stop,
 2 Control lever,
 3 Full-load stop,
 4 Guide block,
 5 Fulcrum lever,
 6 Link fork,
 7 Control rack,
 8 Pump plunger,
 9 Control-rod stop (spring-loaded),
 10 Slider,
 11 Guide pin,
 12 Sliding bolt,
 13 Bell crank,
 14 Governor hub,
 15 Adjusting nut,
 16 Governor spring,
 17 Flyweight,
 18 Camshaft.

three concentrically arranged helical springs: the idle-speed spring and the two maximum-speed springs.

Starting the engine

The accelerator-pedal position for starting the engine is stipulated in the engine's operating instructions.

The fuel-delivery quantity for starting the cold engine at very low outside temperatures is obtained with the accelerator pedal pushed to the floor. Normally, the fuel quantity delivered with the control lever in the low-idle position suffices to start the engine when it is warmed up. In such cases, pushing the pedal to the floor only results in an unnecessary cloud of exhaust smoke (Figs. 4 and 7).

Operating characteristics

Low-idle position

After the engine has started and the control lever (accelerator pedal) has been released, the control lever reverts to the low-idle position. Meanwhile, the governor begins to operate, positioning the control rack at the low-idle position (Fig. 5).

The engine's low idle speed is defined as the lowest speed at which the engine still operates reliably when it is not subjected to any loads beyond those of its own internal friction and those of the permanent ancillaries such as alternator, fuel-injection pump, fan, etc. To overcome this load at idle, the engine requires a given fuel-delivery quantity.

This is provided by the control lever in the position specified as low-idle.

Intermediate speed

The engine is loaded between no load and full load (Fig. 6). As soon as the driver presses the accelerator pedal slightly, the engine accelerates.

As a result the governor flyweights move outwards so that the governor initially attempts to oppose this speed increase. However, as soon as the speed exceeds low idle speed by only a small amount, the flyweights come up against the inner spring seats which are spring-loaded by

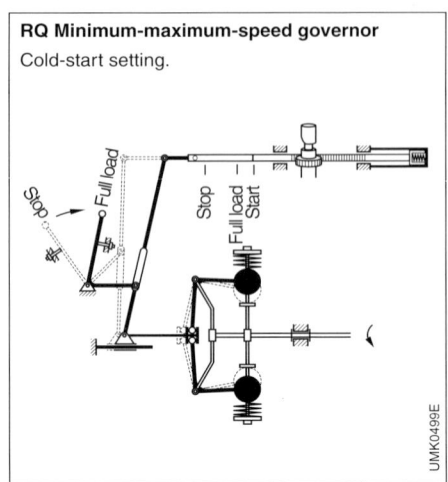

RQ Minimum-maximum-speed governor
Cold-start setting.

Fig. 4

Fig. 5

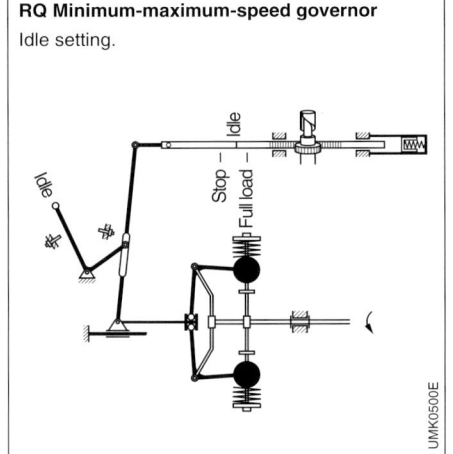

RQ Minimum-maximum-speed governor
Idle setting.

Fig. 6

RQ Minimum-maximum-speed governor
Part-load setting.

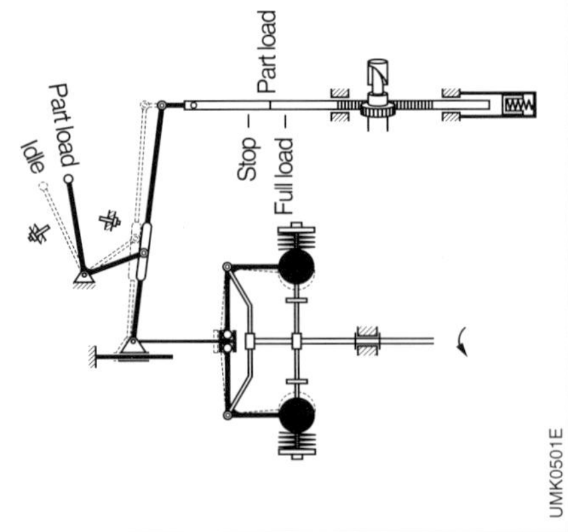

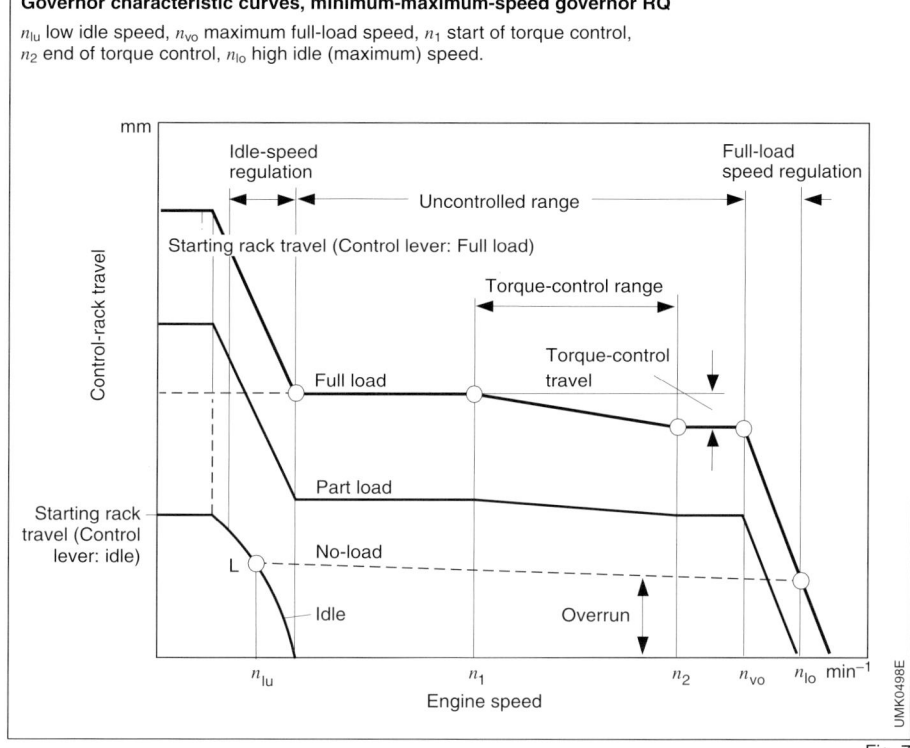

Governor characteristic curves, minimum-maximum-speed governor RQ

n_{lu} low idle speed, n_{vo} maximum full-load speed, n_1 start of torque control,
n_2 end of torque control, n_{lo} high idle (maximum) speed.

Fig. 7

the maximum-speed springs. Due to the fact that the maximum-speed springs do not yield to the flyweight's centrifugal force until the engine attempts to exceed its rated speed, the flyweights remain in this position until the engine reaches its maximum speed.

In other words, the governor is ineffective between the low idle and high idle (maximum) speeds, and the driver alone determines the position of the control rack, and therefore the engine torque, by means of the accelerator pedal.

The torque-control phase which this process traverses is described later on in this book.

Torque-control

The torque-control mechanism is incorporated in the RQ governor's flyweights between the inner spring seats and the maximum-speed springs (Fig. 8). The torque-control spring is held inside a spring retainer against the outside of which the maximum-speed springs are supported.

Fig. 8

Torque-control mechanism for minimum-maximum-speed governor RQ

1 Adjusting nut, **2** Spring seat, **3** Maximum-speed springs, **4** Idle-speed spring, **5** Flyweight, **6** Shim, **7** Torque-control spring, **8** Spring retainer.
a Torque-control travel.

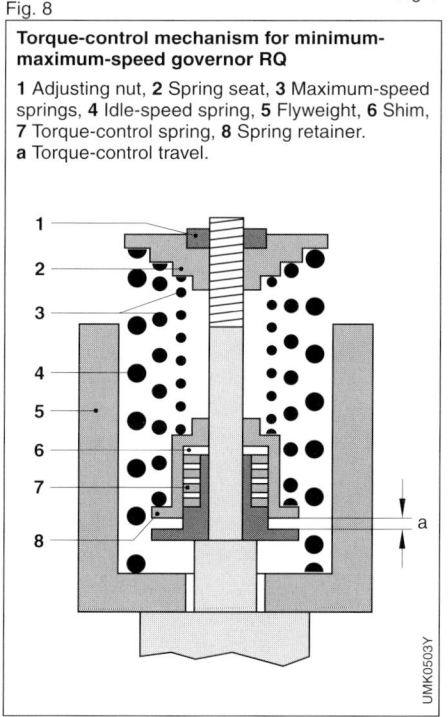

The torque-control mechanism thus comes into operation before the maximum-speed springs. The distance between the inner spring seat and the spring retainer represents the torque-control travel. It is adjusted using shims. The start of torque control n_1 depends upon the engine's fuel-requirement curve. At a speed slightly below the maximum speed, the torque-control spring is compressed to such an extent that the inner spring seat and the spring retainer are pressed together n_2. Without the torque-control spring, the governor is ineffective between low idle and high idle (maximum) speed. As soon as the torque-control springs yield, the flyweights are able to move outwards in the range between n_1 and n_2 representing the torque-control travel range. They shift the control rack in the Stop (shutoff) direction by the required increment (positive torque control).

High idle (maximum) speed

Maximum-speed regulation commences when the engine exceeds its rated speed n_{vo}. This can therefore be anywhere between full and part load, depending upon the control-lever setting (Fig. 9). In other words, as soon as maximum-speed regulation starts, the control-rack position is no longer solely dependent upon the driver's wishes but is also determined by the governor.

The maximum-speed regulation travel of the flyweights is selected to achieve speed regulation from maximum full-load rack travel to the zero-delivery rack travel.

RQU Minimum-maximum-speed governors

Design

The RQU governor is particularly well-suited for regulation at very low speeds. A step-up transmission (between 1:1.5 and 1:3.7, depending upon application) is fitted between the injection-pump camshaft and the governor hub which it drives (Figs. 10 and 20).

The RQU governor was specifically designed for the ZW, P9 and P10 injection pumps which are used for large engines which usually operate at low speeds.

Similar to the RQV governor, the RQU governor has a two-piece linkage lever which is guided in a plate cam.

Operating characteristics

Its function and operating characteristics correspond to those of the RQ governor.

Fig. 9

Minimum-maximum-speed governor RQ

Full-load setting.

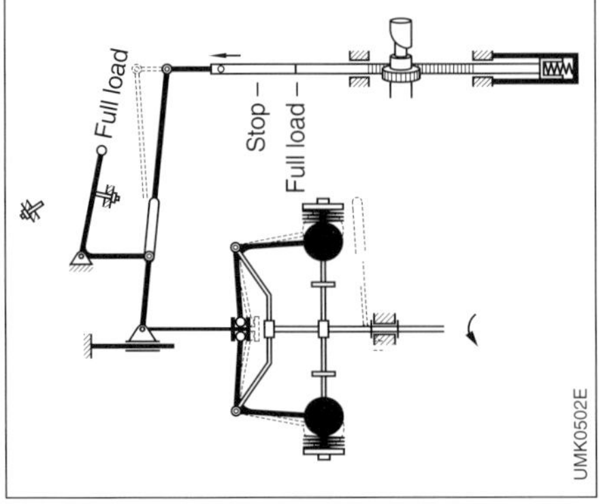

UMK0502E

Fig. 10

Minimum-maximum-speed governor RQU

Stop setting.
1 Full-load stop, **2** Plate cam, **3** Step-up gear.

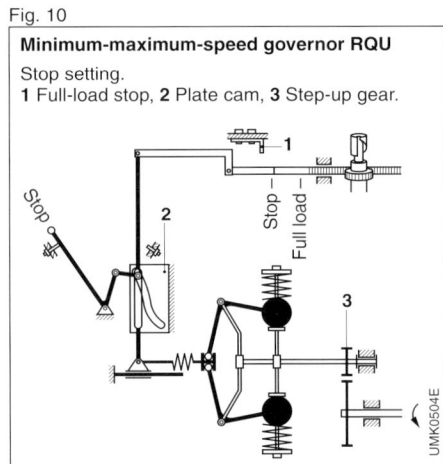

UMK0504E

RQ and RQU Maximum-speed governors

Design
The maximum-speed governor has no idle-speed stage. This is the main difference between it and the minimum-maximum-speed governor.

Operating characteristics
In operation, the maximum-speed governor behaves like the maximum-speed stage of the RQ and RQU minimum-maximum-speed governor.

High-idle (maximum) speeds
The full-load speed regulation starts as soon as the engine exceeds the maximum full-load speed. The flyweight travel for maximum-speed regulation is selected so that the speed-regulation breakaway takes place from maximum full-load rack travel to zero-delivery rack travel.

RQV Variable-speed governors

Design
The RQV governor is similar in design to the RQ governor.

The governor springs are incorporated in the flyweights which, in contrast to those on the RQ governor, continue to move outward within the specified adjustment range as long as the speed increases (Fig. 11). Each control-lever position is allocated a given speed at which speed regulation begins. Control-lever movement is transmitted through the two-piece linkage lever (toggle lever) and the guide block to the variable-fulcrum lever and thus to the control rack. The fulcrum lever's pivot point can shift inside the sliding-block guide. This fact, together with the fulcrum lever also being guided in a plate cam attached to the governor housing, results in a change in the lever ratio of the fulcrum lever. The sliding bolt is the connection element between the flyweight speed-sensing mechanism and the fulcrum lever.

Fig. 11

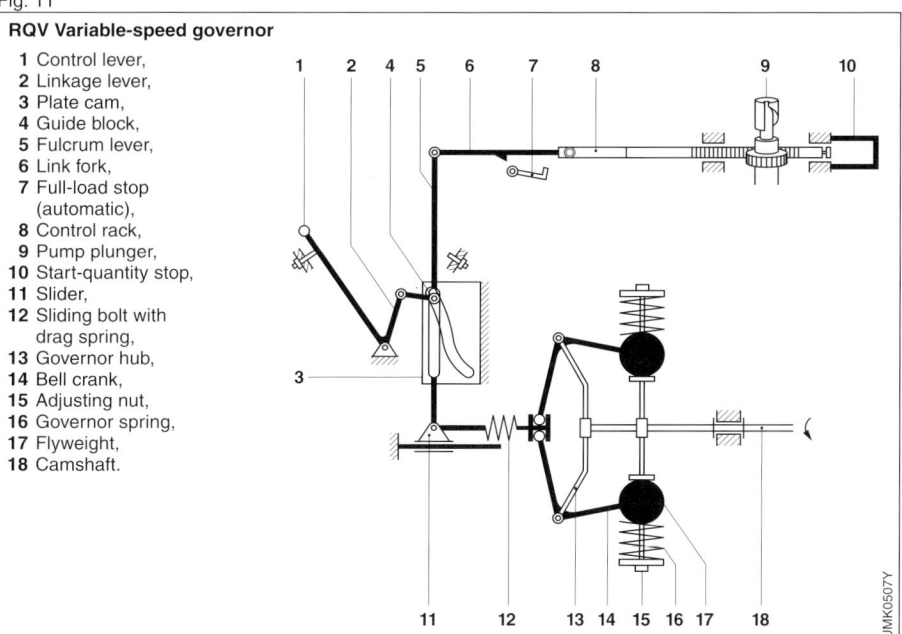

RQV Variable-speed governor

1 Control lever,
2 Linkage lever,
3 Plate cam,
4 Guide block,
5 Fulcrum lever,
6 Link fork,
7 Full-load stop (automatic),
8 Control rack,
9 Pump plunger,
10 Start-quantity stop,
11 Slider,
12 Sliding bolt with drag spring,
13 Governor hub,
14 Bell crank,
15 Adjusting nut,
16 Governor spring,
17 Flyweight,
18 Camshaft.

UMK0507Y

It is spring-loaded for pressure and tension (drag spring). As in the RQ governor, the spring sets fitted in the flyweights are generally comprised of three concentrically arranged helical springs. The outer spring provides low-idle-speed control; it is retained between flyweight and adjusting nut where it serves as a pretensioning element. As it continues outwards, the flyweight initially covers the short distance representing the idle-speed travel range (idle-speed stage). It then contacts the spring seat, and the inner springs between spring seat and adjusting nut also come into effect (Figs. 12 and 13).

Flyweight for RQV variable-speed governor

1 Adjusting nut, **2** Spring seat, **3** Maximum-speed springs, **4** Idle-speed spring, **5** Flyweight, **a** Idle-speed travel.

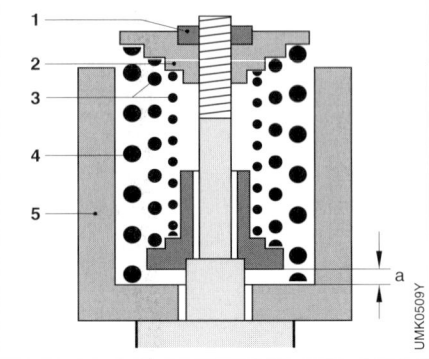

Fig. 12

Fig. 13

RQV Variable-speed governor

1 Control rack, **2** Play-compensating spring, **3** Full-load stop, **4** Adjusting nut, **5** Governor spring, **6** Flyweight, **7** Link fork, **8** Fulcrum lever, **9** Guide block, **10** Linkage lever, **11** Plate cam, **12** Bell crank, **13** Slider, **14** Sliding bolt (spring-loaded).

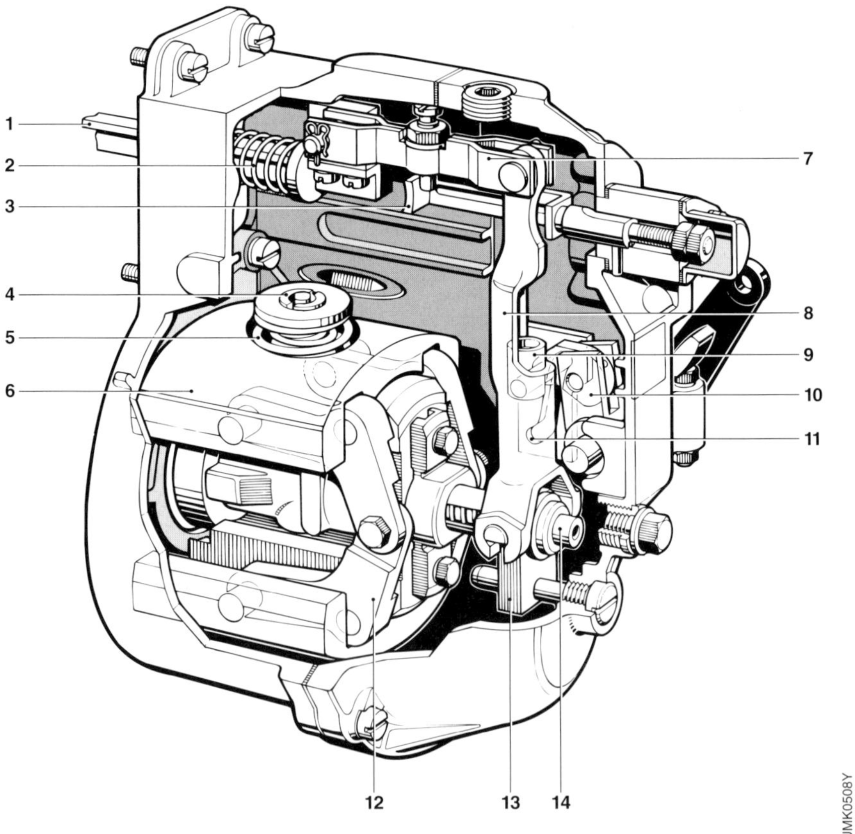

Starting the engine

When controlling the start quantity, the RQV governor differs from the RQ governors as follows:

If the driver presses the accelerator to the floor when the engine runs up to speed for the first time, the control rack remains in its start-quantity position beyond the low idle speed, maintaining this position up to high idle (maximum) speed. This contrasts with the RQ governor, which regulates the speed to the full-load setting as soon as low idle speed is reached. The first speed regulation breakaway must take place before the full-load stop "jumps" to its operating position (Fig. 14).

Operating characteristics

Low-idle setting (Figure 15)

As soon as the engine has started, the driver releases the accelerator pedal and the control lever returns to the low-idle setting.

The control rack also moves back to the low-idle setting as determined by the governor which has since begun to operate (point "L", Fig. 17).

Intermediate speed (Figure 16)

When engine load is increased or decreased at any speed setting determined by the control lever (accelerator pedal), the variable-speed governor increases or decreases the fuel-delivery quantity in order to maintain the set speed within the limits dictated by the speed droop. Example: Using his accelerator pedal, the driver has shifted the control lever from the low-idle position to a position which is to correspond to the vehicle's required speed. The control-lever movement is transferred to the variable-fulcrum lever through the linkage lever. The lever ratio of the fulcrum lever is variable; immediately above the idle range it increases to such an extent that even a relatively small fraction of the total control-lever or flyweight-travel suffices to shift the control rack to the set full-load stop (Fig. 17: Path L – B'). A fixed control-rack stop must therefore be available (never a spring-loaded stop). As the control lever

RQV Variable-speed governor
Cold-start setting.
1 Shutoff stop, **2** Stop for high-idle (maximum) speed.

UMK0511E

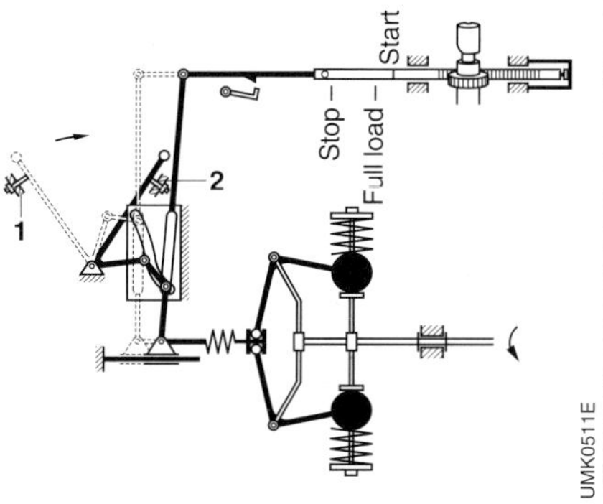

Fig. 14

Fig. 15

RQV Variable-speed governor
Low-idle setting.

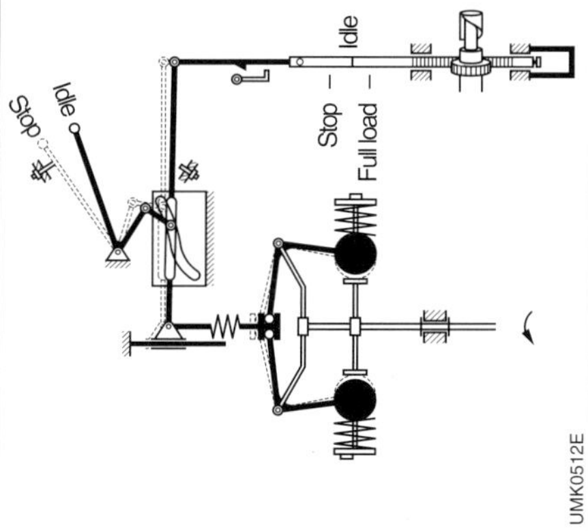

UMK0512E

Fig. 16

RQV Variable-speed governor
Engine loading (part load).
1 Slider, **2** Sliding bolt with drag spring.

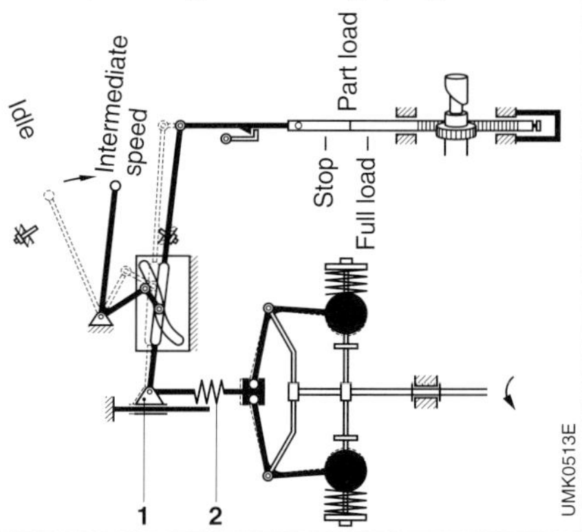

UMK0513E

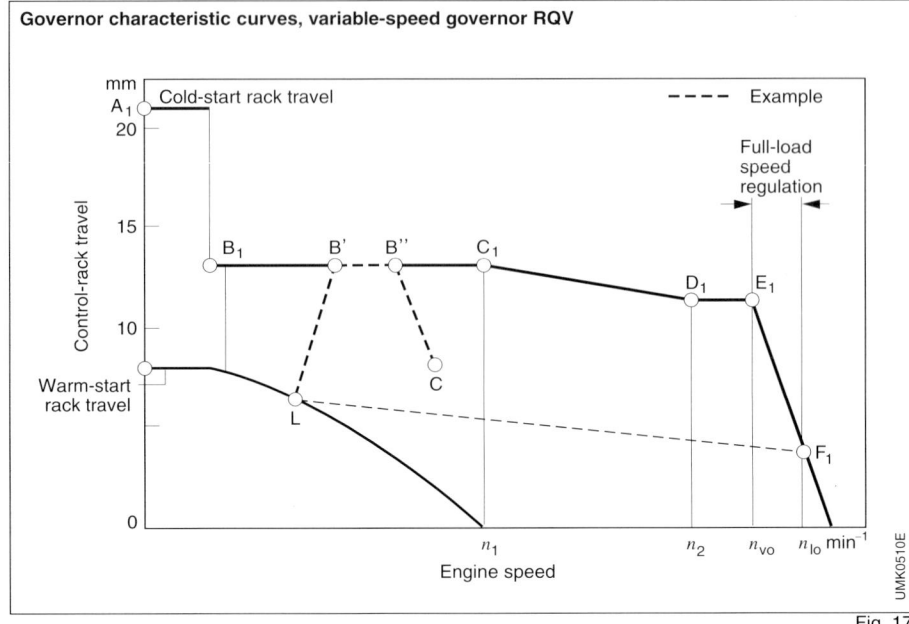

Governor characteristic curves, variable-speed governor RQV

Fig. 17

continues to swivel, it tensions the drag spring. For the time being, the control rack remains up against the full-load stop so that engine speed increases rapidly (Fig. 17: Path B' – B''). In the process, the flyweights move outward but the control rack remains in the full-load position until tension on the drag spring is released. The flyweights are now free to apply force to the fulcrum lever and shift the control rack back toward Stop (shutoff). The delivery quantity is reduced and the engine speed is limited. This engine speed limit corresponds to the control-lever position and the flyweight position (Fig. 17: Path B'' – C).

This means that every control-lever position corresponds to a given engine-speed range as long as the engine is not overloaded or being driven by a vehicle which is rolling downhill (overrun). If engine load now increases slightly, for instance when driving uphill, the engine and governor speeds decrease. As a result, the flyweights move inward and shift the control rack in the full-load direction, thus keeping the engine speed constant at a level determined by the control-lever position and the speed droop. However, if the gradient (or load) increases to

such an extent that engine speed continues to fall even after the control rack has been shifted to the full-load stop, the flyweights respond to this additional reduction in speed by moving even further inward. In doing so, they shift the sliding bolt to the left.

In other words, the flyweights attempt to shift the control rack even further in the full-load delivery direction. However, the fact that the control rack is already up against the full-load stop prevents it moving any further – as a result, tension is applied to the drag spring. This is equivalent to engine overload and the driver must shift to a lower gear. The opposite applies when driving downhill. Here, the engine accelerates due to it being driven by the vehicle (overrun). The flyweights therefore move outward and the control rack is pulled back as far as the stop in the shutoff direction. If engine speed continues to increase, and with the control rack in the Stop (shutoff) position, the drag spring is tensioned in the opposite direction. The governor-response pattern described above is basically valid at all control-lever settings, and operates whenever engine load or speed changes to such an extent that the control rack comes up against either the maximum-fuel-delivery stop or the shutoff stop.

Torque control

Torque control comes into effect between n_1 and n_2 (Fig. 17), and at full load along the line $C_1 - D_1$. With the RQV governor, the torque-control mechanism is incorporated in a special control-rack stop, or in a torque-control strap which takes the place of the normal link fork. (Refer to the Chapter "Control-rack stops".)

High idle (maximum) speed (Fig. 18)

Full-load speed regulation (Fig. 17: Path $E_1 - F_1$) starts as soon as the engine exceeds maximum full-load speed, when the flyweights move outward and the control rack is shifted in the Stop (shutoff) direction. High-idle (maximum) speed n_{lo} is reached when all load is removed from the engine.

RQUV Variable-speed governors

Design

The RQUV variable-speed governor is applied for the control of very low engine speeds such as those which are typical in marine applications.

The RQUV is a variant of the RQV governor. It is available with a variety of different step-up gears which are installed between the injection-pump camshaft and the governor hub (upward conversion ratios between approx. 1:1.5 and 1:3.7) (Fig. 19). As with the RQV governor, the lever ratio of the fulcrum lever is variable.

For this reason, this governor is also equipped with a plate cam (Fig. 20).

The RQUV governor is fitted to ZW, P9, and P10 injection pumps.

Operating characteristics

Operation and operating characteristics correspond to those of the RQV governor, but without start-quantity.

RQV Variable-speed governor
Full-load setting, start of speed regulation.

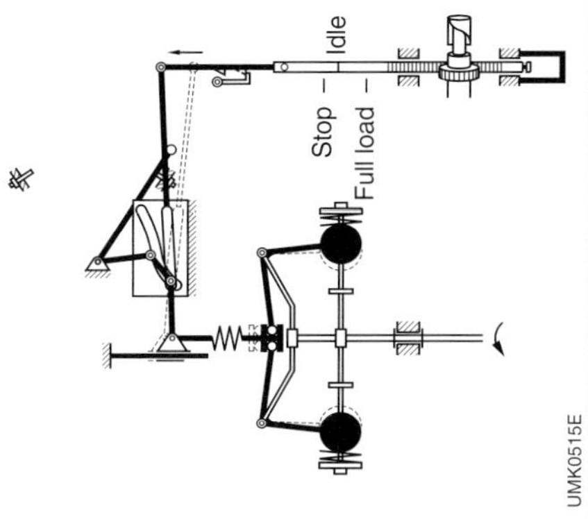

UMK0515E

Fig. 18

Fig. 19

RQUV Variable-speed governor
1 Step-up gearing.

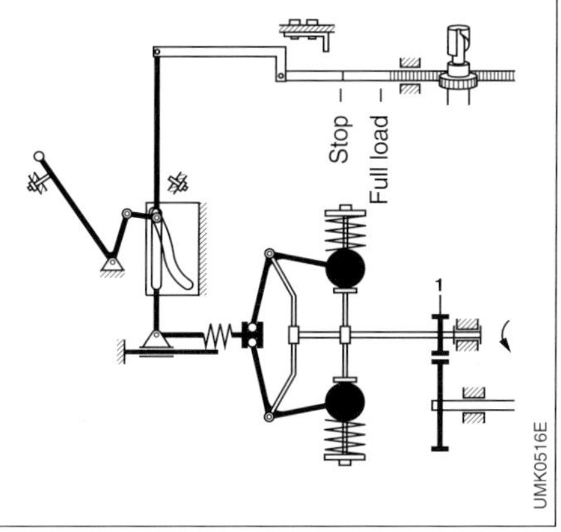

UMK0516E

Fig. 20

RQUV Variable-speed governor
1 Plate cam, **2** Step-up gearing.

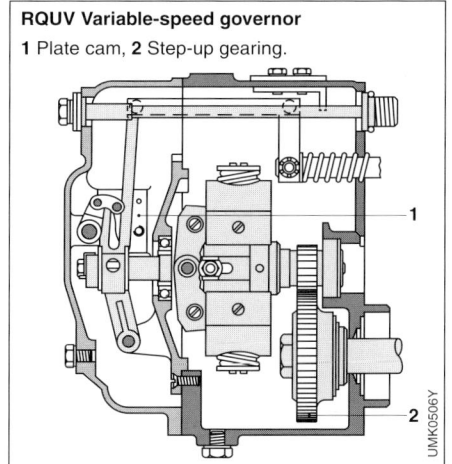

UMK0506Y

RQV..K Variable-speed governors

Design

Basically, the RQV..K governor (Figs. 21 and 22) is fitted with the same flyweight speed-sensing mechanism as the RQV governor, with the governor springs incorporated in the flyweights. The essential difference between the two governor types lies in the torque control. Whereas, with all other governor types, the torque control reduces the delivery quantity if the speed continues to increase at full load; with the RQV..K governor, the full-load delivery quantity can be increased as well as reduced.

Starting the engine

The operating instructions for the engine are to be followed as already described for the RQ governor. If the cold-start quantity is needed, the control lever must be shifted to the high idle (maximum) speed position (Fig. 23).

In the process, the rocker pivots under the full-load stop and the control rack shifts to the start-quantity position A_1 (Diagram, Fig. 26). There is a start-quantity stop on the injection pump. When the starter is actuated, the injection pump injects the start quantity through the nozzle into the engine.

A temperature-sensitive travel stop is provided to adjust the RQV..K governor's start quantity.

Operating characteristics

Idle speed (Fig. 24)

As soon as the engine has started, the control lever moves back to the idle position.

Fig. 21

RQV..K Variable-speed governor

1 Control lever,
2 Linkage lever,
3 Plate cam,
4 Guide block,
5 Variable-fulcrum lever,
6 Rocker,
7 Plate-cam return spring,
8 Strap (spring-loaded for tension),
9 Control rack,
10 Full-load stop,
11 Pump plunger,
12 Start-quantity stop,
13 Slider,
14 Guide lever,
15 Sliding bolt,
16 Bell crank,
17 Adjusting nut,
18 Governor spring,
19 Flyweight,
20 Camshaft.

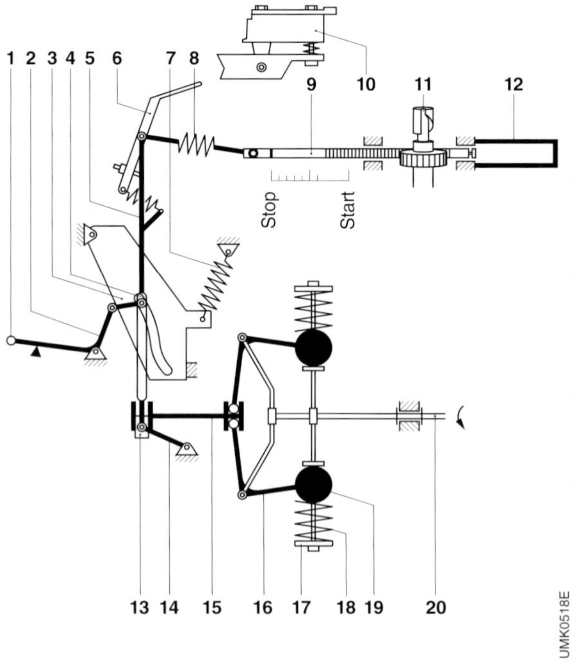

RQV..K Variable-speed governor

1 Full-load stop with rocker guide, **2** Control rack, **3** Adjusting nut, **4** Governor spring, **5** Flyweight, **6** Rocker, **7** Control lever, **8** Plate cam, **9** Guide block, **10** Fulcrum lever, **11** Slider, **12** Bell crank, **13** Sliding bolt, **14** Guide lever.

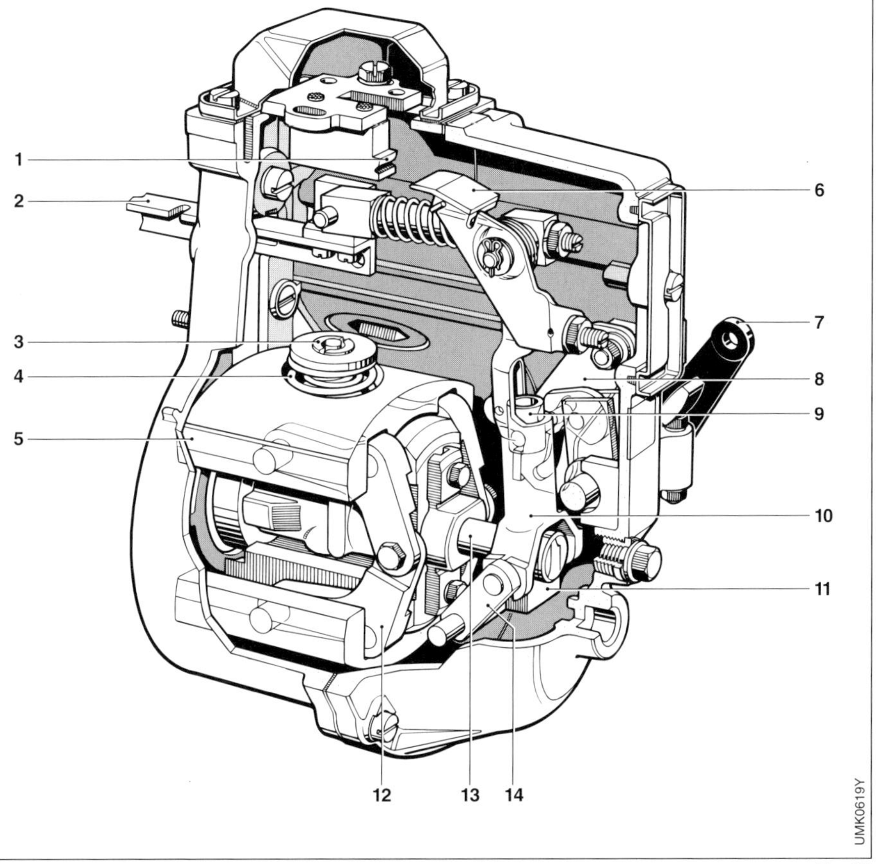

Fig. 22

Fig. 23

RQV..K Variable-speed governor

Cold-start setting.
1 Start-quantity stop.

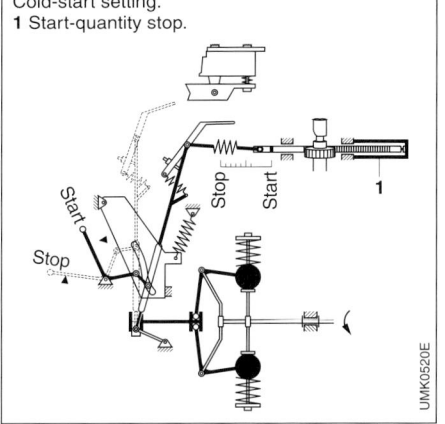

Fig. 24

RQV..K Variable-speed governor

Idle setting.
1 Rocker, **2** Full-load stop with rocker guide (adjustable).

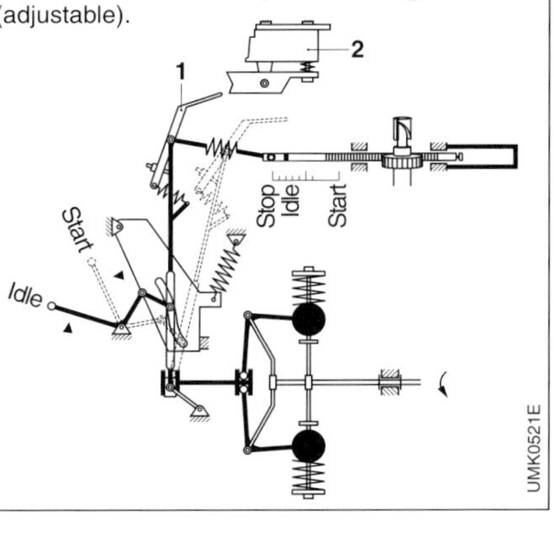

The spring-loaded rocker slides under the full-load stop and back to the idle position. The engine now runs at idle speed.

Intermediate speed
The family of curves demonstrate, for instance at B, the possibilities of speed regulation for intermediate speeds.

Full-load delivery at low engine speeds
Figure 25
If, for instance, the control lever is moved from low idle to high idle (maximum) speed, the guide block follows the rocker guide in the plate cam while at the same time moving downwards in the fulcrum-lever guide. The fulcrum lever swivels to the right around the sliding-block pivot point and, by means of the strap, shifts the control rack in the full-load direction. Fuel delivery increases and the engine speed rises.

The flyweights move outward and the sliding sleeve shifts slightly to the right. This causes a swivelling movement; the guide lever and the variable-fulcrum lever are lifted slightly, so that the rocker slides along the rocker guide of the full-load stop (Fig. 26: A – B on the curve). When the control lever is pushed forward, as soon as the rocker contacts the rocker guide, the plate cam lifts from its stop on the housing against the force of the return spring.

Torque control
The RQV..K governor's torque-control mechanism features a rocker on the end of the fulcrum lever which follows the rocker guide (i.e. a curved path) on the full-load stop. This rocker-guide path is contoured to correspond to the engine's fuel requirements. The strap, in its function as the connection between variable-fulcrum lever and control rack, transfers the rocker movement to the control rack. The result is a full-load delivery quantity which corresponds to the required torque curve. Depending upon the shape of the rocker-guide contour, the delivery quantity can be either increased or decreased. In order to adjust the fuel delivery, the full-load stop can be shifted in the longitudinal direction.

Full-load delivery at medium intermediate speeds with torque control (Figure 27)
If the speed continues to increase, the flyweights move further out and the rocker slides along the rocker guide of the full-load stop. Until the curve changes direction at B, the torque control is characterized by an increase in full-load delivery with increasing speed (negative torque control). Once the direction reverses, the torque control reduces the full-load delivery (positive torque control, Fig. 26: B – C on curve).

High idle (maximum) speed (Figure 28)
When torque control has been completed, and speed regulation starts, the plate cam is again positioned against the housing.

Fig. 25

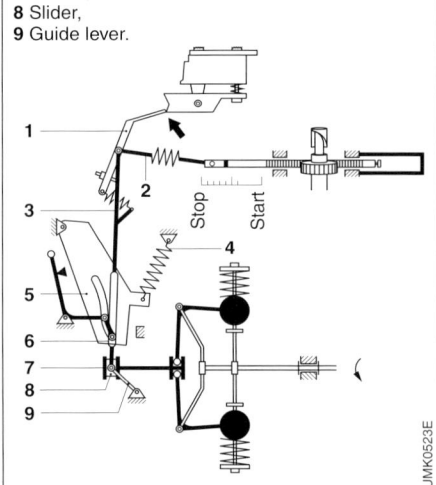

RQV..K Variable-speed governor

Full-load delivery at low speeds. Start of negative torque control.
1 Rocker,
2 Strap,
3 Fulcrum lever,
4 Return spring,
5 Plate cam,
6 Guide block,
7 Sleeve,
8 Slider,
9 Guide lever.

Governor characteristic curves, RQV..K variable-speed governor

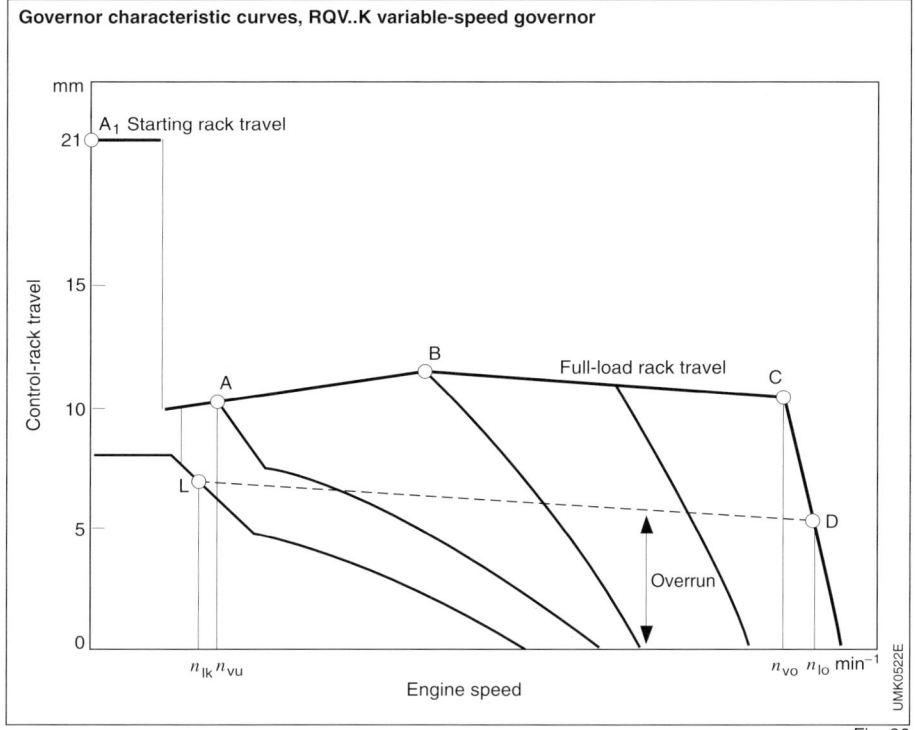

Fig. 26

Fig. 27

RQV..K Variable-speed governor

Full-load delivery at intermediate speed.
Torque-control reversal.

Fig. 28

RQV..K Variable-speed governor

Maximum full-load speed.
End of positive torque control (dashed line: speed regulation).

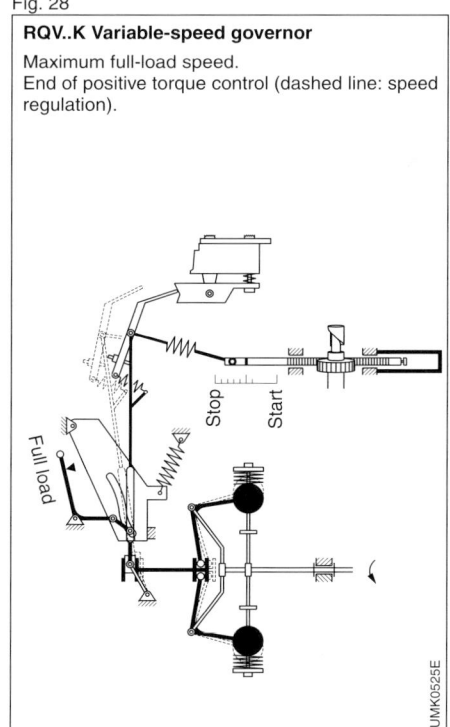

Further increases in speed initiate the regulation of the high idle (maximum) speed. The flyweights move further out, causing the sleeve to move to the right by a corresponding increment. As a result, the top part of the variable-fulcrum lever swivels to the left around the guide-block pivot point. The control rack shifts in the Stop (shutoff) direction (Fig. 26: C – D on the curve).

During operation this means that every control-lever position corresponds to a given engine-speed range as long as the engine is not overloaded or being driven by the vehicle when travelling downhill (overrun). If the load on the engine now increases slightly, for instance when driving uphill, the engine and governor speeds drop. As a result, the flyweights move inward and shift the control rack in the full-load delivery direction, thus maintaining engine speed constant at a level determined by the control-lever position (accelerator pedal) and the speed droop.

However, if the load (or gradient) increases to such an extent that engine speed drops even after the control rack has assumed the full-load stop position, the reduction in speed causes the flyweights to move even further inwards; the sleeve moves further in the full-load delivery direction.

Because the control rack cannot move any further in the full-load delivery direction, the lower section of the fulcrum lever moves to the left against the force of the plate-cam return spring, lifting the plate cam away from its stop.

Downhill driving induces the opposite response: Here, the engine accelerates due to it being driven by the vehicle (overrun).

The flyweights therefore move outward and the control rack is shifted in the Stop (shutoff) direction.

If the engine speed continues to increase, and with the control rack in the Stop (shutoff) position, the spring-loaded strap which connects the fulcrum lever with the control rack gives way. If the driver now brakes slightly, or changes to a higher gear, the bar returns to its normal length.

The above governor behavior applies basically for all control-lever settings, should for any reason the engine load or speed change to such an extent that the control rack comes up against either the full-load stop or the shutoff stop.

RSV Variable-speed governors

Design
The design of the RSV variable-speed governor (Figs. 29 and 30) differs substantially from that of the comparable RQV type. It has only one governor spring, and this is swivel-mounted. When the engine speed is set at the control lever, the spring's position and tension adapt for the required speed, providing a state of mutual balance between the torque at the tensioning lever and that generated by the flyweights.

All settings at the control lever, as well as the flyweight travel, are transferred to the control rack through the governor linkage.

The starting spring hooked into the top end of the variable-fulcrum lever pulls the control rack into the start position, automatically setting the start quantity.

The full-load stop and the torque-control mechanism are integrated in the governor. The auxiliary idle-speed spring and adjusting screw fitted in the governor cover serve for idle stabilization.

One end of the governor spring is hooked into the tensioning lever, and the other end into the rocker. The rocker screw can be adjusted to vary the force which the governor spring exerts at the tensioning lever's pivot point. This makes it possible to adjust the governor's speed droop within certain limits without the governor spring having to be changed. This is one of the advantages of the RSV governor. Lighter flyweights are available for higher rotational speeds.

RSV Variable-speed governor

1 Governor housing, **2** Starting spring, **3** Control rack, **4** Strap, **5** Rocker, **6** Swivelling lever, **7** Control lever, **8** Governor cover, **9** Stop/Idle stop, **10** Tensioning lever, **11** Guide lever, **12** Governor spring, **13** Auxiliary idle-speed spring, **14** Torque-control spring, **15** Flyweight, **16** Guide bushing, **17** Fulcrum lever, **18** Full-load stop.

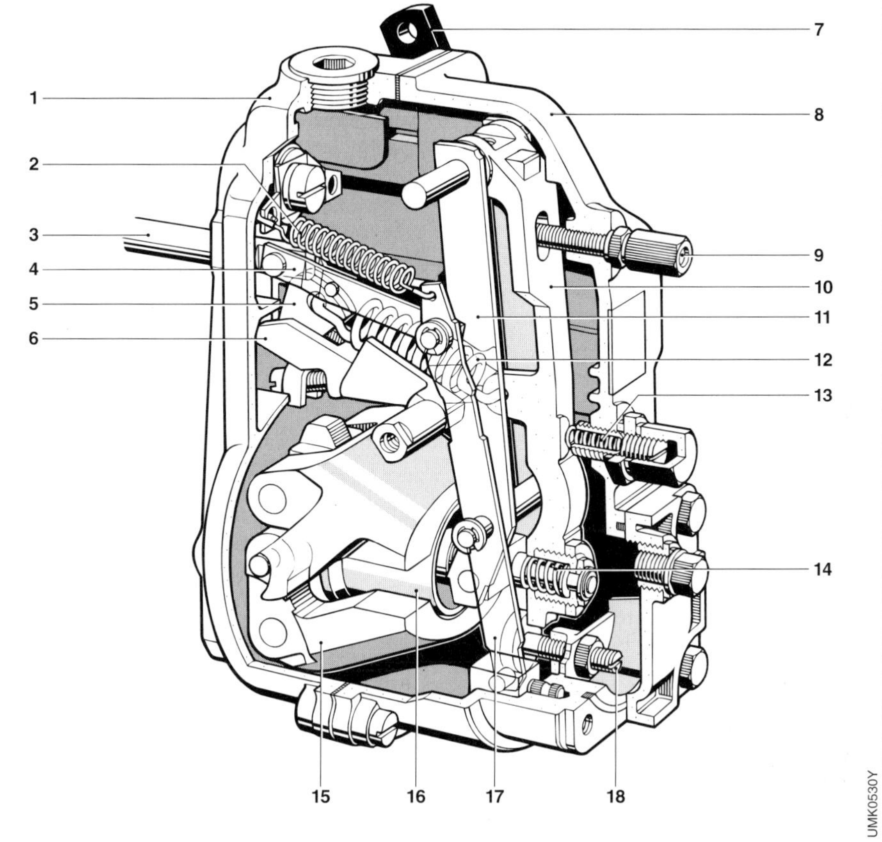

Fig. 29

Fig. 30

RSV Governor (schematic)

1 Pump plunger,
2 Control rack,
3 Stop for high idle (maximum) speed,
4 Control lever,
5 Starting spring,
6 Swivelling lever,
7 Rocker,
8 Camshaft,
9 Governor hub,
10 Flyweight,
11 Sliding bolt,
12 Stop/Idle stop,
13 Guide lever,
14 Fulcrum lever,
15 Governor spring,
16 Auxiliary idle-speed spring,
17 Tensioning lever,
18 Torque-control spring,
19 Full-load stop.

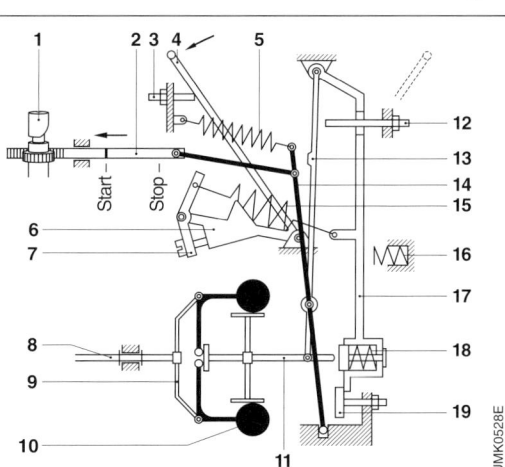

Starting the engine

When the engine is at a standstill, the RSV governor's control rack is always at Point A (Fig. 32) irrespective of the control-lever setting. For this reason, the supplementary TAS unit (For description, see "Temperature-dependent starting stop") is recommended for this type of governor in particular.

Operating characteristics

Low-idle speed (Figure 31)

The control lever is positioned up against the idle stop. As a result, the governor spring is almost completely relaxed and practically vertical. The governor spring has a very weak effect in this position so that the governor springs move outward even at very low speeds. The sliding bolt can then move to the right, causing the guide lever to do the same. This, in turn, pivots the fulcrum lever to the right. The final result is that the control rack moves toward Stop (shutoff), to the idle position L (Fig. 32). The tensioning lever comes up against the auxiliary idle-speed spring which assists the low-idle-speed control.

Low intermediate speeds (Fig. 33)

Only a very small shift of the control lever past its low-idle setting is needed to shift the control rack from its initial position (Fig. 32: Point L) to its full-load position (Fig. 32: Point B′).

The injection pump injects the full-load delivery quantity into the engine cylinders and the engine speed increases (B′ – B″).

As soon as the centrifugal force of the flyweights exceeds the governor-spring tension which corresponds to the control-lever position, the flyweights move outwards and pull the guide bushing, sliding bolt, fulcrum lever, and control rack back to a position for reduced fuel delivery (Fig. 32: Point C). The engine speed stops increasing and is held constant by the governor provided that conditions otherwise remain unchanged.

Torque control

On governors equipped with torque control, the compression force against the torque-control spring increases steadily once the tensioning lever hits the full-load stop (Figure 32: D – E). This causes the guide lever, fulcrum lever, and control rack to move accordingly in the Stop (shutoff) direction and apply "torque control" to the fuel delivery. In other words, fuel delivery is reduced by a quantity corresponding to the torque-control travel.

High-idle (maximum) speed (Figure 34)

Basically, if the control lever is moved to the maximum-speed stop, the governor operates as described above (see "Low intermediate speeds").

The only difference is that the swivelling lever fully tensions the governor spring. This causes the governor spring to apply very high forces to pull the tensioning lever back to the full-load stop, and the control rack to the full-load delivery position.

Fig. 31

RSV Variable-speed governor

Idle setting.
1 Idle stop,
2 Shutoff stop,
3 Governor spring,
4 Fulcrum lever,
5 Auxiliary idle-speed spring,
6 Tensioning lever,
7 Sliding bolt.

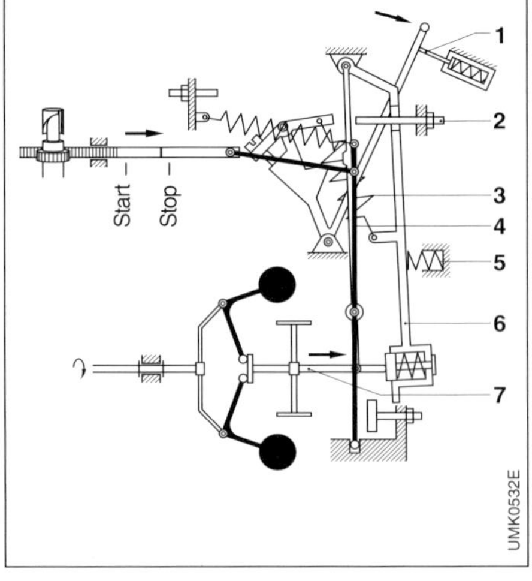

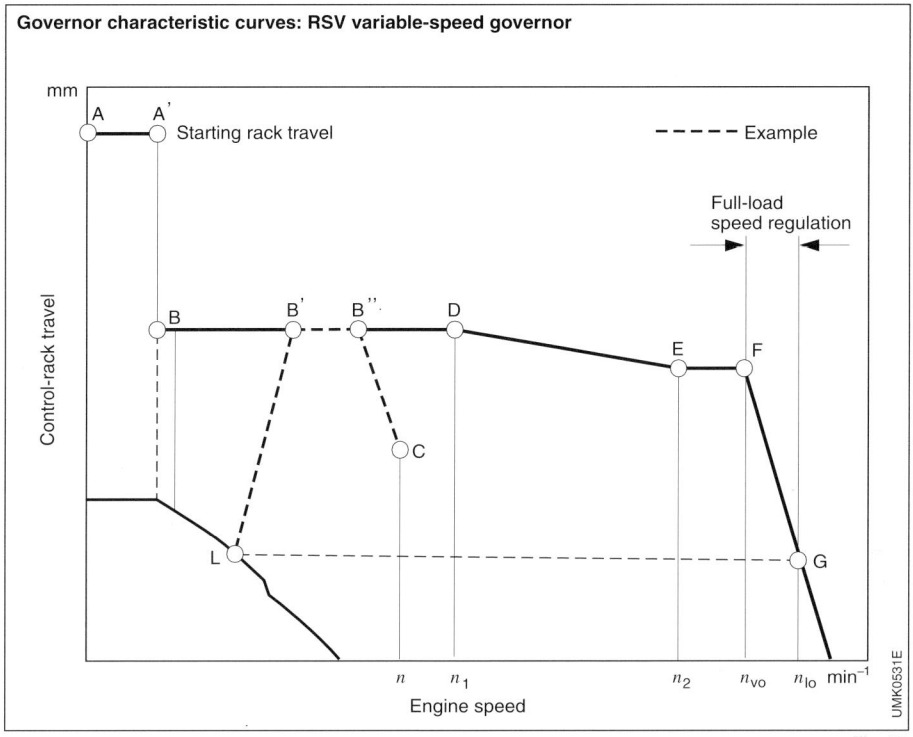

Governor characteristic curves: RSV variable-speed governor

Fig. 32

Fig. 33

RSV Variable-speed governor

Full load at low rotational speed.
Start of torque control.
a Torque-control rack travel.

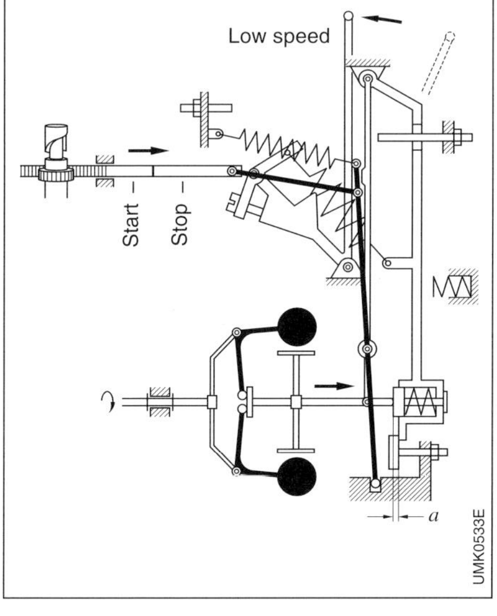

Fig. 34

RSV Variable-speed governor

No-load, regulated from full load.
1 Swivelling lever, **2** Tensioning lever,
3 Guide lever, **4** Fulcrum lever,
5 Torque-control spring.

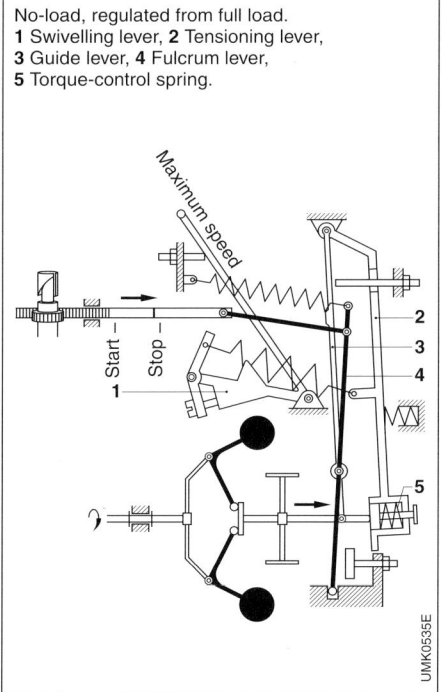

The engine speed increases and the centrifugal force of the flyweights continues to rise. As soon as the maximum full-load speed n_{vo} is reached, the centrifugal force overcomes the governor-spring tension and the tensioning lever is deflected to the right. Sliding bolt, guide lever and control rack (connected through the fulcrum lever) move toward Stop (shutoff) (Fig. 32: F – G).

This adjustment is completed as soon as the injected fuel quantity has been reduced to suit the new loading conditions. The high idle (maximum) speed n_{lo} is reached when the entire load is removed from the engine.

Stopping the engine

Stopping the engine with the control lever (Figure 35)

Engines with governors not equipped with a special shutoff device are stopped by moving the governor control lever to the Stop (shutoff position).

In the process, the lug of the swivelling lever (inclined arrow in Fig. 35, Pos. 1) presses against the guide lever. This then swivels to the right and shifts the variable-fulcrum lever, and with it the control rack, to the Stop (shutoff) position. As the sliding bolt is relieved of the pressure from the governor spring, the flyweights are free to swing outward.

Stopping the engine with the stop lever (Figure 36)

When the governor is equipped with a special shutoff device, the control rack can be shifted to Stop by pushing the stop lever to the Stop (shutoff) position. When the stop lever is shifted to "Stop", the upper section of the control lever then swivels to the right around pivot point C in the guide lever, pulling the control rack to the Stop (shutoff) position via the strap. After it has been released, the stop lever is returned to its initial position by the return spring (not shown).

Fig. 35

RSV Variable-speed governor

Stopping the engine with the governor control lever.
1 Swivelling-lever lug, **2** Shutoff stop.

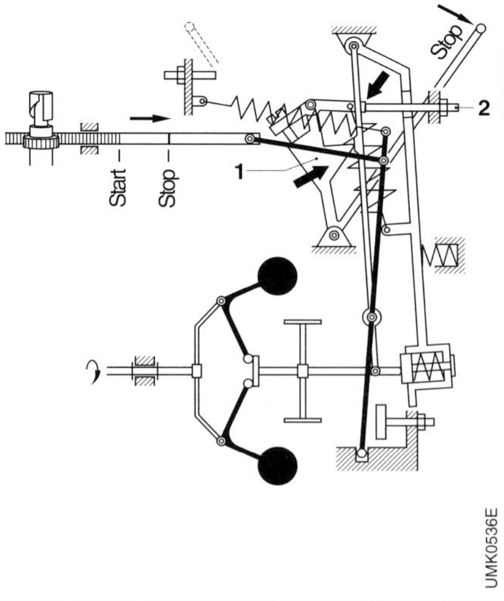

Fig. 36

RSV Variable-speed governor

Stopping the engine with the shutoff lever.
1 Idle-speed stop, **2** Stop (shutoff) lever.

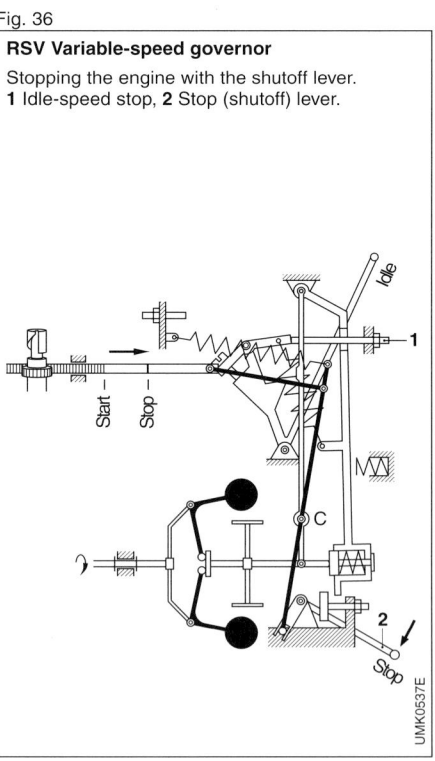

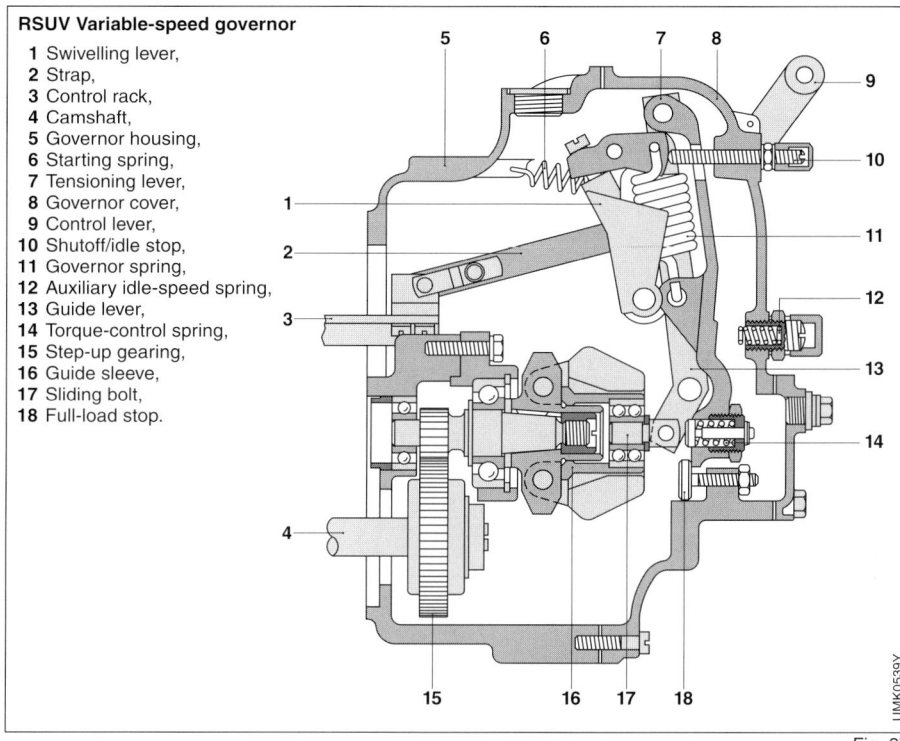

RSUV Variable-speed governor

1 Swivelling lever,
2 Strap,
3 Control rack,
4 Camshaft,
5 Governor housing,
6 Starting spring,
7 Tensioning lever,
8 Governor cover,
9 Control lever,
10 Shutoff/idle stop,
11 Governor spring,
12 Auxiliary idle-speed spring,
13 Guide lever,
14 Torque-control spring,
15 Step-up gearing,
16 Guide sleeve,
17 Sliding bolt,
18 Full-load stop.

Fig. 37

Fig. 38

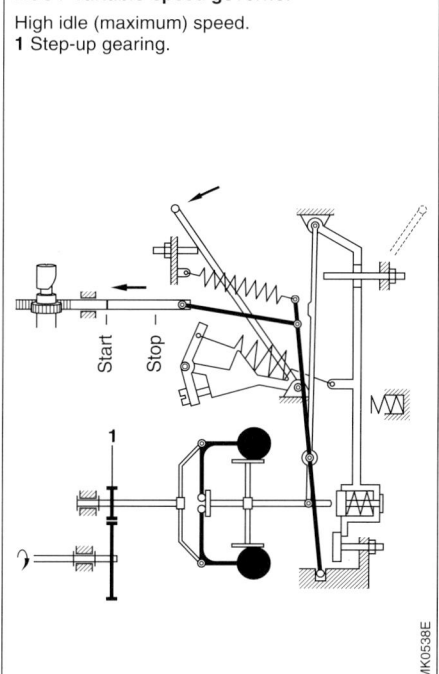

RSUV Variable-speed governor

High idle (maximum) speed.
1 Step-up gearing.

RSUV Variable-speed governors

Design

The RSUV governor is used for regulating the extremely low speeds which are typical, for example, of low-speed marine engines. The major design difference relative to the RSV governor lies in the step-up gearing between the injection pump's camshaft and the RSUV governor hub (Fig. 37).
The RSUV variable-speed governor is used on the Size P in-line fuel-injection pump.

Operating characteristics

Method of operation and operating characteristics correspond to those of the RSV governor. Figure 38 shows an RSUV variable-speed governor in the maximum-speed setting.

127

RS Minimum-maximum-speed governors

Design

The RS minimum-maximum-speed governor is a development of the RSV variable-speed governor, and features low control-lever forces. On the RSV governor, the control lever tensions the swivel spring and therefore sets the speed. On the RS governor, this spring is blocked in the high-idle (maximum) speed position by an adjustable stop on the governor cover. The RS governor also has a facility for setting an intermediate speed, for instance for vehicles with power take-off. On the RS governor, the operating direction of the RSV governor stop lever (Figs. 39 and 40) is reversed to allow use as an accelerator lever.

Starting the engine

To start, the accelerator lever pivots in the full-load direction and pushes the sliding bolt against the spring retainer via the variable-fulcrum lever and the guide lever. The retainer's idle-speed spring shifts the control rack to the start-quantity setting (Fig. 42).

Operating characteristics

Low idle speed

The flyweights start to move out even at very low speeds, and the sliding bolt and guide lever move to the right. The guide lever pivots the variable-fulcrum lever to the right so that the control rack is moved toward Stop to the idle position L (Fig. 41). Also, the sliding bolt presses up against the spring retainer which, in addition to the torque-control spring, has an idle-speed spring for low-idle-speed control. Idle-speed stop screw and auxiliary idle-speed spring as used with the RSV governor are not needed.

High idle (maximum) speed

As soon as engine speed exceeds maximum full-load speed, the control rack moves in the Stop direction (Fig. 41 E – F). With load removed completely, the engine runs at high idle (maximum) speed.

RS Minimum-maximum-speed governor

(External view).
1 Adjusting lever for high-idle (maximum) and intermediate speeds,
2 Accelerator lever (pedal).

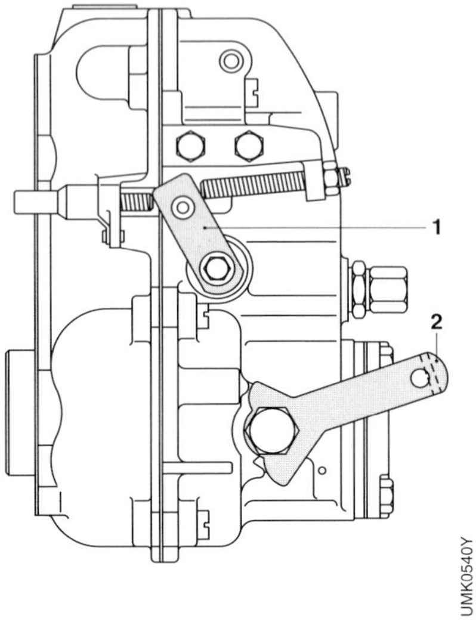

UMK0540Y

Fig. 39

Fig. 40

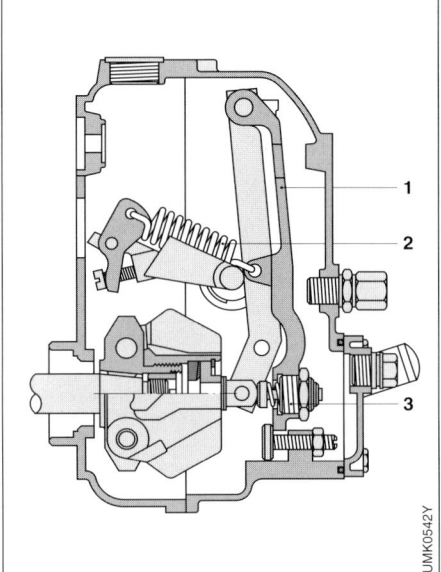

RS Minimum-maximum-speed governor

(Section drawing).
1 Tensioning lever,
2 Governor spring,
3 Spring retainer.

UMK0542Y

Governor characteristic curves, RS minimum-maximum-speed governor

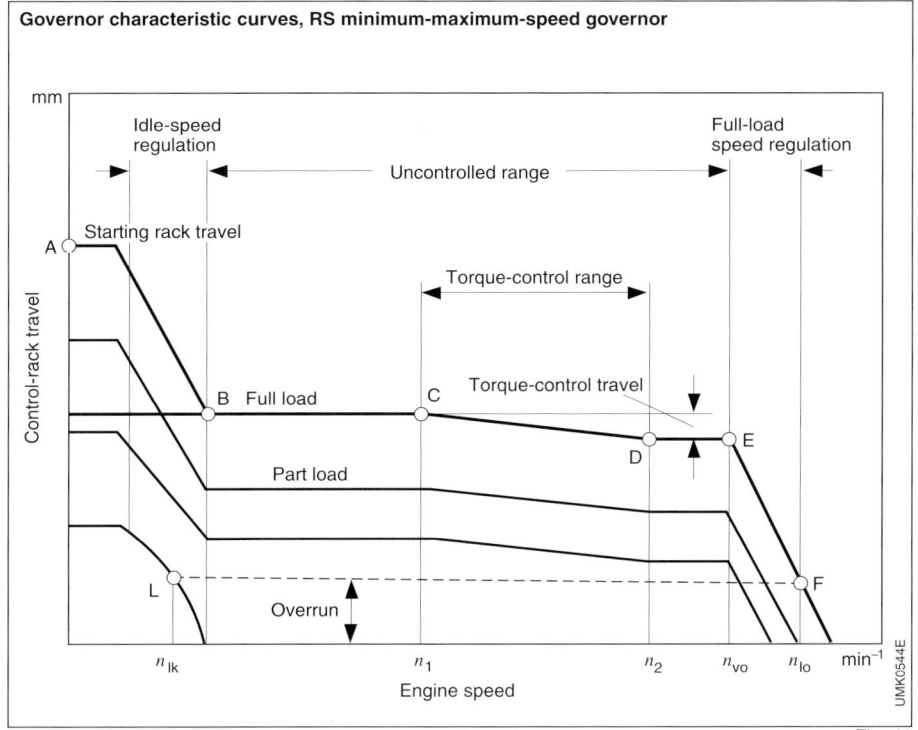

Fig. 41

Fig. 42

RS Minimum-maximum-speed governor

Cold-start setting.
1 Pump plunger,
2 Control rack,
3 Strap,
4 Swivelling lever,
5 Rocker,
6 Control lever,
7 Tensioning lever,
8 Guide lever,
9 Governor spring,
10 Fulcrum lever,
11 Spring retainer,
11.1 Thrust pin,
11.2 Torque-control spring,
11.3 Idle-speed spring,
12 Accelerator lever (pedal),
13 Full-load stop,
14 Sliding bolt,
15 Camshaft.
a Idle stage.

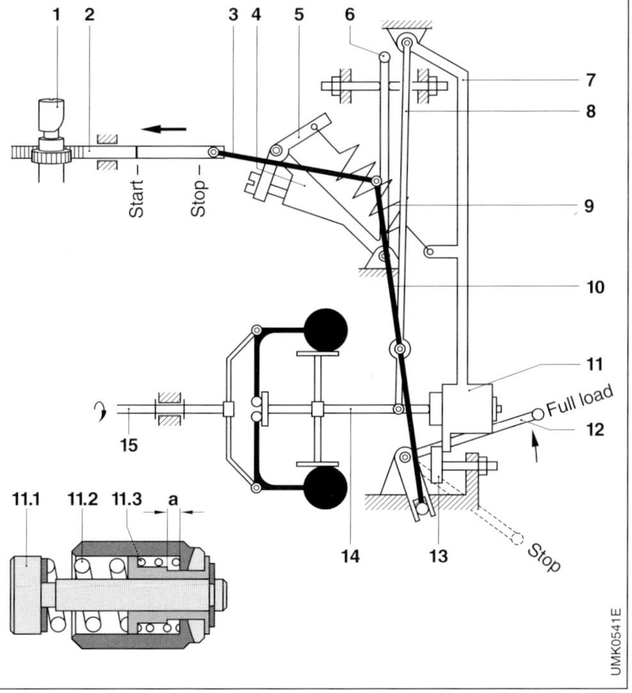

129

RSF Minimum-maximum-speed governors

Design

The RSF mechanical governor was developed specifically as a minimum-maximum-speed governor for vehicle engines fitted with a Size-M in-line fuel-injection pump. It is suitable for use in those on-road vehicles (passenger cars and commercial vehicles) in which control requirements are restricted to low idle and high idle (maximum) speeds. In the uncontrolled range between these two speeds, the driver uses the accelerator pedal to directly adjust the setting of the injection-pump control rack so that the engine develops the right torque (Fig. 43).

The RSF governor complies with today's high standards for control response, operational sophistication, and ride quality. It is mainly used for modern high-speed passenger-car diesel engines. In addition, the RSF governor is particularly suitable for the fitting of add-on adaptation units, as well as being easy to adjust.

Explanation to Fig. 43:

a Idle range (working range of the idle spring), **b** Extended idle range at no-load and minimum part load (working range of the idle spring and the auxiliary idle spring), **c** Uncontrolled range, **d** Torque-control range (working range of the torque-control spring), **e** Torque-control travel, **f** Speed-regulation range (working range of the governor spring), **g** Full-load speed regulation to the high idle speed, **h** Start of the auxiliary idle-spring shutoff, **S** Start setting with accelerator pedal fully depressed (cold start), **S′** Start setting with accelerator pedal released (hot/warm start), **L** Low-idle-speed setting, **O** High-idle-speed setting,

n_{lu} Low idle speed,
n_{lo} High idle speed,
n_{vo} Maximum full-load speed,
n_1 Speed at start of torque control,
n_2 Speed at end of torque control.

The governor's construction can be divided into two main parts: Speed-sensing mechanism and actuator mechanism (Fig. 44).

Fig. 43

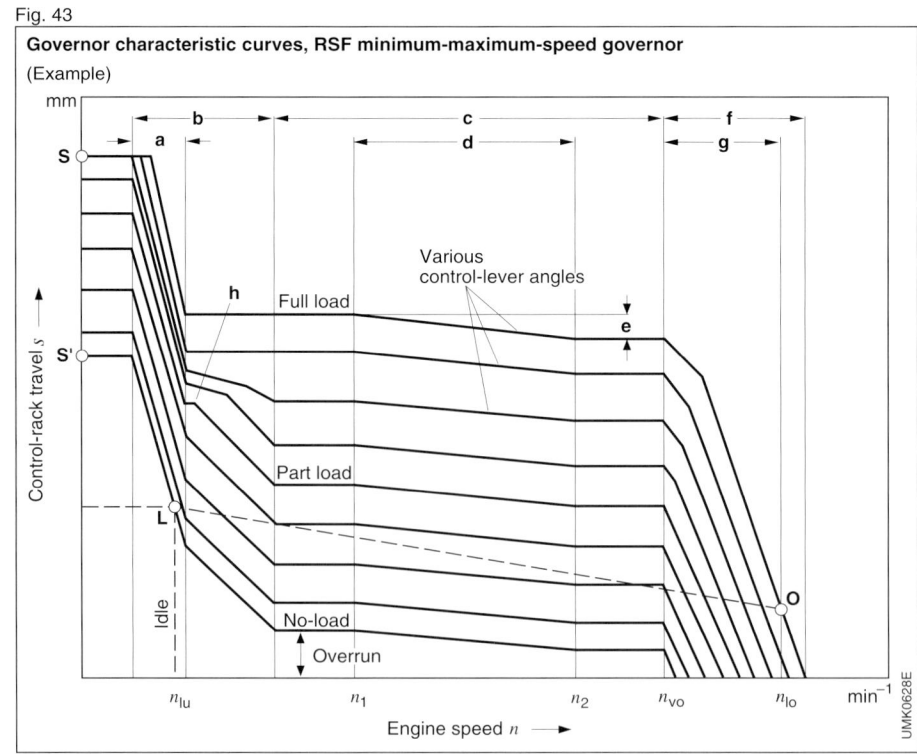

Governor characteristic curves, RSF minimum-maximum-speed governor
(Example)

Control-rack travel s — Engine speed n —

Various control-lever angles

Full load
Part load
No-load
Overrun
Idle

n_{lu} n_1 n_2 n_{vo} n_{lo} min^{-1}

UMK0628E

Speed-sensing mechanism, stage 1 (low idle)

The force path starts at flyweights (22) and travels through sliding sleeve (20) and guide lever (9) to idle spring (12) and auxiliary idle spring (14). Both springs are leaf springs.

Speed-sensing mechanism, stage 2 (until full-load speed regulation)

After completion of idle-speed sleeve travel, the force path is from governor sliding sleeve (20) through torque-control spring retainer (18) and tensioning lever (16) to governor spring (17). The flyweights (22) are fastened directly to the injection-pump camshaft and when they move outward, governor sliding sleeve (20) is moved axially. The sliding sleeve does not move except in the idle, full-load torque-control, and speed-regulation breakaway ranges. The fuel-delivery quantity needed for the required engine torque is selected by the actuator-mechanism control lever.

The guide lever (9) is movably connected at pivot point B with the governor sliding sleeve. In addition, guide lever and tensioning lever (16) also pivot at point A.

Actuator mechanism

Input of the desired value is through control lever (6), linkage lever (5) and reverse-transfer lever (11) to variable-fulcrum lever (13), and from there through strap (2) to injection-pump control rack (4). The strap is spring-loaded, and compensates for the extra movement of the fulcrum lever. Similar to the guide lever, the reverse-transfer lever is also fexibly mounted at pivot point B of the sliding sleeve, and is also attached to the fulcrum lever (13) via a shaft and bushing. Full-load adjusting screw (19) is used to adjust full-load delivery at the fulcrum lever's lower bearing point, which also acts as the spring-loaded yield point for the fulcrum lever for absorbing the sliding sleeve's additional travel in case of excessive speed. The bearing shaft of the stop lever (3) protrudes from the governor housing. Attached to its end is a stop (shutoff) lever (1) which switches off

Fig. 44

RSF Minimum-maximum-speed governor

1 Stop (shutoff) lever,
2 Strap,
3 Stop lever,
4 Control rack,
5 Linkage lever (inner),
6 Control lever (outer),
7 Full-load stop,
8 Adjusting screw for idle speed,
9 Guide lever,
10 Idle stop,
11 Reverse-transfer lever,
12 Idle-speed spring,
13 Variable-fulcrum lever,
14 Auxiliary idle spring,
15 Adjusting screw for auxiliary idle spring,
16 Tensioning lever,
17 Governor spring,
18 Spring retainer (torque control),
19 Full-load adjusting screw,
20 Governor sliding sleeve,
21 Auxiliary-idle-spring switch-off device,
22 Flyweight.

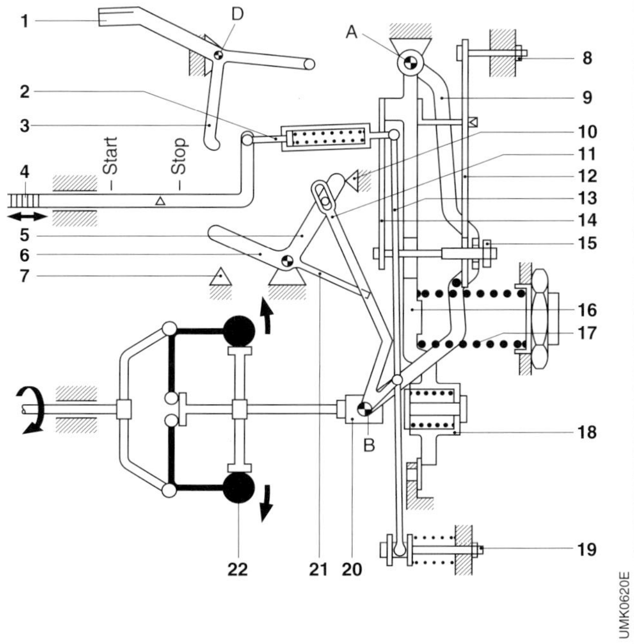

UMK0620E

the engine, whereby the stop lever pulls the control rack in the Stop (shutoff) direction.

Starting the engine

The engine is to be started according to the vehicle manufacturer's operating instructions. Normally, it can be started without pressing the accelerator pedal (Fig. 45). Only with a cold engine at low temperatures is the accelerator pedal to be pushed to the floor so that control lever (6) is shifted up against full-load stop (7) – a fixed stop on the governor housing. Reverse-transfer lever (11) swivels around pivot point B, and in doing so shifts fulcrum lever (13) in the "Start" direction. The result is that the control rack (4) is moved to the start setting and the engine receives the required start quantity. Rapid speed regulation from the governor start setting is made possible by lifting the auxiliary idle spring (14) away from fulcrum lever (9) by a switch-off device (21) when the control lever is in the full-load position.

Operating characteristics

Low idle speed (Fig. 46)

Once the engine starts and the accelerator pedal is released, a return spring pulls control lever (6) back to the low-idle position. Linkage lever (5) is now up against idle-speed stop screw (10).

During warm-up, low idle speed follows the idle-speed control curve and stabilizes at Point L (Fig. 62). When speed increases, flyweights (22) move outward, and shift governor sliding sleeve (20) to the right. During operation within the idle range, sliding sleeve movement is tansmitted through reverse-transfer lever (11) and fulcrum lever (13); moving the control rack (4) toward Stop (shutoff). At the same time, sliding-sleeve movement causes guide lever (9) to swivel about pivot point A, compressing the idle leaf spring (12). This spring's pretension (and thus the low idle speed) is set using adjusting screw (8). At a given speed, the fulcrum lever also comes up against the adjusting nut for auxiliary idle spring (14).

Fig. 45

RSF Minimum-maximum-speed governor

Cold-start setting (only the components concerned in the governing process are shown).
4 Control rack, **6** Control lever (outer),
7 Full-load stop, **9** Guide lever,
11 Reverse-transfer lever,
13 Variable-fulcrum lever,
14 Auxiliary-idle spring,
21 Auxiliary-idle spring switch-off.

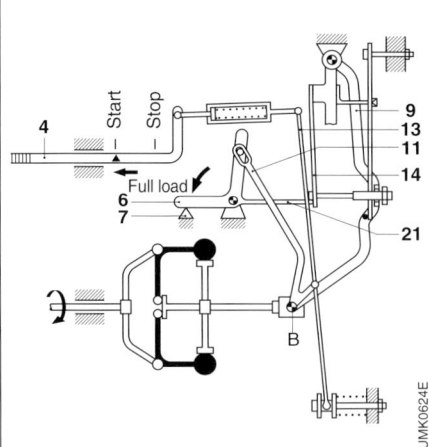

Fig. 46

RSF Minimum-maximum-speed governor

Idle-speed setting (only the components concerned in the governing process are shown).
4 Control rack, **5** Linkage lever (inner),
6 Control lever, **8** Adjusting screw for idle speed,
9 Guide lever, **10** Idle stop,
11 Reverse-transfer lever, **12** Idle-speed spring,
13 Variable-fulcrum lever, **14** Auxiliary idle spring,
15 Adjusting screw for auxiliary idle spring,
18 Spring retainer (torque control),
20 Governor sliding sleeve, **22** Flyweight.

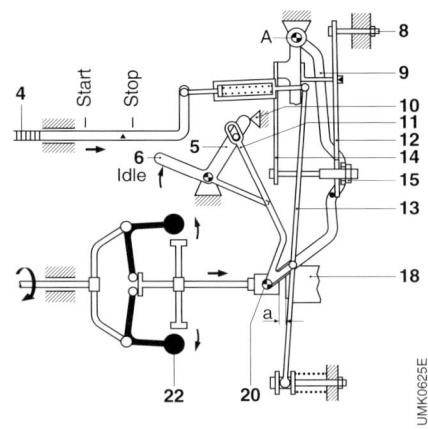

Intermediate speed

After passing through the idle stage (a), governor sliding sleeve (20) and spring retainer (18) for torque control come into contact with each other. In the uncontrolled range between idle and maximum speed, the position of the flyweights (22) remains constant up to high idle (maximum) speed, the only exception being for minor adjustments necessary for torque control. Control-rack setting and therefore delivery quantity are selected directly through control lever (6).

That is, the driver uses the accelerator pedal to select the delivery quantity necessary to increase speed or climb hills (control-lever position between idle stop and full-load stop). If the driver depresses the accelerator pedal fully, the control rack shifts to the full-load delivery setting.

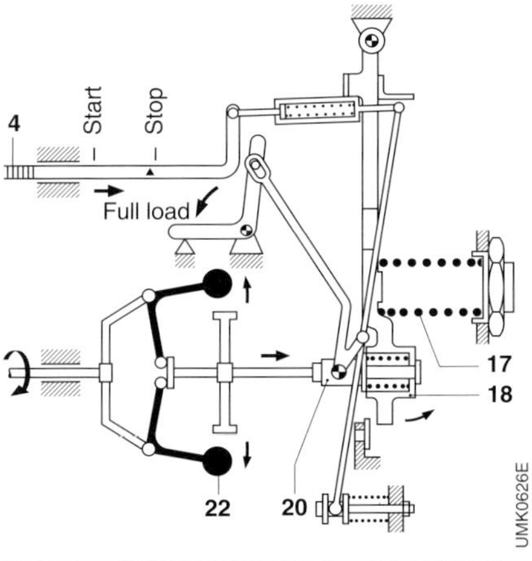

Torque control

If torque control is fitted, full-load delivery quantity is reduced when a given n_1 is exceeded, because centrifugal force acting on sliding sleeve (20) exceeds the force of the torque-control spring fitted in spring retainer (18). The torque-control spring yields to this force. As a result, if speed increases further, control rack (4) shifts by the torque-control travel distance. Torque control ends at speed n_2.

Instead of positive torque control, the RSF governor can have negative torque control. Here, the control-rack setting is controlled by a spring combination.

High idle (maximum) speed (Fig. 47)

With the accelerator pedal pressed to the floor, the full-load quantity is injected until the maximum full-load speed n_{vo} is reached. If engine speed increases beyond full-load speed, the force exerted by the flyweights (22) suffices to overcome the force of governor spring (17) and full-load-speed regulation starts. Engine speed increases slightly, and rack travel is reduced due to the rack being moved towards Stop (shutoff) and the delivery quantity decreases. Breakaway depends upon governor-spring pretension. The engine adjusts to high idle (maximum) speed n_{lo} when all load is removed.

When driving downhill with the accelerator released, the vehicle drives the engine (overrun) and accelerates it. No fuel is injected during this operating mode (overrun fuel shutoff).

Stopping the engine

When Stop (shutoff) lever (1) is moved by hand, stop lever (3) shifts control rack (4) to the Stop (shutoff) position. The fuel supply is interrupted and the engine stops. A pneumatic shutoff device can also be used for engine switchoff.

Fig. 47

RSF Minimum-maximum-speed governor

Full-load setting
(only the components concerned in the governing process are shown).
4 Control rack, **17** Governor spring,
18 Spring retainer (torque control),
20 Governor sliding sleeve, **22** Flyweight.

Add-on modules and shutoff devices

Control-lever stops

Every mechanical governor is equipped with travel stops at both ends of the control lever's travel range, at the minimum and the maximum position. When the driver fully depresses the accelerator pedal, the control lever comes up against an adjustable stop screw. Adjusting this screw causes the following changes:

– On a minimum-maximum-speed governor the rack travel, i.e. the delivery quantity.
– On a variable-speed governor the maximum speed.

This stop screw is always sealed. Any tampering with the screw will automatically invalidate the warranty coverage. The low idle speed is usually adjusted with the other stop screw. This stop can be either rigid or spring-loaded.

Rigid stop

In the case of the rigid stop (Fig. 1), the fuel-injection installation must be equipped with an extra device to switch off the engine.

Spring-loaded stop

In the case of the spring-loaded stop (Fig. 2), the stop position is reached after overcoming the low-idle position.

If required, the bottom stop can also be adjusted to "Stop"; for this the engine must be provided with an idle-speed stop.

Reduced-delivery stop or intermediate-speed stop

Stops for intermediate control-lever positions are available as optional equipment.

Depending upon the governor type, either a reduced-delivery stop can be fitted in order to lower the full-load delivery, or an intermediate-speed stop in order to set a speed lower than the rated speed (Figs. 3 and 4).

Fig. 1

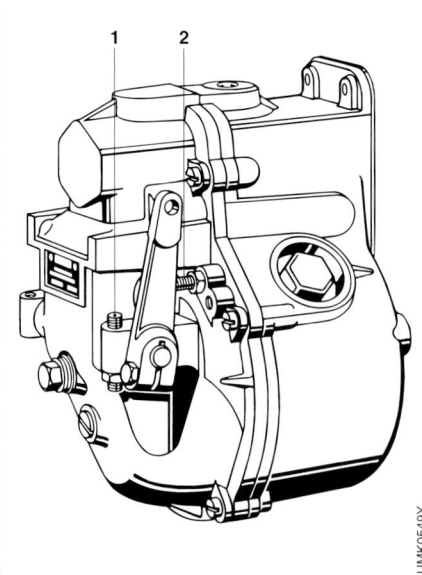

Rigid control-lever stops

1 Stop for low idle speed (or "Stop"),
2 Full-load stop on minimum-maximum-speed governors, or for the rated speed on variable-speed governors.

UMK0549Y

Fig. 2

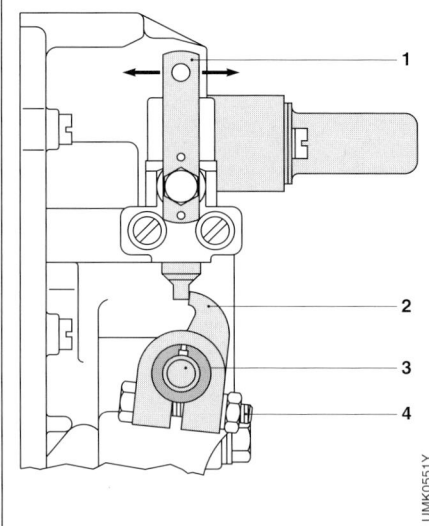

Stops

Reduced-delivery stop for the minimum-maximum-speed governor, or for intermediate speed for the variable-speed governor (outside view).
1 Lever, **2** Stop lever, **3** Control-lever shaft, **4** Clamping screw.

UMK0551Y

Spring-loaded control-lever stop (RQ and RQV)

1 Spring, **2** Threaded sleeve, **3** Bolt, **4** Stop lever, **5** Screw plug, **6** Lock nut, **7** Mounting bracket, **8** Control-lever shaft, **9** Clamping screw.
a Stop (shutoff), b Idle.

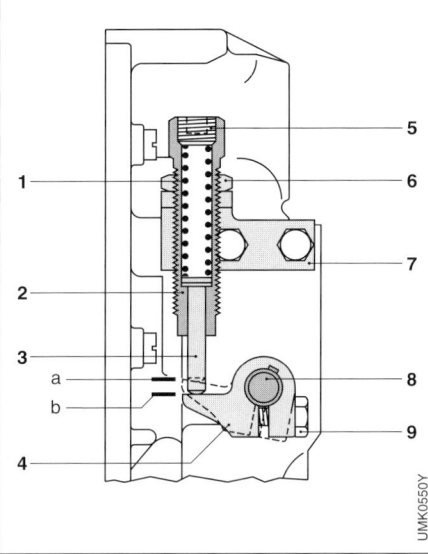

Fig. 3

Control-rack stops

In addition to travel stops for low idle or "Stop", full-load or high idle (maximum) speed (fitted on all mechanical governors to define the rack travel), a special stop is required to limit the rack travel under full-load conditions or when the start quantity is injected.

In addition, full-load stops are available to carry out specific corrective functions. Control-rack stops are mounted either on the injection pump or on the governor itself. Several of these versions will be described in more detail below.

Rigid start-quantity stop

This rigid stop is used mainly on RQ governors for very-low engine idle speeds. As soon as the engine is running, the governor cancels the excess start quantity to prevent undesirable side-effects (such as exhaust smoke).

Fig. 4

Stops

For reduced-delivery and intermediate speed.
a Blocked,
b Released.
1 Lever, **2** Housing, **3** Spring, **4** Control shaft, **5** Bolt, **6** Control-lever shaft.

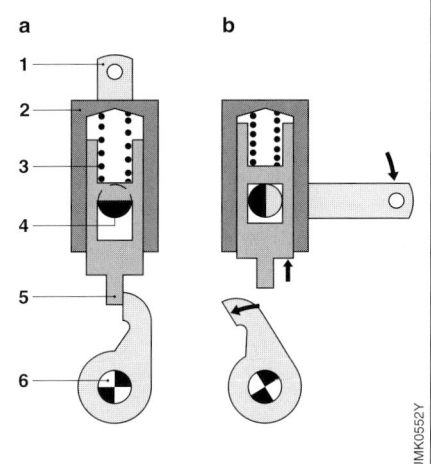

Fig. 5

Rigid control-rack stop

For start-quantity limitation on RQ governor.
1 Adjustment of the excess fuel for starting,
2 Stop pin,
3 Stop lug,
4 Start-quantity limitation,
5 Link fork.

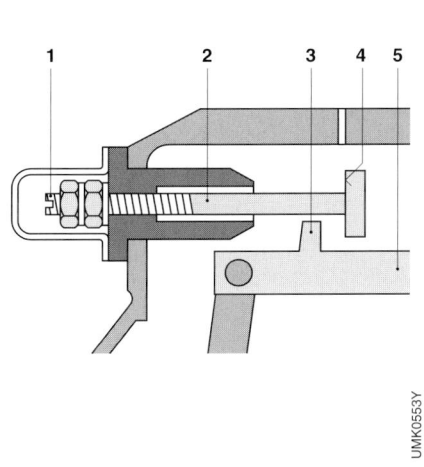

135

Spring-loaded start-quantity stop for RQ governors

When the engine is started with the accelerator pressed to the floor, the stop pin overcomes the resistance of the spring and moves to the set start-quantity position. The spring fitted inside the stop opposes the idle-speed spring and in doing so causes the early return of the control rack back from the start setting. In other words, when the accelerator is suddenly pressed to the floor from idle, this prevents the control rack moving to the start-quantity setting (Fig. 6).

Automatic full-load stop

With the engine at a standstill, the flyweight governor springs acting through the sliding bolt overcome the rocker-arm spring, with the result that the rocker arm forces the stop strap (with the full-load stop lug) downward (arrows, Fig. 8a).

This means that the control rack can shift to the start position when the accelerator pedal is fully depressed during starting (Fig. 8a). Once the engine has started (Fig. 8b), the centrifugal force causes the sliding bolt to move away from the rocker arm. And for the same reason, the control rack shifts back from the start quantity to a lower delivery quantity.

As a result, the rocker-arm spring forces the rocker arm with its long lever arm upward and the lug on the stop strap again limits the control-rack travel to full-load delivery at the stop piece on the link fork.

Control-rack stop with external torque-control mechanism (for RQV governors)

This external stop can be used to set full-load rack travel and torque-control (torque-control start, torque-control characteristic, and torque-control travel). Torque control itself results from the interplay between governor drag spring and the torque-control spring, whereby both these springs must be accurately matched to each other (Fig. 7).

If a draw lever for provision of the start quantity is fitted to this control-rack stop, the rocker lever is not fitted, i.e., the start-quantity release is no longer speed-controlled (Fig. 9).

Fig. 6

> **Spring-loaded control-rack stop**
>
> For RQ governor with start-quantity limitation.
> **1** Spring,
> **2** Governor cover,
> **3** Governor housing,
> **4** Control-rack link fork,
> a Start-quantity stop travel.

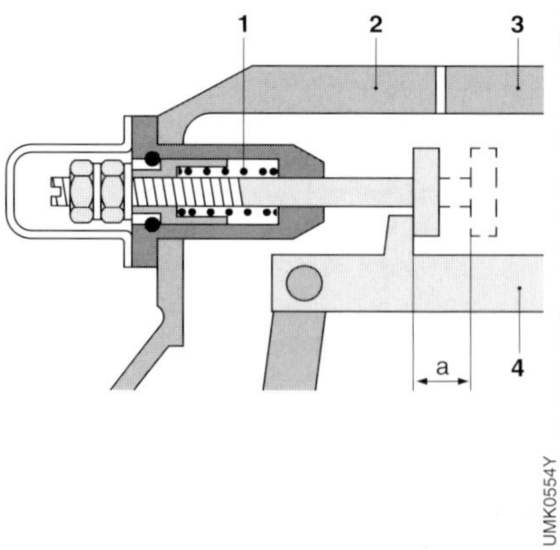

UMK0554Y

Fig. 7

> **Control-rack stop**
>
> For RQV governor with torque control.
> Torque-control spring overcomes the drag spring.
> **1** Torque-control spring,
> **2** Control rack,
> **3** Drag spring,
> a Torque-control travel.

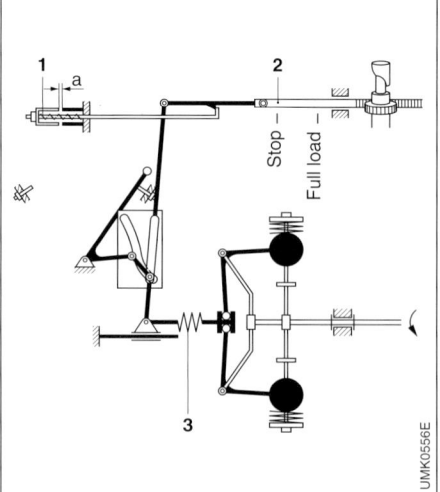

UMK0556E

Automatic full-load control-rack stop

For RQV governor.
a Provision of the start quantity,
b Full-load delivery position.
1 Full-load delivery adjustment,
2 Governor cover,
3 Governor housing,
4 Full-load stop,
5 Link fork,
6 Control rack,
7 Rocker arm,
8 Variable-fulcrum lever,
9 Control-lever shaft,
10 Rocker-arm spring,
11 Sliding bolt,
a   Start-quantity stop travel.

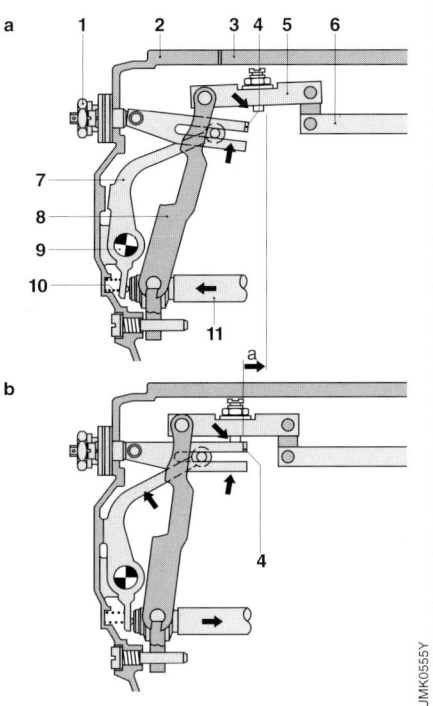

Fig. 8

Fig. 9

Control-rack stop

For RQV governor, with draw lever
for excess fuel for starting, and torque
control.
a Start-quantity position,
b Full-load delivery with torque control.
1 Locking bolt,
2 Governor cover,
3 Fork lever,
4 Variable-fulcrum lever,
5 Draw lever,
6 Draw-lever return spring,
7 Threaded sleeve,
8 Torque-control spring,
9 Adjusting screw,
a Torque-control travel,
b Excess fuel for starting.

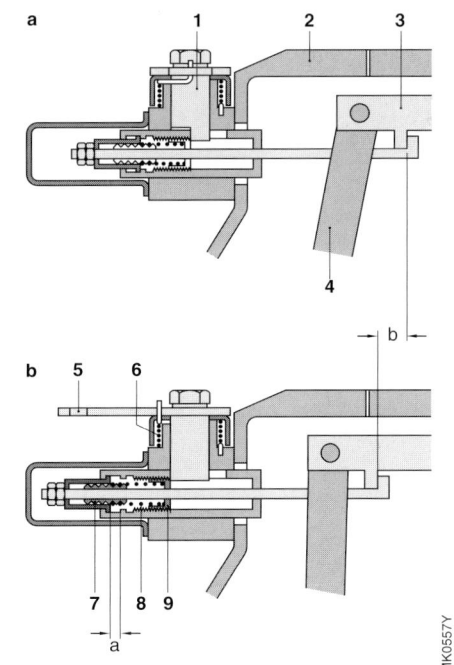

Control-rack stop with internal torque-control mechanism (for RQV governors)

In those cases in which there is not enough room for a rack stop with external torque-control mechanism, a stop with internal mechanism is installed instead because it projects only 25% as far as the "external" version. Torque-control start and torque-control travel can be set, but not the torque-control characteristic (Fig. 10).

Pump-side control-rack stops

The full-load quantity is usually set at the governor. However, there are also rigid and spring-loaded control-rack stops available which are mounted on the pump's drive end. Normally, these stops are used to set maximum start quantity, and in some rare cases full-load quantity.

Rigid version

A rigid stop adjusted for excess start quantity (Fig. 11) can be used in place of the governor-side stop shown in Fig. 5. A rigid stop, set to full load, never permits excess start quantity.

Spring-loaded version

A spring-loaded control-rack stop, as in Figure 12, can replace the pump-side stop shown in Figure 6. They operate identically.

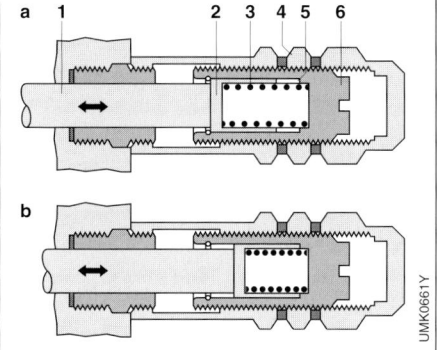

Rigid control-rack stop
1 Control rack, 2 Screw plug,
3 Adjusting screw, 4 Stop surface.

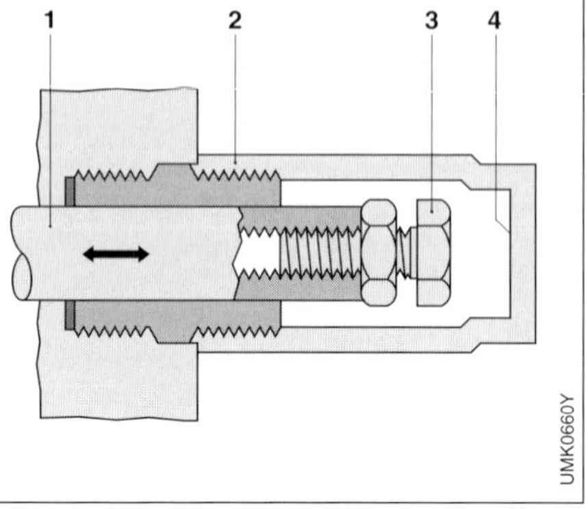

UMK0660Y

Fig. 11

Fig. 12

Spring-loaded control-rack stop
a Full-load setting, b Start setting.
1 Control rack, 2 Stop bushing, 3 Spring,
4 Lock nut, 5 Stop, 6 Adjusting bushing.

a

b

UMK0661Y

Fig. 10

RQV Governor

With internal torque-control mechanism.
1 Strap with torque-control mechanism,
2 Adjusting screw for torque-control start,
3 Torque-control spring,
4 Adjusting spring for torque-control travel,
5 Full-load stop,
6 Control rack,
a Torque-control travel.

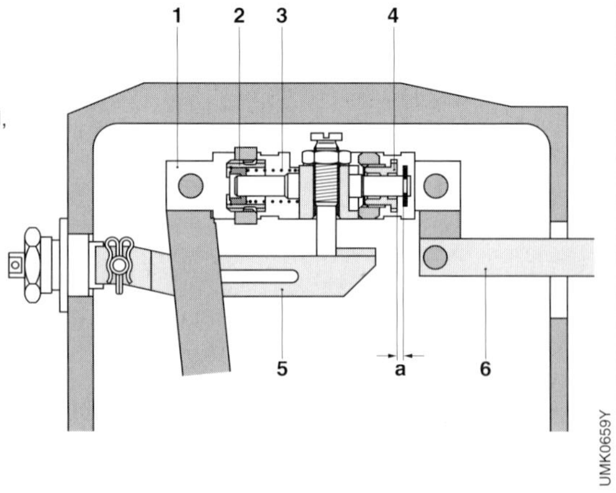

UMK0659Y

Manifold-pressure compensator (LDA)

Application

On pressure-charged engines, the full-load delivery is matched to the charge-air pressure. At lower speeds, however, the charge-air pressure is lower than at higher speeds, and as a result the air charge drawn in by the engine cylinders also weighs less. This means that the full-load delivery must be adapted to correspond to the reduced air weight. This is the task of the manifold-pressure compensator (LDA). Above a given (selectable) charge-air pressure, it limits the full-load delivery in the lower speed range. There are LDA versions available for mounting on the injection pump, and on either the rear or the top of the governor. The version described below is intended for mounting on the RSV governor (Figs. 13, 14 and 16).

Design and function

The construction of all such control-rack stops is basically the same.

A hermetically-sealed diaphragm is clamped between a housing bolted onto the governor and a mating cover. The cover contains a fitting for the charge-air pressure, and a compression spring acts against the diaphragm from below. The other end of the spring is supported on a guide bushing which is screwed into the

Fig. 13

Manifold-pressure compensator (LDA)

For RSV governor.
1 Threaded pin, **2** Plate washer, **3** Diaphragm, **4** Spring, **5** Guide bushing, **6** Bell-crank shaft, **7** Bell crank, **8** Strap, **9** Control rack, **10** Governor housing, **11** Starting spring, **12** Governor cover, **13** Variable-fulcrum lever, p_L charge-air pressure.

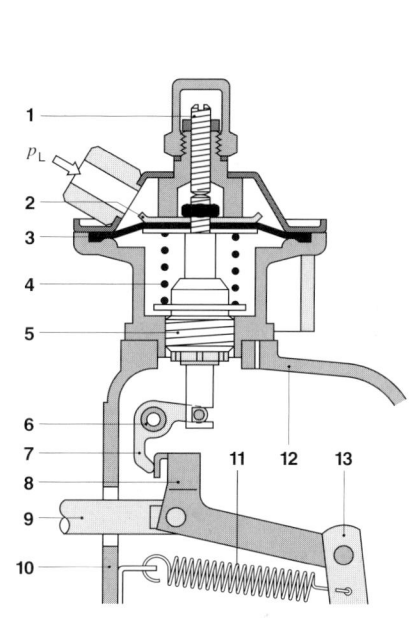

Fig. 14

Manifold-pressure compensator (LDA)

For RSV governor.
a Operating position, **b** Strap starting position referred to bell crank.
1 Strap, **2** Bell-crank shaft, **3** Bell crank.

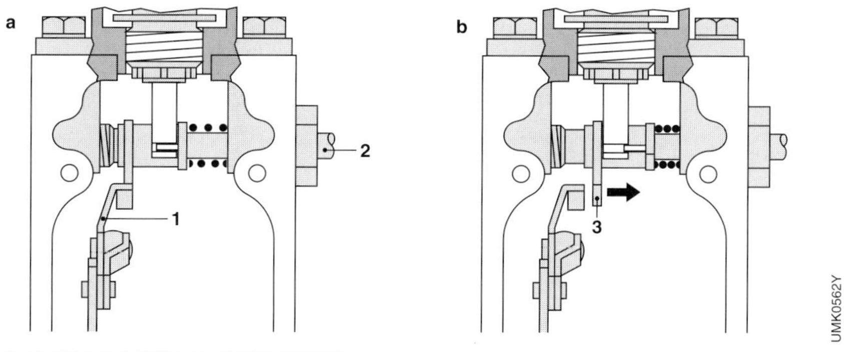

housing, and with which it is possible to vary the compression spring's pretension within certain limits. A threaded stud with a lateral slot at its lower extremity is screwed into the diaphragm. Extending into this slot is a pin, the other end of which is attached to the bell crank. The threaded stud can be used to make adjustments after the LDA has been installed. As soon as charge-air pressure is applied to the diaphragm, the threaded stud moves downward to compress the spring.

Maximum charge-air pressure results in maximum stud travel.

The threaded stud exerts pressure against a pivoting bell crank mounted in the governor housing; the movement is transferred via the strap to the injection pump's control rack.

When the charge-air pressure drops, the control rack is moved back toward Stop (shutoff) (Fig. 16).

Figure 15 shows a version of the LDA for use with the RQV governor.

So that the control rack can be shifted to the start-quantity position for starting the engine, the bell crank can be disengaged from the strap by moving its shaft to the side. The shaft is moved either manually, with a linkage mechanism or control cable, or by means of a solenoid which is only effective during starting.

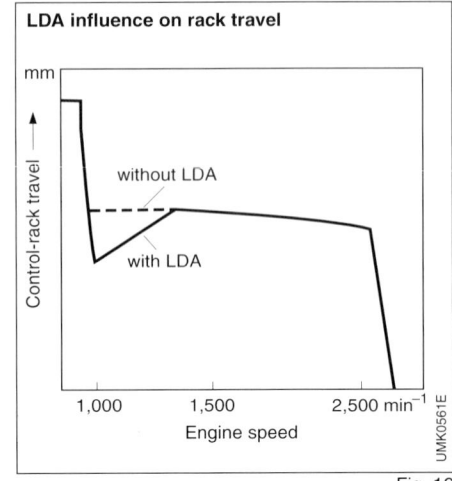

LDA influence on rack travel

Fig. 16

A temperature-sensitive element can be employed to interrupt the current to the solenoid when the engine temperature makes the start quantity unnecessary. Another option is the hydraulic start-quantity locking device (HSV), which operates using the engine-oil pressure. After the engine has started, the oil pressure increases and cancels the excess start quantity. The HSV unit is bolted to the side of the governor housing.

Fig. 15

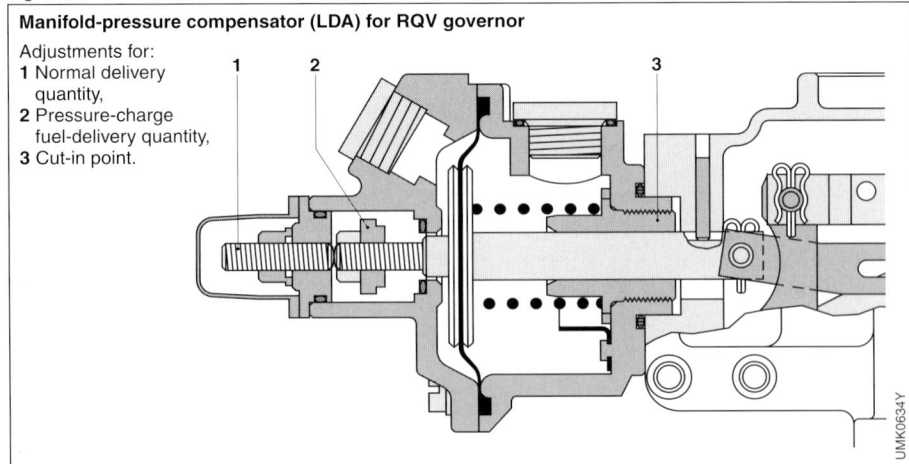

Manifold-pressure compensator (LDA) for RQV governor

Adjustments for:
1 Normal delivery quantity,
2 Pressure-charge fuel-delivery quantity,
3 Cut-in point.

Altitude-pressure compensator (ADA)

Application

In those countries or regions in which traffic is subjected to considerable variations in altitude, the injected fuel quantity must be adapted above a given altitude to the inefficient cylinder charge resulting from the reduced intake-air density.

The altitude-pressure compensator (ADA, Figs. 17 and 18) reduces the injected fuel quantity along with increasing altitude (or reduced air pressure). It is mounted on the governor cover of the RQ(V) and RSF governors.

Design and function

The ADA used with the RQV governor comprises an aneroid capsule fitted vertically inside a housing. Using an adjusting screw and a spring-loaded threaded pin, the aneroid capsule can be set to a given altitude above which it starts to expand.

Changes in the capsule's length are transferred to the swivelling plate cam through the spring-loaded threaded pin on the base of the capsule and by the fork on the end of the threaded pin. The plate cam acts upon the bolt which is attached to the plate strap.

Expansion of the capsule causes the plate cam to swivel downwards, the pin attached to the stop strap pulls the control rack in the Stop (shutoff) direction, and the delivery quantity is reduced.

When the capsule contracts (with lower altitudes), the delivery quantity increases. An adjusting screw is provided to adjust the full-load delivery quantity by shifting the cam plate horizontally.

With the RSF governor, design and configuration are similar although here a spring-loaded threaded pin together with an attached lever transfers the changes in altitude to the pump's control rack (Fig. 18). From the constructional viewpoint, the version for the RQ governor is similar.

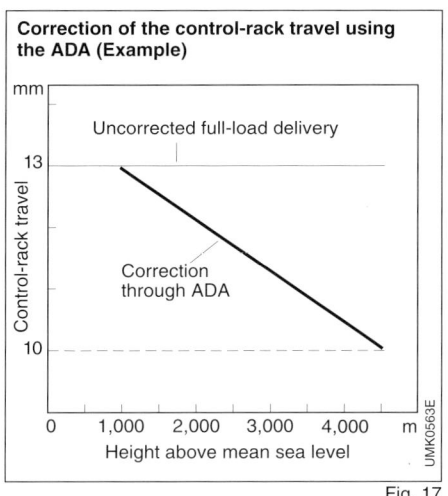

Correction of the control-rack travel using the ADA (Example)

Uncorrected full-load delivery

Correction through ADA

Control-rack travel (mm) vs. Height above mean sea level (m)

UMK0563E

Fig. 17

Fig. 18

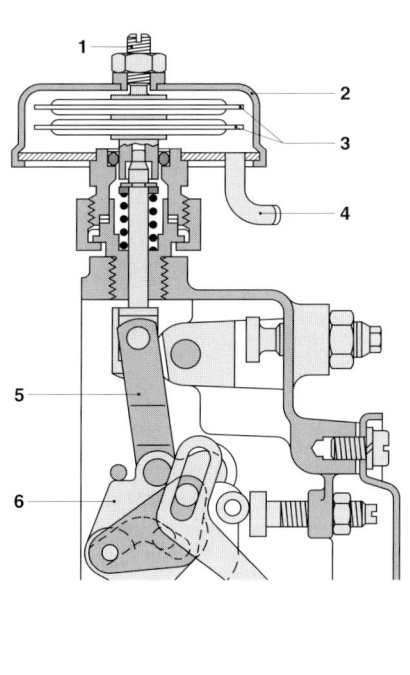

Altitude-pressure compensator (ADA)

1 Adjusting screw,
2 Cap,
3 Aneroid capsule,
4 Opening to outside,
  (for registering the absolute pressure),
5 Correction linkage,
6 Template.

UMK0633Y

Manifold-pressure compensator, absolute metering (ALDA)

Application

On a pressure-charged engine, the charge-air pressure is based upon the prevailing atmospheric pressure whose influence is particularly marked when driving in regions with considerable changes in altitude.

Atmospheric pressure + charge-air pressure = absolute pressure.

Design and function

The RSF governor's (absolute metering) manifold-pressure compensator is equipped with aneroid capsules (adjustable for a given altitude) to which intake-manifold pressure is applied through a fitting.

The capsules change their length in response to every change in pressure, and in doing so adapt the injected fuel quantity through a linkage and the control rack (Fig. 19).

Fig. 19

Absolute-metering manifold-pressure compensator (ALDA)

1 Connection to engine intake manifold (registration of absolute pressure), **2** Adjusting screw, **3** Aneroid capsule, **4** Barometric cells, **5** Correction linkage, **6** Template.

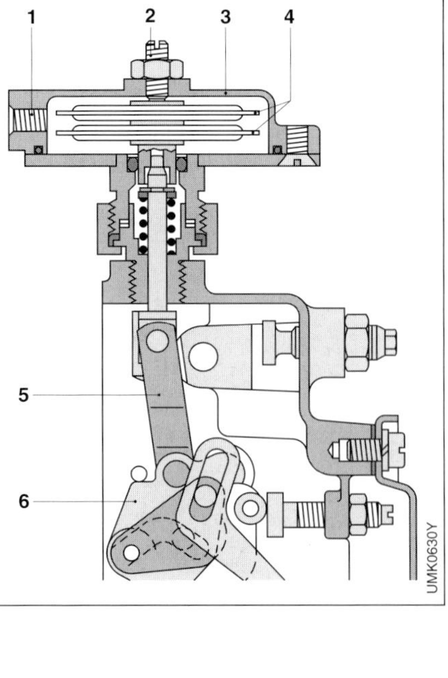

Pneumatic idle-speed increase (PLA)

Application

The injected fuel quantity needed by a diesel engine at idle decreases along with increasing engine temperature.

When the engine is cold, the temperature-dependent idle-speed increase fitted to the RSF governor raises the idle speed accordingly, thereby improving the engine's warm-up behaviour. It also prevents the cold engine stalling when additional loads are switched, such as power-assisted steering, air-conditioner, etc.

The PLA ceases to function above a certain temperature (Fig. 20).

Design and function

Depending upon temperature, pressure is applied to the aneroid capsule's diaphragm which in turn shifts a sliding bolt to change the pretension on the idle spring. The control rack now moves towards "increased fuel quantity" via the governor linkage and the control rack.

Fig. 20

Pneumatic idle-speed increase (PLA)

1 Idle-speed spring, **2** Sliding bolt, **3** Diaphragm, **4** Pressure connection, **5** Aneroid capsule, **6** Spring.

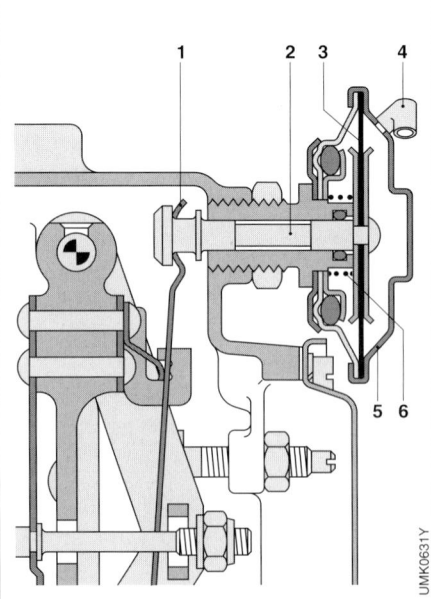

Electronic idle-speed control (ELR)

Application
Instead of the pneumatic idle-speed increase (PLA) usually fitted, the idle speed can be controlled electronically to comply with increased demands regarding comfort.

Design and function
The electronic idle-speed control is comprised of:
– ECU, and
– Electrical actuator solenoid.

The ECU triggers the actuator solenoid to adjust the correct idle speed for all temperature and load conditions. As shown in Fig. 21, the solenoid actuator is arranged on the RSF-governor cover so that when voltage is applied, the solenoid armature increases the force exerted by the idle spring, thus raising the idle speed.

Anti-bucking device (ARD)

Application
On vehicles with higher comfort standards, an anti-bucking device (ARD) can be fitted which virtually eliminates this type of bucking oscillation.

Design and function
The anti-bucking device is comprised of:
– Speed sensor,
– ECU, and
– Electrical actuator solenoid.

The signals from the speed sensor are registered by the ECU and processed. In order to prevent bucking, the ECU triggers the actuator solenoid (Fig. 21), which is mounted on the RSF governor cover, so that the lower suspension point of the control lever is shifted to reduced rack travel, and therefore reduced injected fuel quantity. Rack-travel reduction is synchronized to the bucking oscillations but is 180° out of phase with them.

Fig. 21

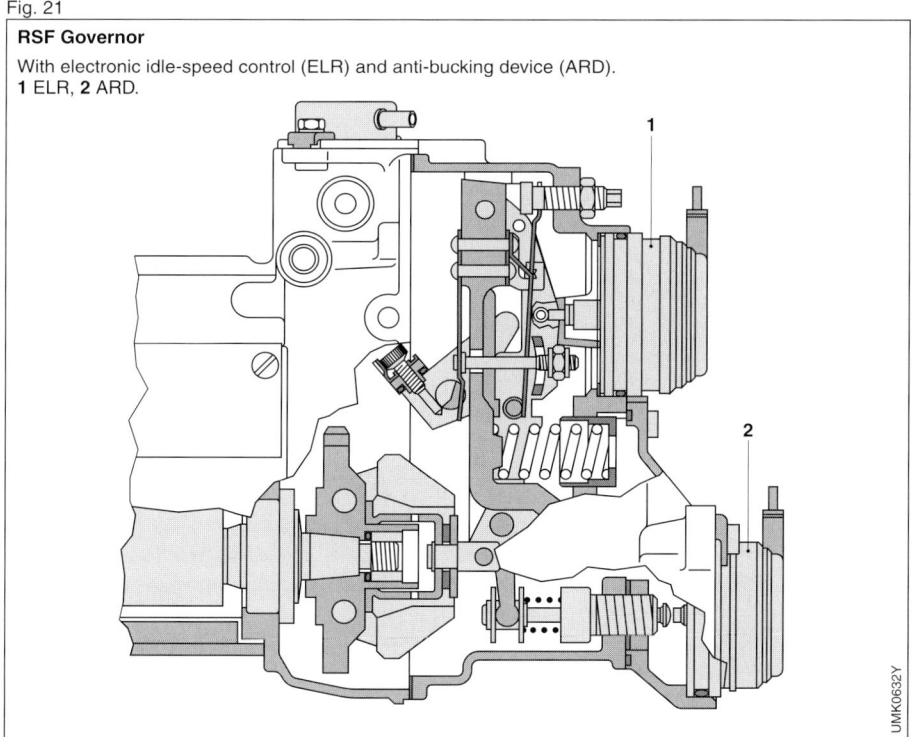

RSF Governor
With electronic idle-speed control (ELR) and anti-bucking device (ARD).
1 ELR, **2** ARD.

UMK0632Y

Port-closing sensor (FBG)

Application

The precise synchronization of the start of injection (port closing) to the engine's crankshaft is imperative if the demands made on a modern-day diesel engine are to be complied with. These include:
– Optimum combustion process for compliance with increasingly stringent emission-control legislation,
– Reduction of noise,
– Optimization of fuel economy.
Using the port-closing adjustment method, the injection pump's start of delivery can be more precisely adjusted and at lower cost than with traditional methods.

Design and function

The governor's flyweight mount is provided with a tooth-shaped signal mark which is precisely in the center of the mounting bore at the "Start-of-delivery (port closing), Cylinder 1" setting.
Figure 22 shows the corresponding start-of-delivery timing device for the RSF governor.

A similar method is used for the RQ and RQV governors, whereby the signal mark is on a pointer which is rigidly attached to the governor hub.
Attached to the governor housing is an adjustable sliding flange with mounting bore into which the following devices can be screwed in order to scan the signal mark:
– Inductive-type sensor for measurements with the engine running.
This permits the timing device to be checked as a function of engine speed (Fig. 22a).
– Photosensor for determining the start-of-delivery point with the engine at standstill (Fig. 22b).
– Blocking pin to lock the pump camshaft in the start-of-delivery position. This permits the pump to be mounted on the engine more quickly and more accurately.
The pin engages with the signal mark and thus prevents the pump's camshaft being turned inadvertently (Fig. 22c).
– A screw plug for closing off the mounting bore following completion of the start-of-delivery timing (Fig. 22d).

Fig. 22

Port-closing sensor (FBG)

Shown for
the RSF governor
(other port-closing
sensors have a
sliding flange).
a Measuring with
the inductive-type sensor,
b Measuring with the
light-signal sensor,
c Blocking the camshaft
with the blocking pin,
d Sensor hole closed
with the screw plug.

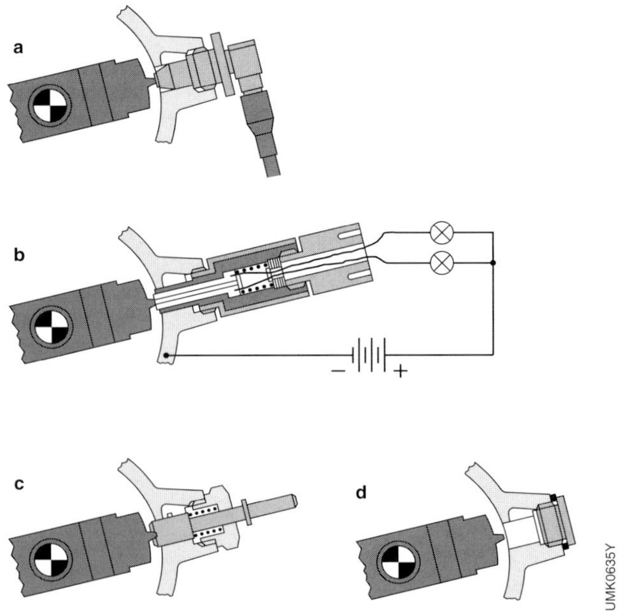

UMK0635Y

Rack-travel sensor (RWG)

Application
The rack-travel sensor is used to monitor the position of the control rack. It provides a signal for a variety of functions on the engine and in the vehicle:
– Shift-point signal for hydraulic transmissions,
– Load signal as shift indication for manual transmissions,
– Signal for fuel-consumption measurement,
– Actuating signal for exhaust-gas recirculation (EGR), and
– Diagnosis signal.

Design
The sensor comprises a magnetically-soft laminated core with two outer limbs. A measuring coil is fastened to one of the limbs, and a reference coil to the other (Fig. 23). The short-circuiting ring for registering the rack travel is fastened to the control rack and fits over the lower limb without contacting it. The upper limb with the fixed reference short-circuiting ring forms the reference unit. The rack-travel sensor is connected to the electronic processing circuit through a 3-core cable with a plug.

Operation
The measuring principle is based on the fact that the short-circuiting ring surrounding the laminated-core limb shields the alternating magnetic field generated by the coil. The extension of the magnetic field is limited to the area between the coil and the short-circuiting ring. This means that the inductance depends upon the position of the short-circuiting ring (Fig. 23). An evaluation circuit converts the ratio of measuring-coil inductance to reference-coil inductance into a voltage ratio which is proportional to the rack travel (Fig. 24).

Installation
On the P and MW in-line injection pumps, the rack-travel sensor is mounted in the governor housing at the same height and on the same side as the control rack. The

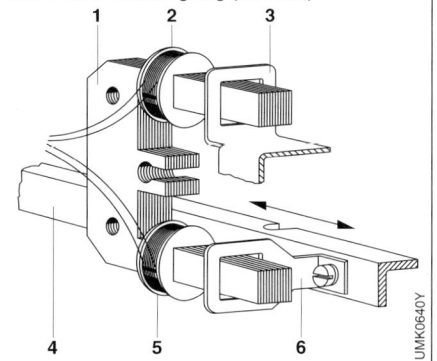

Rack-travel sensor (RWG) (Functional principle)
1 Laminated iron core, **2** Reference coil, **3** Short-circuiting ring (fixed), **4** Control rack, **5** Measuring coil, **6** Short-circuiting ring (movable).

Fig. 23

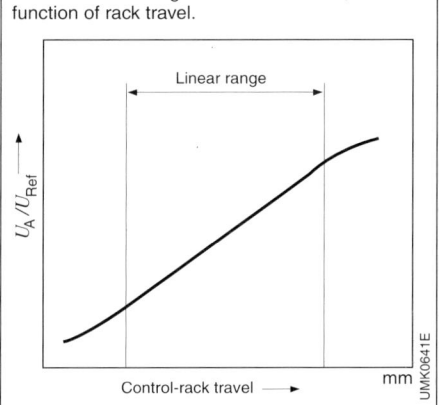

Voltage ratio
Between measuring and reference coils, as a function of rack travel.

U_A / U_{Ref}

Linear range

Control-rack travel

Fig. 24

in-line injection pump can also be delivered already modified for the subsequent mounting of the rack-travel sensor.
The mounting flange on the pump housing is closed-off with a special cover. With the RSF governor, the rack-travel sensor is already integrated in the governor housing.

145

Temperature-dependent starting stop (TAS)

Application

On modern-day engines an increased start quantity is only needed with a cold engine at extremely low outside temperatures. Environmental considerations dictate that the unnecessary injection of an increased start quantity (excess fuel for starting) be avoided.

To this end, the governor is fitted with a device which permits only the required rack travel as specified by the engine manufacturer. This device is the "Temperature-dependent starting stop" (TAS). It is available for practically all governor models.

Design and function

An expansion element, which reacts to the surrounding temperature, or a temperature-controlled solenoid, limits the warm-start fuel delivery by means of corrections to the rack travel. The warm-start fuel delivery therefore becomes a function of the engine's temperature. The following options are available for mounting the expansion element or solenoid; the selection is made according to the specific installation configuration of the injection pump and the governor:

1. If there is space available on the injection pump's drive end, the expansion element is mounted to act directly on the control rack (Fig. 25).

Fig. 25

TAS

With expansion element, acting directly on the control rack.
1 Control rack, **2** Stop pin, **3** Expansion element.

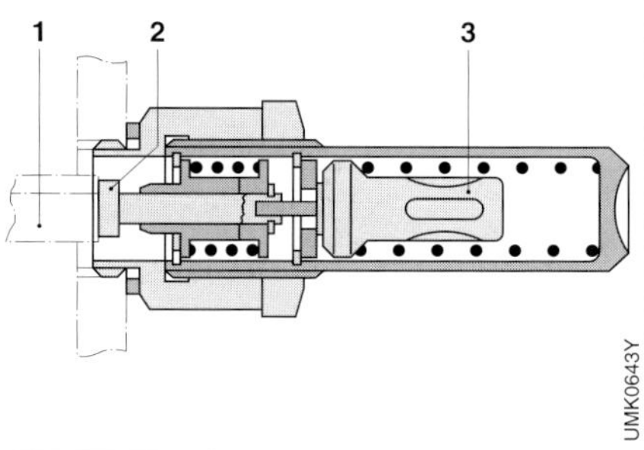

UMK0643Y

Fig. 27

Electromagnetic start-quantity release

With temperature-dependent control for the RQ-/RQV governors.
1 Latch.

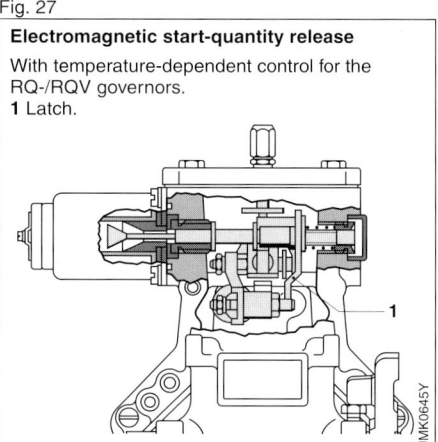

UMK0645Y

Fig. 26

TAS

Mounted on the governor-side, with expansion element for RQ governor.

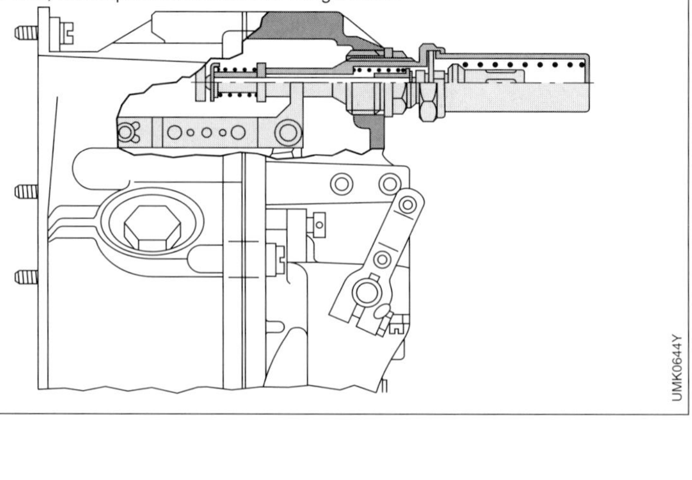

UMK0644Y

Electromagnetic start-quantity release device

With temperature-dependent control for RSV governor, with control-rack stop or LDA.
1 Latch.

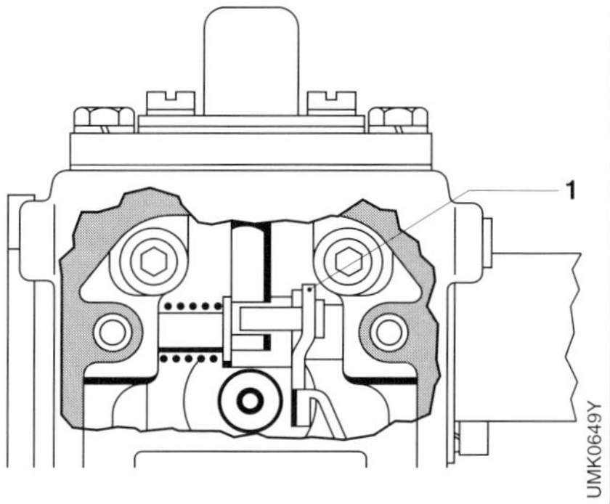

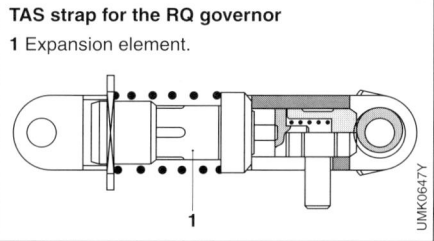

UMK0649Y

Fig. 28

Fig. 29

TAS strap for the RQ governor

1 Expansion element.

UMK0647Y

Fig. 30

Mechanical start-quantity locking device

With expansion element for RQ/RQV governor with LDA.

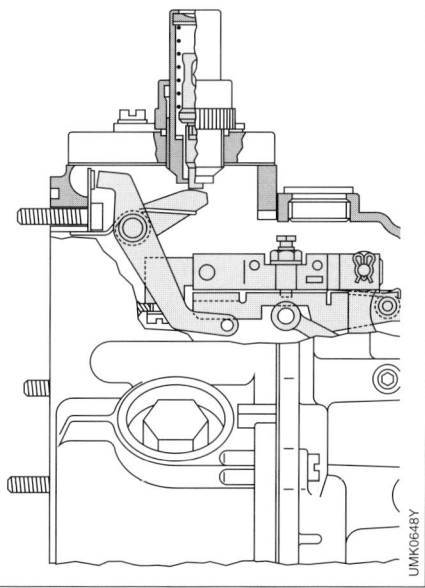

UMK0648Y

The warm-start setting is shown in Fig. 25, i.e. the stop pin is pushed against a spring by the expansion-element adjusting pin to limit rack travel. With this layout, warm-start rack travel is equal to or greater than full-load rack travel.

2. If fitted at the governor end (Fig. 26), as with the RQ governor, the expansion element is installed on the governor cover. Together with the spring force, the expansion element reduces the starting rack travel when starting the warm engine. Here, the warm-start rack travel is equal to or greater than full-load rack travel.

3. The following rule applies for cold and warm starts in the case of plunger-and-barrel assemblies with a starting groove:
Cold start: with excess fuel for starting and with retarded start of injection.
Warm start: without excess fuel for starting and without retarded start of injection.

On RQ/RQV governors, the release device for excess starting fuel is standard, in the form of a solenoid which is triggered as a function of temperature. For cold starting, the latch is shifted to permit starting rack travel. With the engine warm, the solenoid is switched off and the latch blocks the control rack so that warm-start delivery is equal to full-load delivery (Fig. 27).

4. On RQ/RQV governors equipped with a governor-side LDA, an expansion element and a lever arrangement in the governor are responsible for the temperature-dependent control-rack starting travel.
This permits a start-quantity rack travel or a normal fuel-delivery rack travel for cold start and warm start respectively (Fig. 30).

5. In special cases, the RQ governor can be equipped with an expansion cartridge (TAS strap, Fig. 29).

6. With the RSV governor with control-rack stop or LDA, the excess fuel for starting can be released through a temperature-controlled solenoid. During cold-start, this moves the latch and releases the control rack for delivery of the excess fuel for starting (Fig. 28).

Stabilizer

Application

The stabilizer is used mainly on governors for engine-generator sets for stabilizing systems which easily become unstable, or which operate at their instability limit, or for the reduction of the speed droop on stable systems. A stabilizer is not suitable for shortening the settling time, nor for reducing the dynamic speed droop.

Design and function

The stabilizer is hydraulic and consists of a plunger which is precisely fitted into a housing with a minimum of clearance. The housing is bolted to the governor cover, and the plunger chamber is connected to an oil reservoir via an adjustable throttle bore. A spring is hooked to the plunger. On the RSV governor, this spring is connected to the tensioning lever and on the RQV governor to the sliding bolt.

The connection is free from backlash. The oil reservoir is connected to the engine lube-oil circuit and designed so that under normal tilt conditions no air can enter the plunger chamber (Figs. 31 and 32).

Principle

When speed fluctuations or speed oscillations occur, flyweight movement is damped by a spring which is switched in temporarily. The spring causes an increase of the dynamic speed droop. After settling to the new operating conditions, the spring is switched out again. When the governor flyweights move outward (or inward), the spring is stretched or compressed. Its tensile inertia acts together with that of the governor-spring, causing a temporary increase in speed droop. This has a stabilizing effect on the entire control circuit. Since the spring's other end is hooked into the hydraulic plunger, the plunger is shifted in its housing until the spring force drops to zero. The stabilizer damping effect depends upon the spring constants of the spring (various spring versions), and the setting of the

Fig. 31

RSV Governor with stabilizer

1 Throttle screw, **2** Oil-input line, **3** Governor cover, **4** Hollow screw with input throttle, **5** Oil reservoir, **6** Housing, **7** Plunger, **8** Retaining pin, **9** Stabilizer spring, **10** Screw plug, **11** Threaded bushing, **12** Hexagon nut, **13** Full-load adjustment screw.

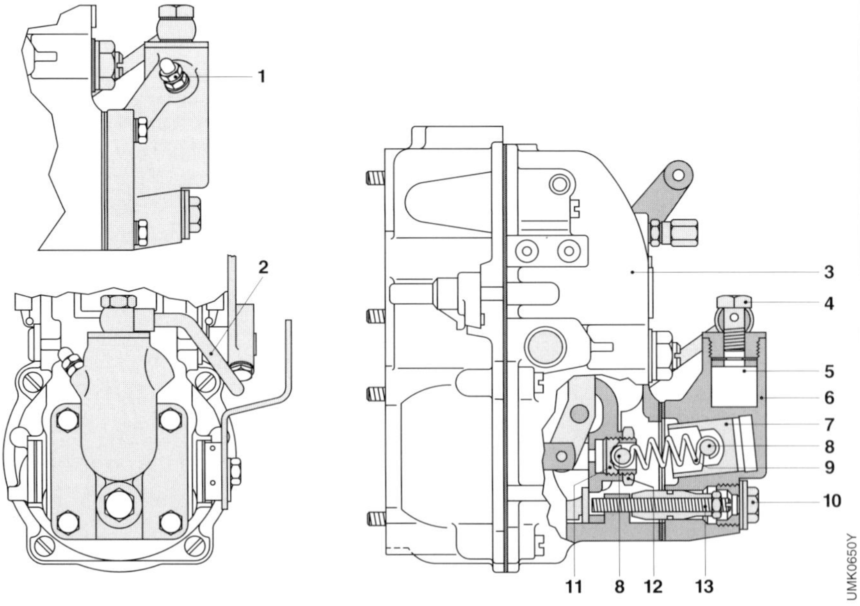

throttle screw between plunger chamber and oil reservoir. The plunger chamber must be completely free of air. This is a prerequisite for efficient stabilizer operation. The stabilizer has been designed so that air is bled off automatically. When operated for the first time, or after being out of operation for a considerable period, a short settling period is required before the device becomes fully effective.

Pneumatic shutoff device (PNAB)

Application
To shut off the engine, the key is turned to "Stop".

Design and function
When the key is turned, vacuum is applied to the diaphragm of the RSF governor's shutoff-device. The control rack is pulled to Stop (shutoff). For this function, the vehicle must be equipped with a vacuum pump (Fig. 33).

Fig. 32

Pneumatic shutoff device (PNAB)

1 Pneumatic shutoff device,
2 Stop (shutoff) lever for manual actuation,
3 Stop lever, 4 Spring-loaded strap,
5 Control rack.

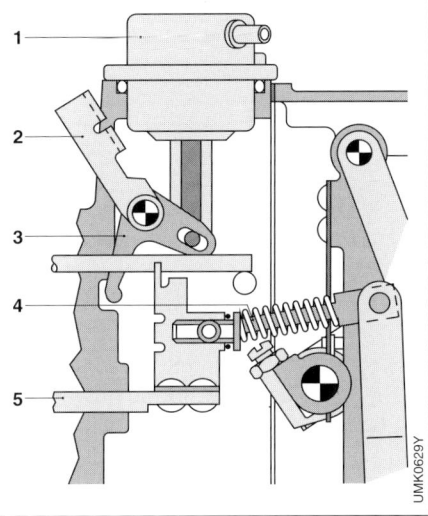

Fig. 33

RQ-/RQV Governor with stabilizer

1 Throttle bore, 2 Oil-inlet bore, 3 Adjustable throttle screw, 4 Oil overflow, 5 Connecting strap.

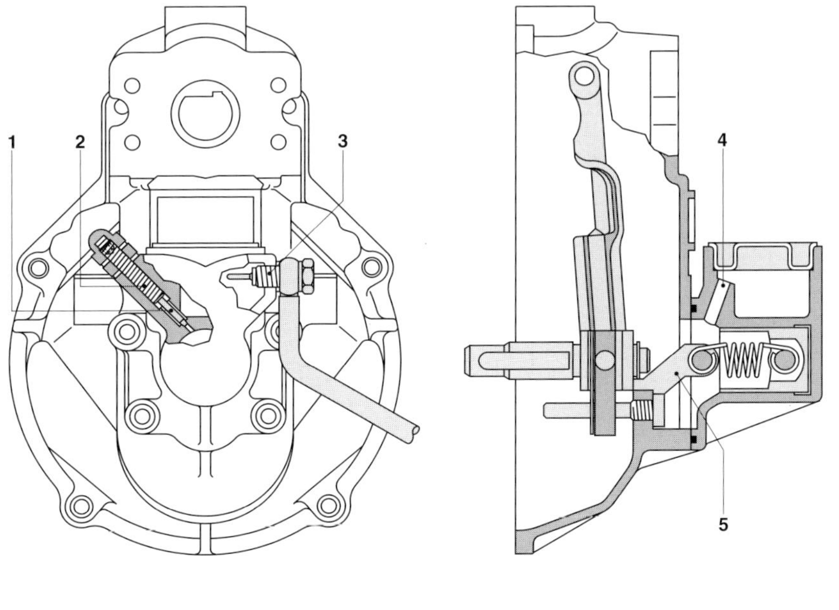

Governor adjustment and testing

The governor has already been assembled and preliminary adjustments made at the factory. In order to carry out final adjustments and to check its mechanical functioning, the governor is mounted on an in-line fuel-injection pump to form a unit which is then adjusted on an injection-pump test bench (Fig. 1).

When carrying out adjustment and testing work, there are a number of basic differences between mechanical governors and the electronic diesel control (EDC) which must be taken into account:

Considering the mechanical governor, all assemblies which directly affect the governor characteristics (flyweight assembly, spring sets, correction devices such as torque-control spring retainer, stops etc.) are all mounted inside the governor housing.

This means that these assemblies must also be matched individually to the particular governor characteristics, even though they have been correctly selected for the version in question.

During adjustment and testing, the following tests are therefore performed at a number of different loading stages in accordance with the relevant rack-travel map (Example: RSF minimum-maximum-speed governor):
– Rack-travel measurement and measurement of sliding-sleeve position,
– Adjustment of the rotational-speed stages, delivery quantities and control-lever positions,
– Testing the torque-control characteristic, the auxiliary idle-speed spring and their switch off, together with the load take-up,
– Functioning of the shutoff device.

The fuel-injection pump and the governor must be filled with lube-oil before testing takes place (Example: Size M in-line fuel-

Fig. 1

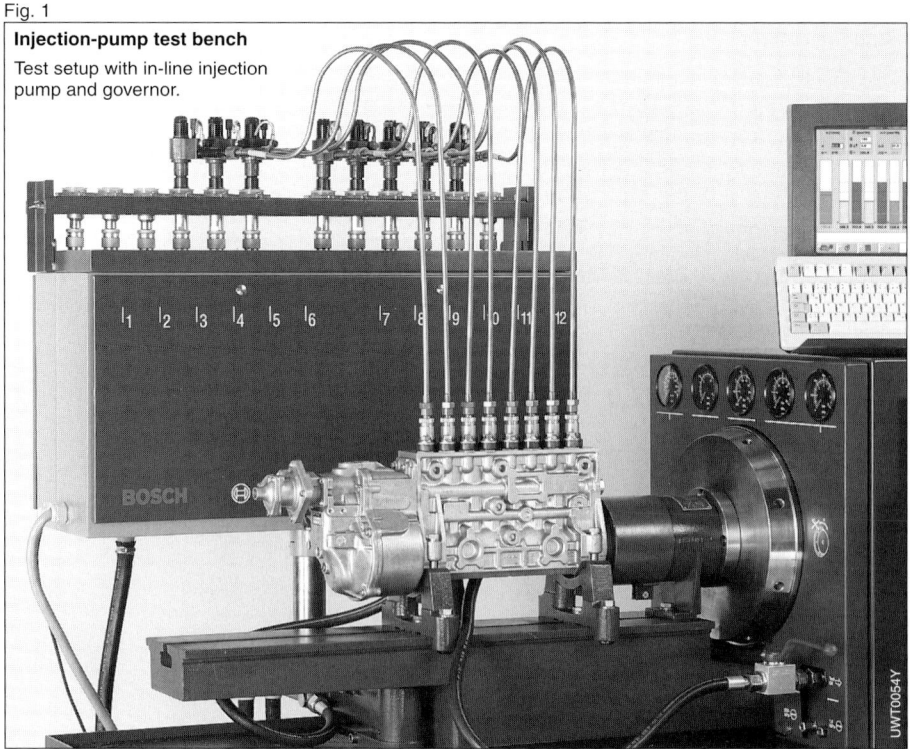

Injection-pump test bench

Test setup with in-line injection pump and governor.

injection pump and RSF minimum-maximum-speed governor). When actually operating in the field, the injection pump and the governor are both connected into the engine's lube-oil circuit. Oil supply is then via the camshaft chamber or via the pump housing.

The port-closing sensor system described above is suitable for testing and adjustment of the start-of-delivery. Here, with the engine running, the inductive-type sensor serves for setting the dynamic start of delivery. With the engine stopped, the light-signal sensor is used to set the static start of delivery. Both sensors are screwed into a special mounting hole in the governor housing. A TDC sensor provides the necessary reference signal for crankshaft position. The same sensors can be used in customer service. The pulse generated in the sensor when the pump is rotating is converted by a special adapter unit so that the commercially available Bosch Motortester can be used to check the specified differential angle between crankshaft

position and the pump's start of delivery (Fig. 2).

In the case of electronic control (governing), the same actuator mechanism is always used independent of the governor characteristics. The maps, and other corrections, are always determined by the ECU. They are a software component and as such are to be checked separately in a specialized workshop.

Maps and parameters are programmed into the ECU at the end of the production line (EOL) by Bosch, or by the engine or vehicle manufacturer.

This means that the same ECU can be used for a variety of different engine and vehicle variants.

The actuator mechanism is not provided with a special oil filling. In addition to the linear solenoid and sensors, it is equipped with an oil supply pump which pumps the lube oil which has entered the actuator back to the injection pump's camshaft chamber.

Fig. 2

Start-of-delivery timing

In-line injection pump with governor, using the port-closing sensor system.
1 Motortester, **2** Adapter unit, **3** In-line pump with governor, **4** Port-closing sensor, 5 TDC sensor.

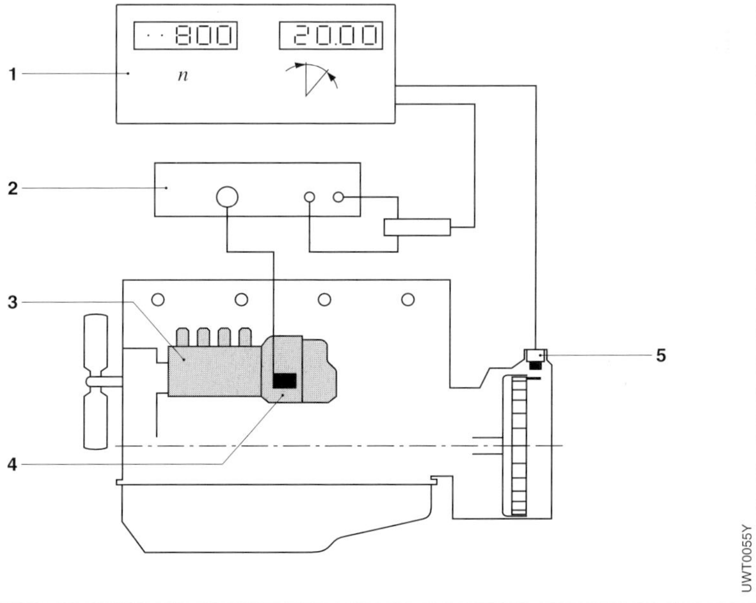

Mechanically-controlled (governed) axial-piston distributor fuel-Injection pumps VE

Fuel-injection systems

Assignments

The fuel-injection system is responsible for supplying the diesel engine with fuel. To do so, the injection pump generates the pressure required for fuel injection. The fuel under pressure is forced through the high-pressure fuel-injection tubing to the injection nozzle which then injects it into the combustion chamber.

The fuel-injection system (Fig. 1) includes the following components and assemblies: The fuel tank, the fuel filter, the fuel-supply pump, the injection nozzles, the high-pressure injection tubing, the governor, and the timing device (if required).

The combustion processes in the diesel engine depend to a large degree upon the quantity of fuel which is injected and upon the method of introducing this fuel to the combustion chamber.

The most important criteria in this respect are the fuel-injection timing and the duration of injection, the fuel's distribution in the combustion chamber, the moment in time when combustion starts, the amount of fuel metered to the engine per degree crankshaft, and the total injected fuel quantity in accordance with the engine loading. The optimum interplay of all these parameters is decisive for the faultless functioning of the diesel engine and of the fuel-injection system.

Fig. 1

Fuel-injection system with mechanically-controlled (governed) distributor injection pump
1 Fuel tank, **2** Fuel filter, **3** Distributor fuel-injection pump, **4** Nozzle holder with nozzle, **5** Fuel return line, **6** Sheathed-element glow plug (GSK) **7** Battery, **8** Glow-plug and starter switch, **9** Glow control unit (GZS).

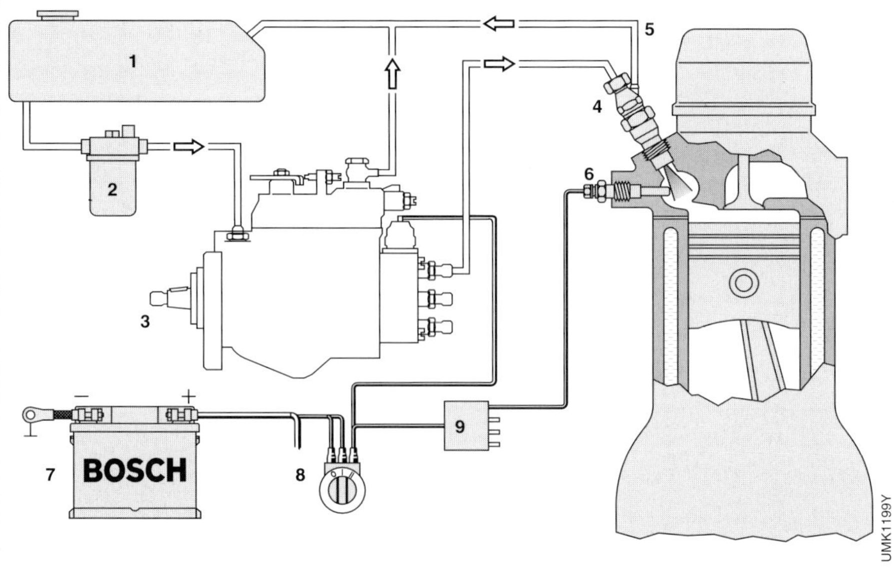

Types

The increasing demands placed upon the diesel fuel-injection system made it necessary to continually develop and improve the fuel-injection pump.

Following systems comply with the present state-of-the-art:
- In-line fuel-injection pump (PE) with mechanical (flyweight) governor or Electronic Diesel Control (EDC) and, if required, attached timing device,
- Control-sleeve in-line fuel-injection pump (PE), with Electronic Diesel Control (EDC) and infinitely variable start of delivery (without attached timing device),
- Single-plunger fuel-injection pump (PF),
- Distributor fuel-injection pump (VE) with mechanical (flyweight) governor or Electronic Diesel Control (EDC). With integral timing device,
- Radial-piston distributor injection pump (VR),
- Common Rail accumulator injection system (CRS),
- Unit-injector system (UIS),
- Unit-pump system (UPS).

Fuel-injection techniques

Fields of application

Small high-speed diesel engines demand a lightweight and compact fuel-injection installation. The VE distributor fuel-injection pump (Fig. 2) fulfills these stipulations by combining
- Fuel-supply pump,
- High-pressure pump,
- Governor, and
- Timing device,

in a small, compact unit. The diesel engine's rated speed, its power output, and its configuration determine the parameters for the particular distributor pump.

Distributor pumps are used in passenger cars, commercial vehicles, agricultural tractors and stationary engines.

Fig. 2: **VE distributor pump fitted to a 4-cylinder diesel engine**

Subassemblies

In contrast to the in-line injection pump, the VE distributor pump has only <u>one</u> pump cylinder and <u>one</u> plunger, even for multi-cylinder engines. The fuel delivered by the pump plunger is apportioned by a distributor groove to the outlet ports as determined by the engine's number of cylinders. The distributor pump's closed housing contains the following functional groups:

– High-pressure pump with distributor,
– Mechanical (flyweight) governor,
– Hydraulic timing device,
– Vane-type fuel-supply pump,
– Shutoff device, and
– Engine-specific add-on modules.

Fig. 3 shows the functional groups and their assignments. The add-on modules facilitate adaptation to the specific requirements of the diesel engine in question.

Design and construction

The distributor pump's drive shaft runs in bearings in the pump housing and drives the vane-type fuel-supply pump. The roller ring is located inside the pump at the end of the drive shaft although it is not connected to it. A rotating-reciprocating movement is imparted to the distributor plunger by way of the cam plate which is driven by the input shaft and rides on the rollers of the roller ring. The plunger moves inside the distributor head which is bolted to the pump housing. Installed in the distributor head are the electrical fuel shutoff device, the screw plug with vent screw, and the delivery valves with their

Fig. 3

The subassemblies and their functions

1 Vane-type fuel-supply pump with pressure regulating valve: Draws in fuel and generates pressure inside the pump.
2 High-pressure pump with distributor: Generates injection pressure, delivers and distributes fuel.
3 Mechanical (flyweight) governor: Controls the pump speed and varies the delivery quantity within the control range.
4 Electromagnetic fuel shutoff valve: Interrupts the fuel supply.
5 Timing device: Adjusts the start of delivery (port closing) as a function of the pump speed and in part as a function of the load.

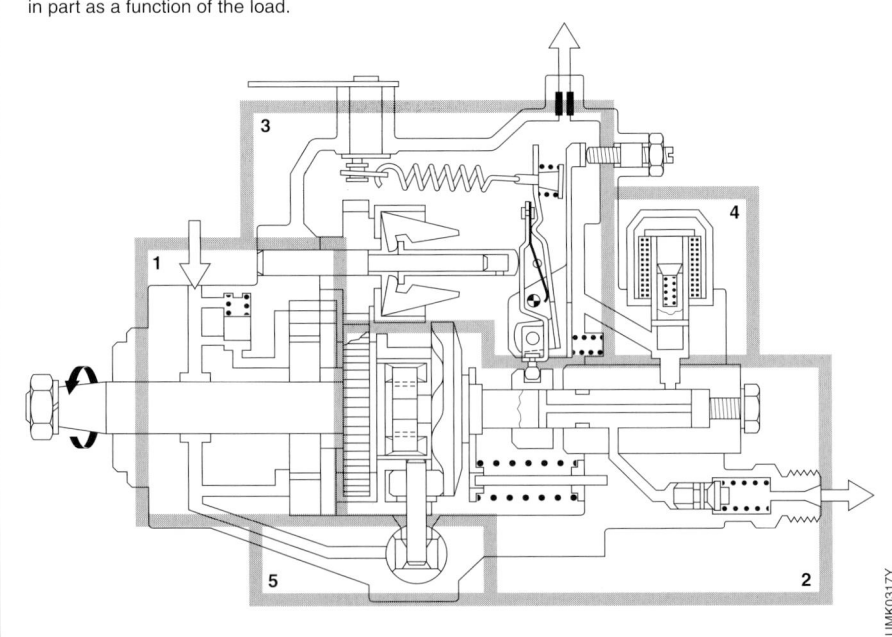

holders. If the distributor pump is also equipped with a mechanical fuel shutoff device this is mounted in the governor cover.

The governor assembly comprising the flyweights and the control sleeve is driven by the drive shaft (gear with rubber damper) via a gear pair. The governor linkage mechanism which consists of the control, starting, and tensioning levers, can pivot in the housing.

The governor shifts the position of the control collar on the pump plunger. On the governor mechanism's top side is the governor spring which engages with the external control lever through the control-lever shaft which is held in bearings in the governor cover.

The control lever is used to control pump function. The governor cover forms the top of the distributor pump, and also contains the full-load adjusting screw, the overflow restriction or the overflow valve, and the engine-speed adjusting screw. The hydraulic injection timing device is located at the bottom of the pump at right angles to the pump's longitudinal axis. Its operation is influenced by the pump's internal pressure which in turn is defined by the vane-type fuel-supply pump and by the pressure-regulating valve. The timing device is closed off by a cover on each side of the pump (Fig. 4).

Fig. 4

The subassemblies and their configuration

1 Pressure-control valve, **2** Governor assembly, **3** Overflow restriction,
4 Distributor head with high-pressure pump, **5** Vane-type fuel-supply pump, **6** Timing device,
7 Cam plate, **8** Electromagnetic shutoff valve.

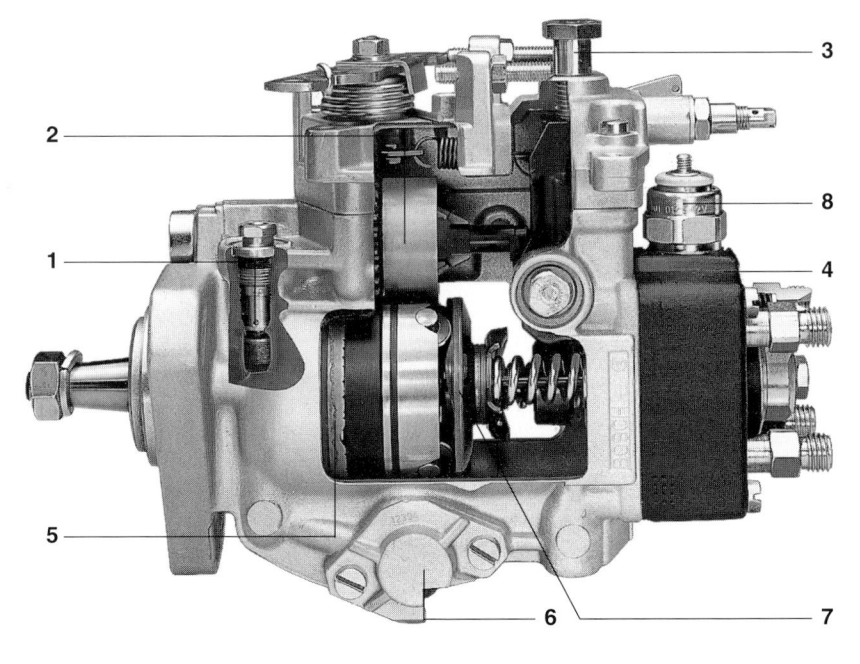

UMK0319Y

Pump drive

The distributor injection pump is driven by the diesel engine through a special drive unit. For 4-stroke engines, the pump is driven at exactly half the engine crankshaft speed, in other words at camshaft speed. The VE pump must be positively driven so that it's drive shaft is synchronized to the engine's piston movement.

This positive drive is implemented by means of either toothed belts, pinion, gear wheel or chain. Distributor pumps are available for clockwise and for counter-clockwise rotation, whereby the injection sequence differs depending upon the direction of rotation.

The fuel outlets though are always supplied with fuel in their geometric sequence, and are identified with the letters A, B, C etc. to avoid confusion with the engine-cylinder numbering. Distributor pumps are suitable for engines with up to max. 6 cylinders.

Fuel supply and delivery

Considering an injection system with distributor injection pump, fuel supply and delivery is divided into low-pressure and high-pressure delivery (Fig. 1).

Low-pressure stage

Low-pressure delivery

The low-pressure stage of a distributor-pump fuel-injection installation comprises the fuel tank, fuel lines, fuel filter, vane-type fuel-supply pump, pressure-control valve, and overflow restriction.

The vane-type fuel-supply pump draws fuel from the fuel tank. It delivers a virtually constant flow of fuel per revolution to the interior of the injection pump. A pressure-control valve is fitted to ensure that a defined injection-pump interior pressure is maintained as a function of supply-pump speed. Using this valve, it is possible to set a defined pressure for a given speed. The pump's

Fig. 1

Fuel supply and delivery in a distributor-pump fuel-injection system

1 Fuel tank, **2** Fuel line (suction pressure), **3** Fuel filter, **4** Distributor injection pump,
5 High-pressure fuel-injection line, **6** Injection nozzle, **7** Fuel-return line (pressureless),
8 Sheathed-element glow plug.

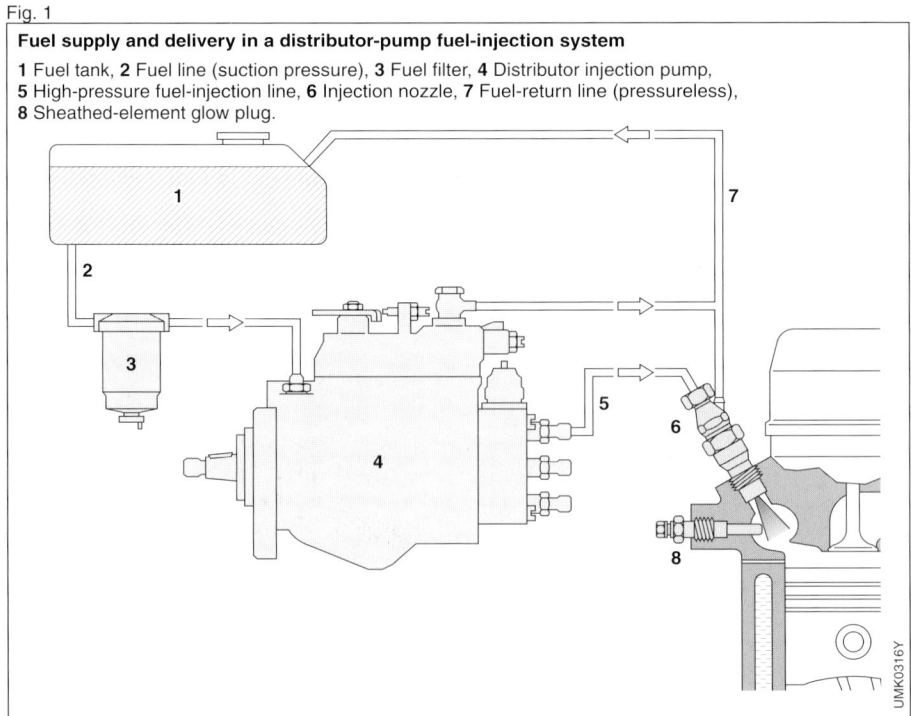

interior pressure then increases in proportion to the speed (in other words, the higher the pump speed the higher the pump interior pressure). Some of the fuel flows through the pressure-regulating valve and returns to the suction side. Some fuel also flows through the overflow restriction and back to the fuel tank in order to provide cooling and self-venting for the injection pump (Fig. 2). An overflow valve can be fitted instead of the overflow restriction.

Fuel-line configuration

For the injection pump to function efficiently it is necessary that its high-pressure stage is continually provided with pressurized fuel which is free of vapor bubbles. Normally, in the case of passenger cars and light commercial vehicles, the difference in height between the fuel tank and the fuel-injection equipment is negligible. Furthermore, the fuel lines are not too long and they have adequate internal diameters. As a result, the vane-type supply pump in the injection pump is powerful enough to draw the fuel out of the fuel tank and to build up sufficient pressure in the interior of the injection pump.

In those cases in which the difference in height between fuel tank and injection pump is excessive and (or) the fuel line between tank and pump is too long, a pre-supply pump must be installed. This overcomes the resistances in the fuel line and the fuel filter. Gravity-feed tanks are mainly used on stationary engines.

Fuel tank

The fuel tank must be of noncorroding material, and must remain free of leaks at double the operating pressure and in any case at 0.3 bar. Suitable openings or safety valves must be provided, or similar measures taken, in order to permit excess pressure to escape of its own accord. Fuel must not leak past the filler cap or through pressure-compensation devices. This applies when the vehicle is subjected to minor mechanical shocks, as well as when

Fig. 2

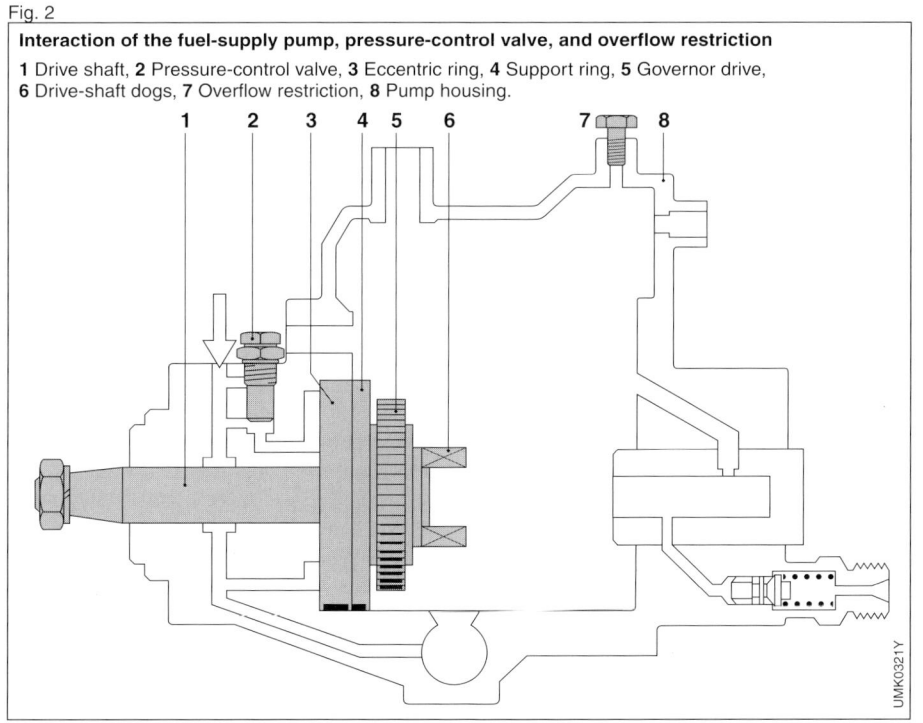

Interaction of the fuel-supply pump, pressure-control valve, and overflow restriction
1 Drive shaft, **2** Pressure-control valve, **3** Eccentric ring, **4** Support ring, **5** Governor drive, **6** Drive-shaft dogs, **7** Overflow restriction, **8** Pump housing.

UMK0321Y

cornering, and when standing or driving on an incline. The fuel tank and the engine must be so far apart from each other that in case of an accident there is no danger of fire. In addition, special regulations concerning the height of the fuel tank and its protective shielding apply to vehicles with open cabins, as well as to tractors and buses

Fuel lines

As an alternative to steel pipes, flame-inhibiting, steel-braid-armored flexible fuel lines can be used for the low-pressure stage. These must be routed to ensure that they cannot be damaged mechanically, and fuel which has dripped or evaporated must not be able to accumulate nor must it be able to ignite.

Fuel filter

The injection pump's high-pressure stage and the injection nozzle are manufactured with accuracies of several thousandths of a millimeter. As a result,

---

Fig. 3: **Vane-type fuel-supply pump with impeller on the drive shaft**

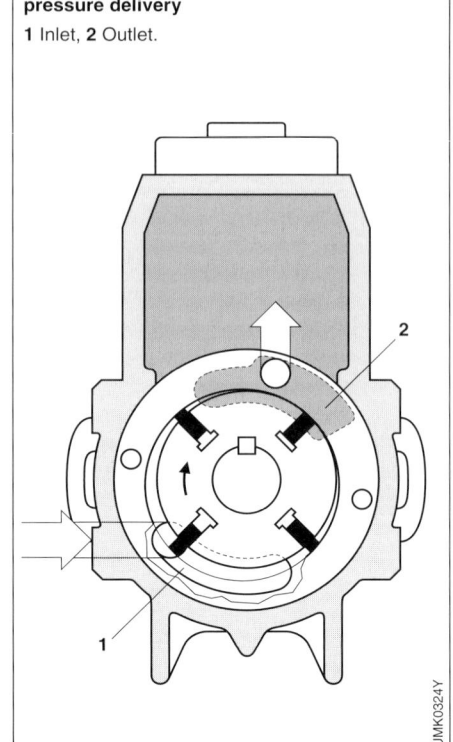

Vane-type fuel-supply pump for low-pressure delivery
1 Inlet, **2** Outlet.

UMK0324Y

Fig. 4

UMK0320Y

contaminants in the fuel can lead to malfunctions, and inefficient filtering can cause damage to the pump components, delivery valves, and injector nozzles. This means that a fuel filter specifically aligned to the requirements of the fuel-injection system is absolutely imperative if trouble-free operation and a long service life are to be achieved. Fuel can contain water in bound form (emulsion) or unbound form (e.g., condensation due to temperature changes). If this water gets into the injection pump, corrosion damage can be the result. Distributor pumps must therefore be equipped with a fuel filter incorporating a water accumulator from which the water must be drained off at regular intervals. The increasing popularity of the diesel engine in the passenger car has led to the development of an automatic water-warning device which indicates by means of a warning lamp when water must be drained.

Vane-type fuel supply pump

The vane-type pump (Figs. 3 and 4) is located around the injection pump's drive shaft. Its impeller is concentric with the shaft and connected to it with a Woodruff key and runs inside an eccentric ring mounted in the pump housing.

When the drive shaft rotates, centrifugal force pushes the impeller's four vanes outward against the inside of the eccentric ring. The fuel between the vanes' undersides and the impeller serves to support the outward movement of the vanes. The fuel enters through the inlet passage and a kidney-shaped recess in the pump's housing, and fills the space formed by the impeller, the vane, and the inside of the eccentric ring. The rotary motion causes the fuel between adjacent vanes to be forced into the upper (outlet) kidney-shaped recess and through a passage into the interior of the pump. At the same time, some of the fuel flows through a second passage to the pressure-control valve.

Pressure-control valve

The pressure-control valve (Fig. 5) is connected through a passage to the upper (outlet) kidney-shaped recess, and is mounted in the immediate vicinity of the fuel-supply pump. It is a spring-loaded spool-type valve with which the pump's internal pressure can be varied as a function of the quantity of fuel being delivered. If fuel pressure increases beyond a given value, the valve spool opens the return passage so that the fuel can flow back to the supply pump's suction side. If the fuel pressure is too low, the return passage is closed by the spring.

Fig. 5

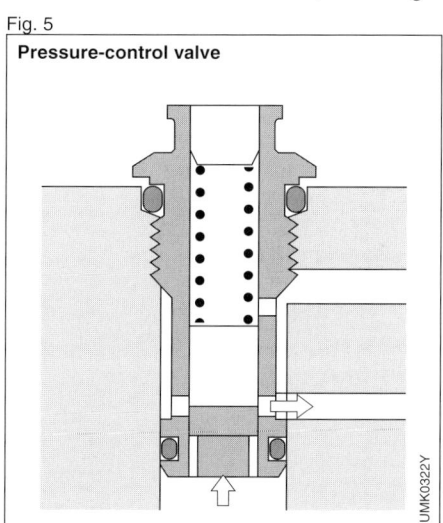

Pressure-control valve

UMK0322Y

Fig. 6

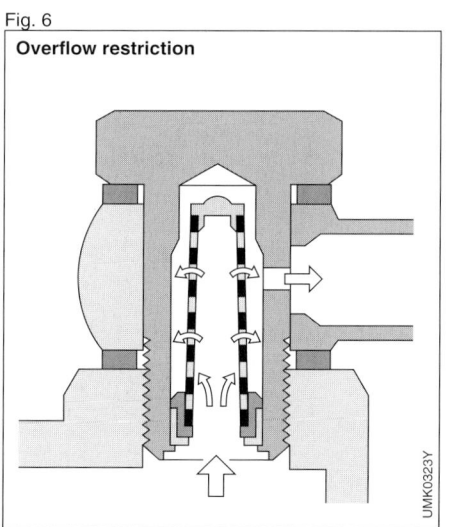

Overflow restriction

UMK0323Y

The spring's initial tension can be adjusted to set the valve opening pressure.

Overflow restriction

The overflow restriction (Figure 6) is screwed into the injection pump's governor cover and connected to the pump's interior. It permits a variable amount of fuel to return to the fuel tank through a narrow passage. For this fuel, the restriction represents a flow resistance that assists in maintaining the pressure inside the injection pump. Being as inside the pump a precisely defined pressure is required as a function of pump speed, the overflow restriction and the flow-control valve are precisely matched to each other.

High-pressure stage

The fuel pressure needed for fuel injection is generated in the injection pump's high-pressure stage. The pressurized fuel then travels to the injection nozzles through the delivery valves and the fuel-injection tubing.

Distributor-plunger drive

The rotary movement of the drive shaft is transferred to the distributor plunger via a coupling unit (Fig. 7), whereby the dogs on cam plate and drive shaft engage with the recesses in the yoke, which is located between the end of the drive shaft and the cam plate. The cam plate is forced against the roller ring by a spring, and when it rotates the cam lobes riding on the ring's rollers convert the purely rotational movement of the drive shaft into a rotating-reciprocating movement of the cam plate.

The distributor plunger is held in the cam plate by its cylindrical fitting piece and is locked into position relative to the cam

Fig. 7

Pump assembly for generation and delivery of high pressure in the distributor-pump interior

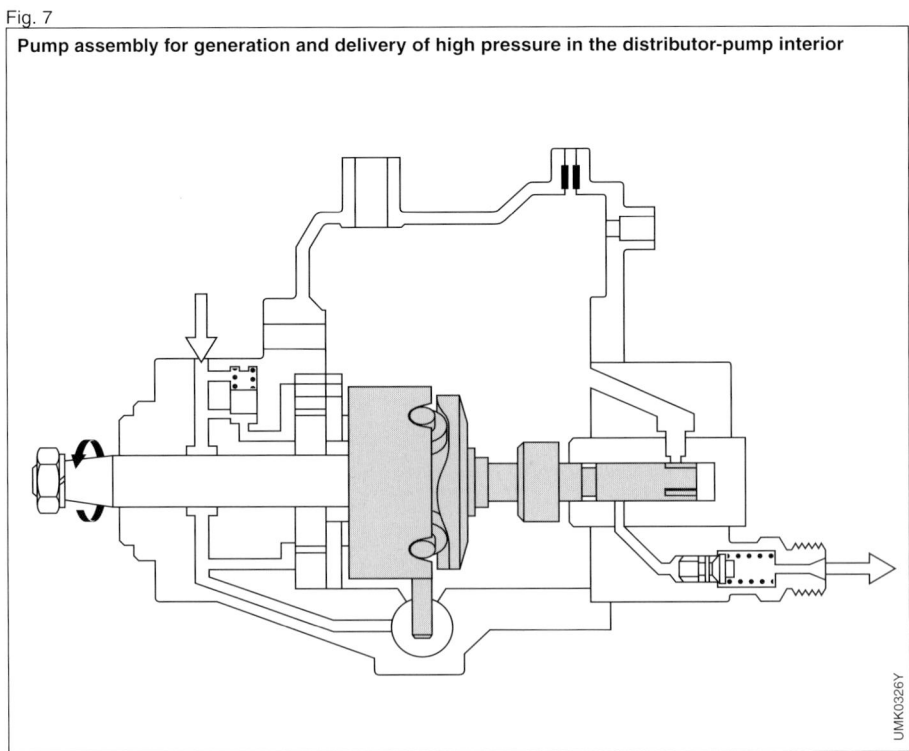

UMK0326Y

Pump assembly with distributor head

Generates the high pressure and distributes the fuel to the respective fuel injector.
1 Yoke, **2** Roller ring, **3** Cam plate, **4** Distributor-plunger foot, **5** Distributor plunger, **6** Link element,
7 Control collar, **8** Distributor-head flange, **9** Delivery-valve holder, **10** Plunger-return spring,
4...8 Distributor head.

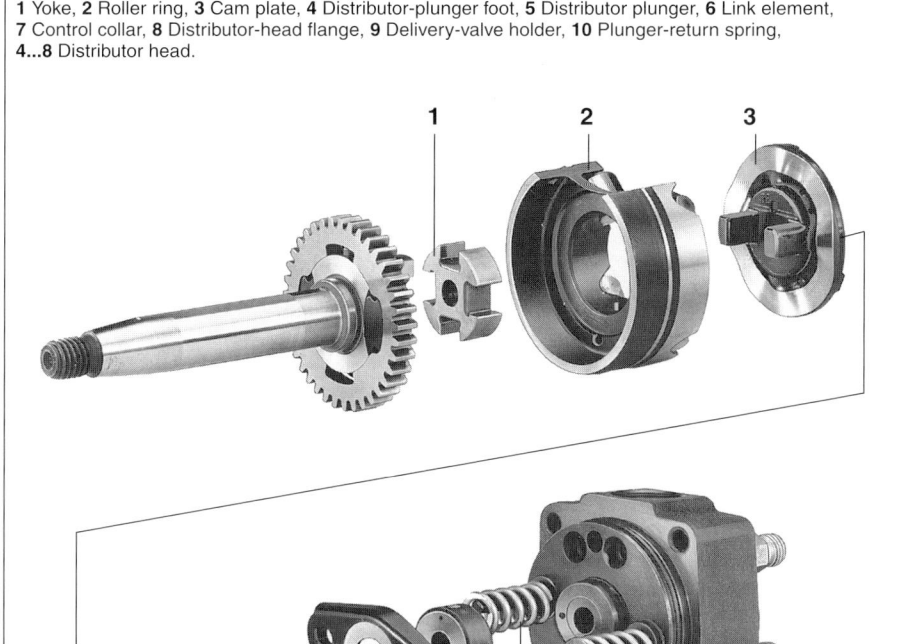

Fig. 8

plate by a pin. The distributor plunger is forced upwards to its TDC position by the cams on the cam plate, and the two symmetrically arranged plunger-return springs force it back down again to its BDC position.

The plunger-return springs abut at one end against the distributor head and at the other their force is directed to the plunger through a link element. These springs also prevent the cam plate jumping off the rollers during harsh acceleration. The lengths of the return springs are carefully matched to each other so that the plunger is not displaced from its centered position (Fig. 8).

Cam plates and cam contours

The cam plate and its cam contour influence the fuel-injection pressure and the injection duration, whereby cam stroke and plunger-lift velocity are the decisive criteria. Considering the different combustion-chamber configurations and combustion systems used in the various engine types, it becomes imperative that the fuel-injection factors are individually tailored to each other. For this reason, a special cam-plate surface is generated for each engine type and machined into the cam-plate face. This defined cam plate is then assembled in the corresponding distributor pump. Since the cam-plate surface is specific to a given engine type, the cam plates are not interchangeable between the different VE-pump variants.

Distributor head

The distributor plunger, the distributor-head bushing and the control collar are so precisely fitted (lapped) into the distributor head (Fig. 8), that they seal even at very high pressures. Small leakage losses are nevertheless unavoidable, as well as being desirable for plunger lubrication. For this reason, the distributor head is only to be replaced as a complete assembly, and never the plunger, control collar, or distributor flange alone.

Fuel metering

The fuel delivery from a fuel-injection pump is a dynamic process comprising several stroke phases (Fig. 9). The pressure required for the actual fuel injection is generated by the high-pressure pump. The distributor plunger's stroke and delivery phases (Fig. 10) show the metering of fuel to an engine cylinder. For a 4-cylinder engine the distributor plunger rotates through 90° for a stroke from BDC to TDC and back again. In the case of a 6-cylinder engine, the plunger must have completed these movements within 60° of plunger rotation.

As the distributor plunger moves from TDC to BDC, fuel flows through the open inlet passage and into the high-pressure chamber above the plunger. At BDC, the plunger's rotating movement then closes the inlet passage and opens the distributor slot for a given outlet port (Fig. 10a). The plunger now reverses its direction of movement and moves upwards, the working stroke begins. The pressure that builds up in the high-pressure chamber above the plunger and in the outlet-port passage suffices to open the delivery valve in question and the fuel is forced through the high-pressure line to the injector nozzle (Fig. 10b). The working stroke is completed as soon as the plunger's transverse cutoff bore reaches the control edge of the control collar and pressure collapses. From this point on, no more fuel is delivered to the injector and the delivery valve closes the high-pressure line.

Fig. 9: **The cam plate rotates against the roller ring, whereby its cam track follows the rollers causing it to lift (for TDC) and drop back again (for BDC)**

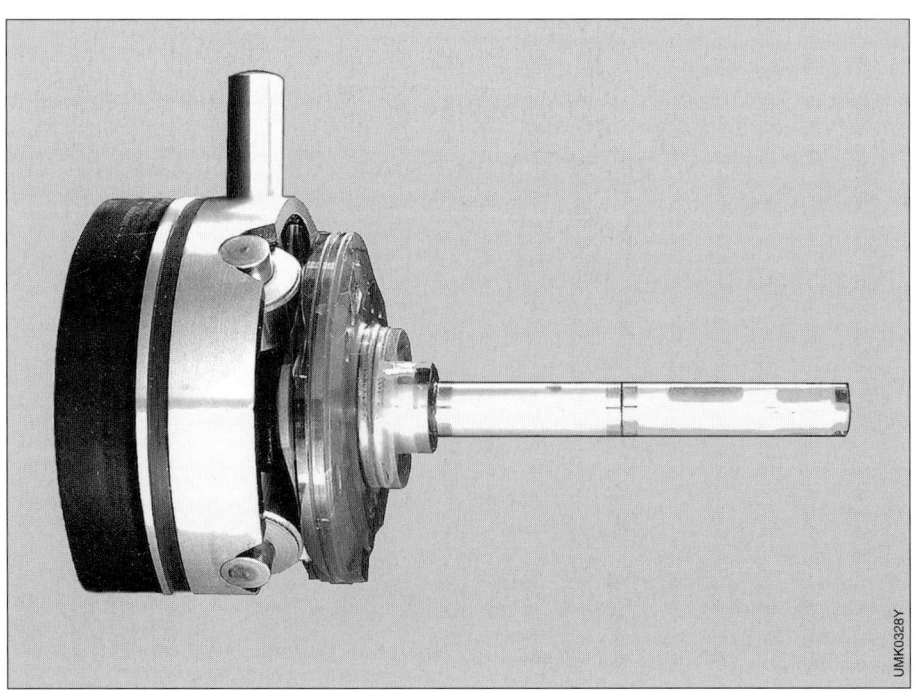

Fig. 10

Distributor plunger with stroke and delivery phases

a Inlet passage closes.
At BDC, the metering slot (**1**) closes the inlet passage, and the distributor slot (**2**) opens the outlet port.

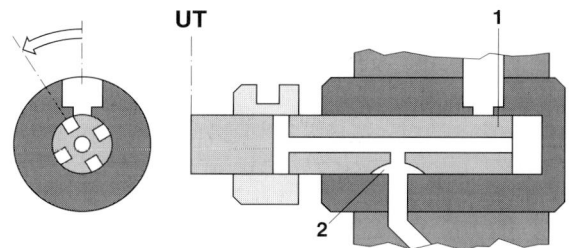

b Fuel delivery.
During the plunger stroke towards TDC (working stroke), the plunger pressurizes the fuel in the high-pressure chamber (**3**). The fuel travels through the outlet-port passage (**4**) to the injection nozzle.

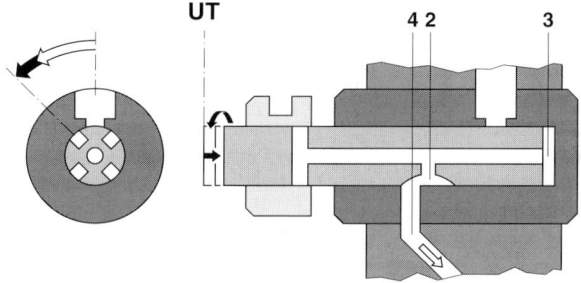

c End of delivery.
Fuel delivery ceases as soon as the control collar (**5**) opens the transverse cutoff bore (**6**).

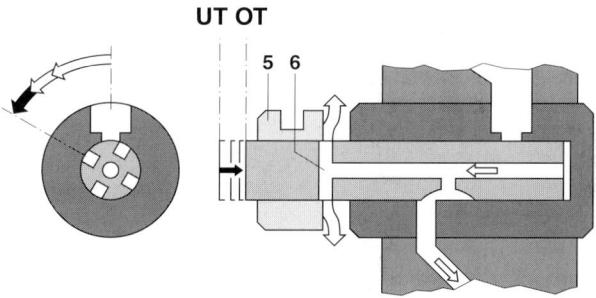

d Entry of fuel.
Shortly before TDC, the inlet passage is opened. During the plunger's return stroke to BDC, the high-pressure chamber is filled with fuel and the transverse cutoff bore is closed again. The outlet-port passage is also closed at this point.

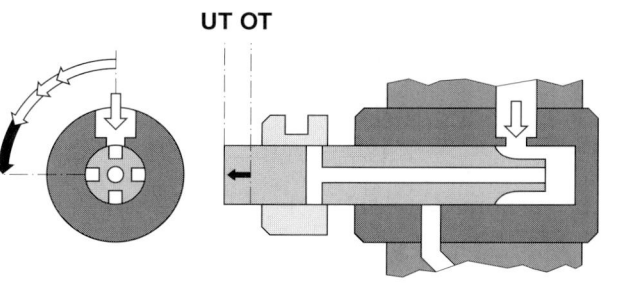

OT = TDC
UT = BDC

UMK0329Y

During the plunger's continued movement to TDC, fuel returns through the cutoff bore to the pump interior. During this phase, the inlet passage is opened again for the plunger's next working cycle (Fig. 10c).

During the plunger's return stroke, its transverse cutoff bore is closed by the plunger's rotating stroke movement, and the high-pressure chamber above the plunger is again filled with fuel through the open inlet passage (Fig. 10d).

Delivery valve

The delivery valve closes off the high-pressure line from the pump. It has the job of relieving the pressure in the line by removing a defined volume of fuel upon completion of the delivery phase. This ensures precise closing of the injection nozzle at the end of the injection process. At the same time, stable pressure conditions between injection pulses are created in the high-pressure lines, regardless of the quantity of fuel being injected at a particular time.

The delivery valve is a plunger-type valve. It is opened by the injection pressure and closed by its return spring.

Between the plunger's individual delivery strokes for a given cylinder, the delivery valve in question remains closed. This separates the high-pressure line and the distributor head's outlet-port passage. During delivery, the pressure generated in the high-pressure chamber above the plunger causes the delivery valve to open. Fuel then flows via longitudinal slots, into a ring-shaped groove and through the delivery-valve holder, the high-pressure line and the nozzle holder to the injection nozzle.

As soon as delivery ceases (transverse cutoff bore opened), the pressure in the high-pressure chamber above the plunger and in the highpressure lines drops to that of the pump interior, and the delivery-valve spring together with the static pressure in the line force the delivery-valve plunger back onto its seat again (Fig. 11).

Fig. 11

Distributor head with high-pressure chamber

1 Control collar, **2** Distributor head, **3** Distributor plunger, **4** Delivery-valve holder, **5** Delivery-valve.

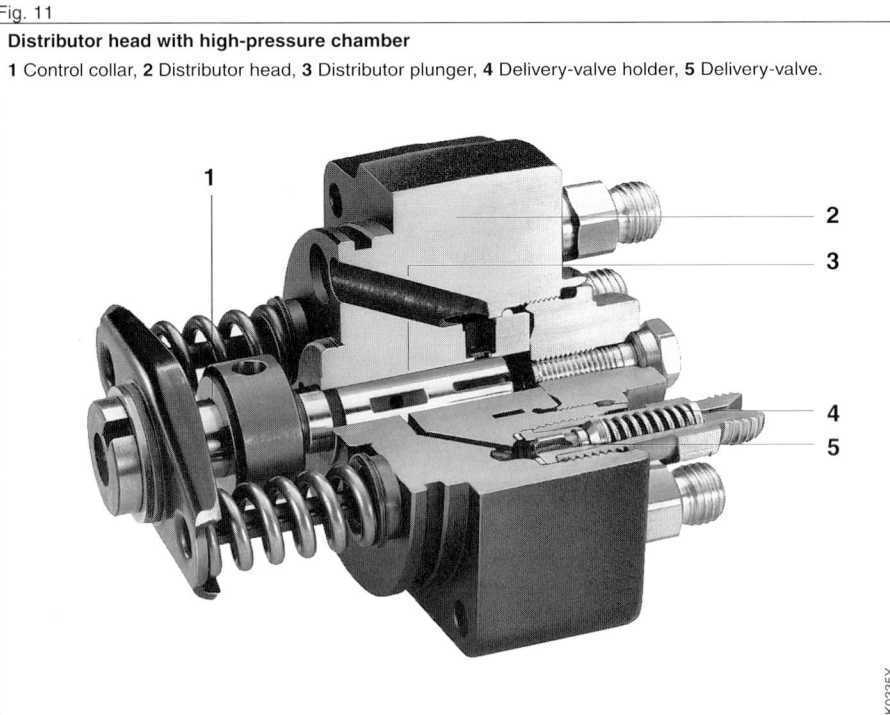

UMK0335Y

Delivery valve with return-flow restriction

Precise pressure relief in the lines is necessary at the end of injection. This though generates pressure waves which are reflected at the delivery valve. These cause the delivery valve to open again, or cause vacuum phases in the high-pressure line. These processes result in post-injection of fuel with attendant increases in exhaust emissions or cavitation and wear in the injection line or at the nozzle. To prevent such harmful reflections, the delivery valve is provided with a restriction bore which is only effective in the direction of return flow. This return-flow restriction comprises a valve plate and a pressure spring so arranged that the restriction is ineffective in the delivery direction, whereas in the return direction damping comes into effect (Fig. 12).

Constant-pressure valve

With high-speed direct-injection (DI) engines, it is often the case that the "retraction volume" resulting from the retraction piston on the delivery-valve plunger does not suffice to reliably prevent cavitation, secondary injection, and combustion-gas blowback into the nozzle-and-holder assembly. Here, constant-pressure valves are fitted which relieve the high-pressure system (injection line and nozzle-and-holder assembly) by means of a single-acting non-return valve which can be set to a given pressure, e.g., 60 bar (Fig. 13).

High-pressure lines

The pressure lines installed in the fuel-injection system have been matched precisely to the rate-of-discharge curve and must not be tampered with during service and repair work. The high-pressure lines connect the injection pump to the injection nozzles and are routed so that they have no sharp bends. In automotive applications, the high-pressure lines are normally secured with special clamps at specific intervals, and are made of seamless steel tubing.

Fig. 12

Delivery valve with return-flow restriction

1 Delivery-valve holder, **2** Return-flow restriction, **3** Delivery-valve spring, **4** Valve holder, **5** Piston shaft, **6** Retraction piston.

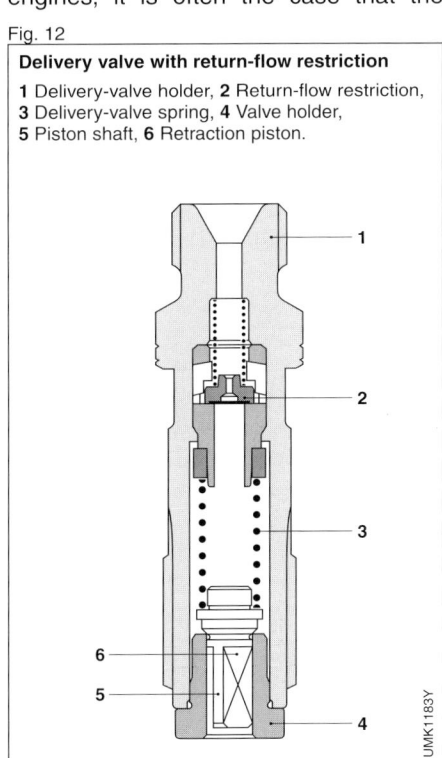

UMK1183Y

Fig. 13

Constant-pressure valve

1 Delivery-valve holder, **2** Filler piece with spring locator, **3** Delivery-valve spring, **4** Delivery-valve plunger, **5** Constant-pressure valve, **6** Spring seat, **7** Valve spring (constant-pressure valve), **8** Setting sleeve, **9** Valve holder, **10** Shims.

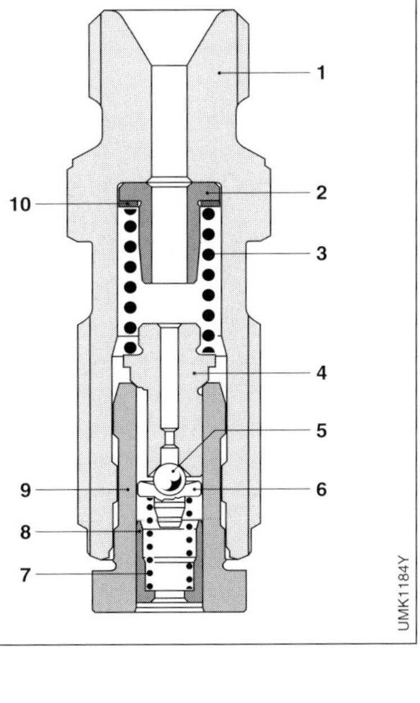

UMK1184Y

Mechanical engine-speed control (governing)

Application

The driveability of a diesel-powered vehicle can be said to be satisfactory when its engine immediately responds to driver inputs from the accelerator pedal. Apart from this, upon driving off the engine must not tend to stall. The engine must respond to accelerator-pedal changes by accelerating or decelerating smoothly and without hesitation. On the flat, or on a constant gradient, with the accelerator pedal held in a given position, the vehicle speed should also remain constant. When the pedal is released the engine must brake the vehicle. On the diesel engine, it is the injection pump's governor that ensures that these stipulations are complied with. The governor assembly comprises the mechanical (flyweight) governor and the lever assembly. It is a sensitive control device which determines the position of the control collar, thereby defining the delivery stroke and with it the injected fuel quantity. It is possible to adapt the governor's response to setpoint changes by varying the design of the lever assembly (Fig. 1).

Governor functions

The basic function of all governors is the limitation of the engine's maximum speed. Depending upon type, the governor is also responsible for keeping certain engine speeds constant, such as idle speed, or the minimum and maximum engine speeds of a stipulated engine-speed range, or of the complete speed range, between idle and maximum speed. The different governor types are a direct result of the variety of governor assignments (Fig. 2):

– Low-idle-speed governing: The diesel engine's low-idle speed is controlled by the injection-pump governor.

Fig. 1

Distributor injection pump with governor assembly, comprising flyweight governor and lever assembly

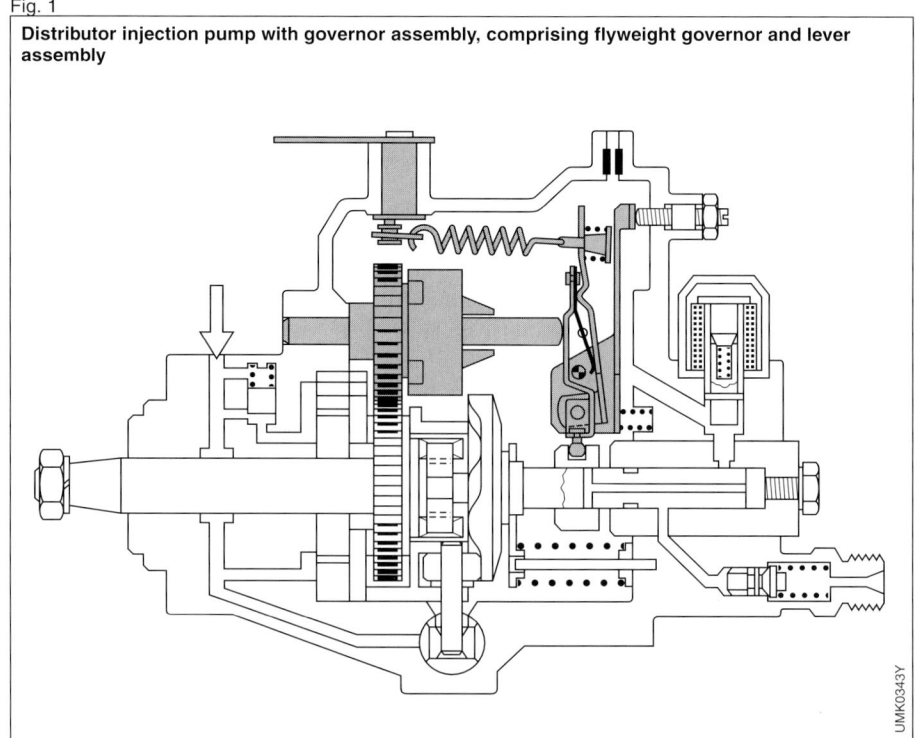

UMK0343Y

– Maximum-speed governing: With the accelerator pedal fully depressed, the maximum full-load speed must not increase to more than high idle speed (maximum speed) when the load is removed. Here, the governor responds by shifting the control collar back towards the "Stop" position, and the supply of fuel to the engine is reduced.

– Intermediate-speed governing: Variable-speed governors incorporate intermediate-speed governing. Within certain limits, these governors can also maintain the engine speeds between idle and maximum constant. This means that depending upon load, the engine speed n varies inside the engine's power range only between n_{VT} (a given speed on the full-load curve) and n_{LT} (with no load on the engine).

Other control functions are performed by the governor in addition to its governing responsibilities:

– Releasing or blocking of the extra fuel required for starting,

– Changing the full-load delivery as a function of engine speed (torque control). In some cases, add-on modules are necessary for these extra assignments.

Speed-control (governing) accuracy

The parameter used as the measure for the governor's accuracy in controlling engine speed when load is removed is the so-called speed droop (P-degree). This is the engine-speed increase, expressed as a percentage, that occurs when the diesel engine's load is removed with the control-lever (accelerator) position unchanged. Within the speed-control range, the increase in engine speed is not to exceed a given figure. This is stipulated as the high idle speed. This is the engine speed which results when the diesel engine, starting at its maximum speed under full load, is relieved of all load. The speed increase is proportional to the change in load, and increases along with it.

$$\delta = \frac{n_{lo} - n_{vo}}{n_{vo}}$$

or expressed in %:

$$\delta = \frac{n_{lo} - n_{vo}}{n_{vo}} \cdot 100\%$$

where
δ = Speed droop
n_{lo} = High idle (maximum) speed
n_{vo} = Maximum full-load speed

The required speed droop depends on engine application. For instance, on an engine used to power an electrical generator set, a small speed droop is required so that load changes result in only minor speed changes and therefore minimal frequency changes. On the other hand, for automotive applications large speed droops are preferable because these result in more stable control in case of only slight load changes (acceleration or deceleration) and lead to better driveability. A low-value speed droop would lead to rough, jerking operation when the load changes.

Fig. 2

Governor characteristics

a Minimum-maximum-speed governor,
b Variable-speed governor.
1 Start quantity, **2** Full-load delivery,
3 Torque control (positive),
4 Full-load speed regulation, **5** Idle.

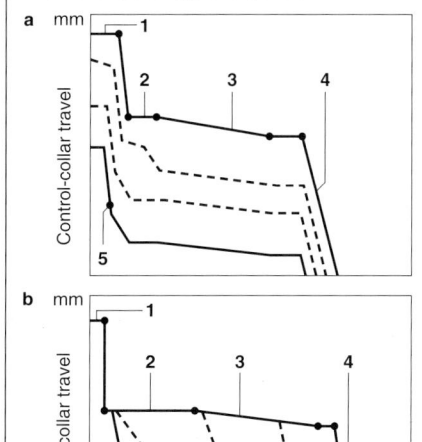

Variable-speed governor

The variable-speed governor controls all engine speeds between start and high idle (maximum). The variable-speed governor also controls the idle speed and the maximum full-load speed, as well as the engine-speed range in between. Here, any engine speed can be selected by the accelerator pedal and, depending upon the speed droop, maintained practically constant (Fig. 4).

This is necessary for instance when ancillary units (winches, fire-fighting pumps, cranes etc.) are mounted on the vehicle. The variable-speed governor is also often fitted in commercial and agricultural vehicles (tractors and combine harvesters).

Design and construction

The governor assembly is driven by the drive shaft and comprises the flyweight housing complete with flyweights.

The governor assembly is attached to the governor shaft which is fixed in the governor housing, and is free to rotate around it. When the flyweights rotate they pivot outwards due to centrifugal force and their radial movement is converted to an axial movement of the sliding sleeve. The sliding-sleeve travel and the force developed by the sleeve influence the governor lever assembly. This comprises the starting lever, tensioning lever, and adjusting lever (not shown). The interaction of spring forces and sliding-sleeve force defines the setting of the governor lever assembly, variations of which are transferred to the control collar and result in adjustments to the injected fuel quantity.

Starting

With the engine at standstill, the flyweights and the sliding sleeve are in their initial position (Fig. 3a). The starting lever has been pushed to the start position by the starting spring and has pivoted around its fulcrum M_2. At the same time the control collar on the distributor plunger has been shifted to its

Fig. 3

Variable-speed governor. Start and idle positions

a Start position, **b** Idle position.
1 Flyweights, **2** Sliding sleeve, **3** Tensioning lever, **4** Starting lever, **5** Starting spring, **6** Control collar, **7** Distributor-plunger cutoff port, **8** Distributor plunger, **9** Idle-speed adjusting screw, **10** Engine-speed control lever, **11** Control lever, **12** Control-lever shaft, **13** Governor spring, **14** Retaining pin, **15** Idle spring.
a Starting-spring travel, c Idle-spring travel, h_1 max. working stroke (start); h_2 min. working stroke (idle): M_2 fulcrum for 4 and 5.

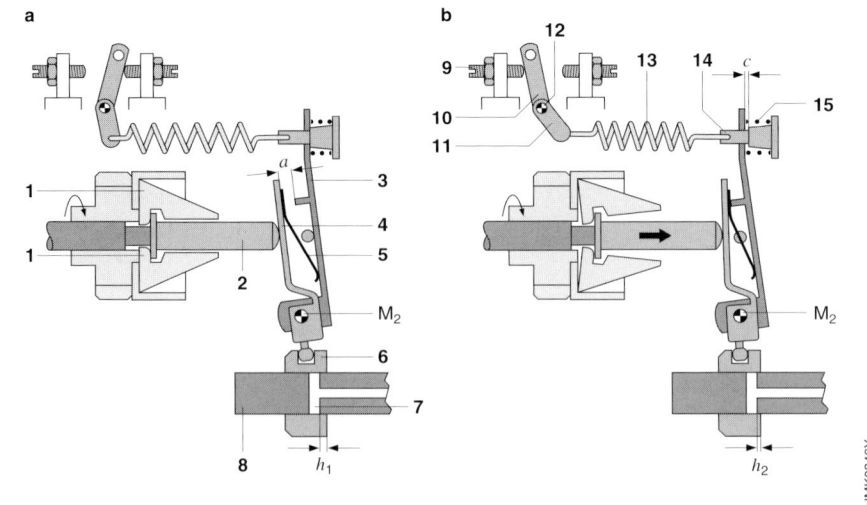

start-quantity position by the ball pin on the starting lever. This means that when the engine is cranked the distributor plunger must travel through a complete working stroke (= maximum delivery quantity) before the cutoff bore is opened and delivery ceases. Thus the start quantity (= maximum delivery quantity) is automatically made available when the engine is cranked.

The adjusting lever is held in the pump housing so that it can rotate. It can be shifted by the fuel-delivery adjusting screw (not shown in Figure 3). Similarly, the start lever and tensioning lever are also able to rotate in the adjusting lever. A ball pin which engages in the control collar is attached to the underside of the start lever, and the start spring to its upper section. The idle spring is attached to a retaining pin at the top end of the tensioning lever. Also attached to this pin is the governor spring. The connection to the engine-speed control lever is through a lever and the control-lever shaft.

It only needs a very low speed for the sliding sleeve to shift against the soft start spring by the amount a. In the process, the start lever pivots around fulcrum M_2 and the start quantity is automatically reduced to the idle quantity.

Low-idle-speed control

With the engine running, and the accelerator pedal released, the engine-speed control lever shifts to the idle position (Figure 3b) up against the idle-speed adjusting screw. The idle speed is selected so that the engine still runs reliably and smoothly when unloaded or only slightly loaded. The actual control is by means of the idle spring on the retaining pin which counteracts the force generated by the flyweights.

This balance of forces determines the sliding-sleeve's position relative to the distributor plunger's cutoff bore, and with it the working stroke. At speeds above idle, the spring has been compressed by the amount c and is no longer effective. Using the special idle spring attached to the governor housing,

Characteristic curves of the variable-speed governor

A: Start position of the control collar,
S: Engine starts with start quantity,
S–L: Start quantity reduces to idle quantity,
L: Idle speed n_{LN} following engine start-up (no-load),
L–B: Engine acceleration phase after shifting the engine-speed control lever from idle to a given required speed n_c,
B–B': The control collar remains briefly in the full-load position and causes a rapid increase in engine speed,
B'–C: Control collar moves back (less injected fuel quantity, higher engine speed). In accordance with the speed droop, the vehicle maintains the required speed or speed n_c in the part-load range,
E: Engine speed n_{LT}, after removal of load from the engine with the position of the engine-speed control-lever remaining unchanged.

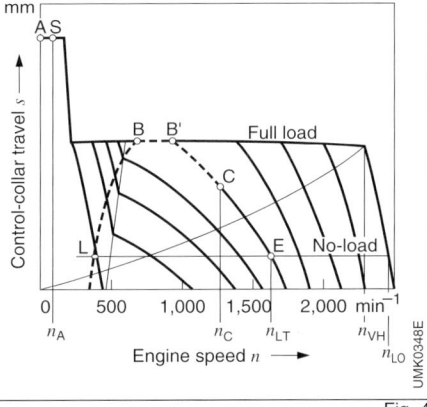

Fig. 4

this means that idle speed can be adjusted independent of the accelerator-pedal setting, and can be increased or decreased as a function of temperature or load.

Operation under load

During actual operation, depending upon the required engine speed or vehicle speed, the engine-speed control lever is in a given position within its pivot range. This is stipulated by the driver through a given setting of the accelerator pedal. At engine speeds above idle, start spring and idle spring have been compressed completely and have no further effect on governor action. This is taken over by the governor spring.

Example (Fig. 5):
Using the accelerator pedal, the driver sets the engine-speed control lever to a specific position corresponding to a desired (higher) speed. As a result of this adjustment of the control-lever position, the governor spring is tensioned by a given amount, with the result that the governor-spring force exceeds the centrifugal force of the flyweights and causes the start lever and the tensioning lever to pivot around fulcrum M_2. Due to the mechanical transmission ratio designed into the system, the control collar shifts in the "Full-load" direction. As a result, the delivery quantity is increased and the engine speed rises. This causes the flyweights to generate more force which, through the sliding sleeve, opposes the governor-spring force.

The control collar remains in the "Full-load" position until a torque balance occurs. If the engine speed continues to increase, the flyweights separate even further, the sliding-sleeve force prevails, and as a result the start and tensioning levers pivot around M_2 and push the control collar in the "Stop" direction so that the control port is opened sooner. It is possible to reduce the delivery quantity to "zero" which ensures that engine-speed limitation takes place. This means that during operation, and as long as the engine is not overloaded, every position of the engine-speed control lever is allocated to a specific speed range between full-load and zero. The result is that within the limits set by its speed droop, the governor maintains the desired speed (Fig. 4).

If the load increases to such an extent (for instance on a gradient) that even though the control collar is in the full-load position the engine speed continues to drop, this indicates that it is impossible to increase fuel delivery any further. This means that the engine is overloaded and the driver must change down to a lower gear.

Fig. 5

Fig. 5: Variable-speed governor, operation under load

a Governor function with increasing engine speed, **b** with falling engine speed.
1 Flyweights, **2** Engine-speed control lever, **3** Idle-speed adjusting screw, **4** Governor spring,
5 Idle spring, **6** Start lever, **7** Tensioning lever, **8** Tensioning-lever stop, **9** Starting spring,
10 Control collar, **11** Adjusting screw for high idle (maximum) speed, **12** Sliding sleeve,
13 Distributor-plunger cutoff bore, **14** Distributor plunger.
h_1 Working stroke, idle, h_2 Working stroke, full-load, M_2 fulcrum for 6 and 7.

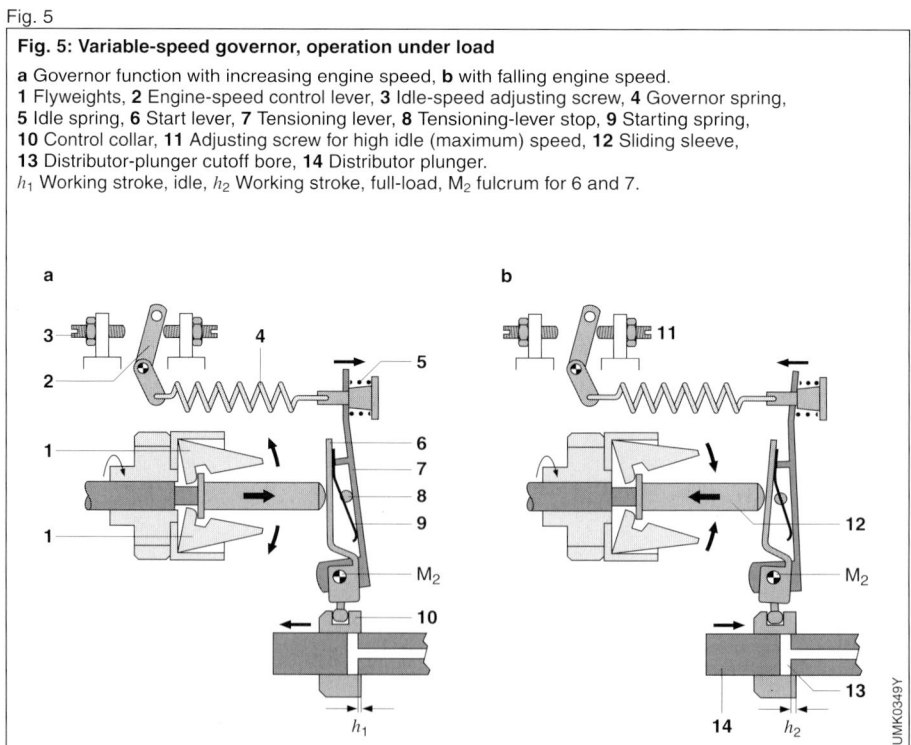

Overrun (engine braking)

During downhill operation the engine is "driven" by the vehicle, and engine speed tends to increase. This causes the flyweights to move outwards so that the sliding sleeve presses against the tensioning and start levers. Both levers change their position and push the control collar in the direction of less fuel delivery until a reduced fuel-delivery figure is reached which corresponds to the new loading level. At the extreme, the delivery figure is zero. Basically, with the variable-speed governor, this process applies for all settings of the engine-speed control lever, when the engine load or engine speed changes to such an extent that the control collar shifts to either its full-load or stop position.

Fig. 6

Characteristic curves of the minimum-maximum-speed governor with idle spring and intermediate spring

a Starting-spring range,
b Range of starting and idle spring,
d Intermediate-spring range,
f Governor-spring range.

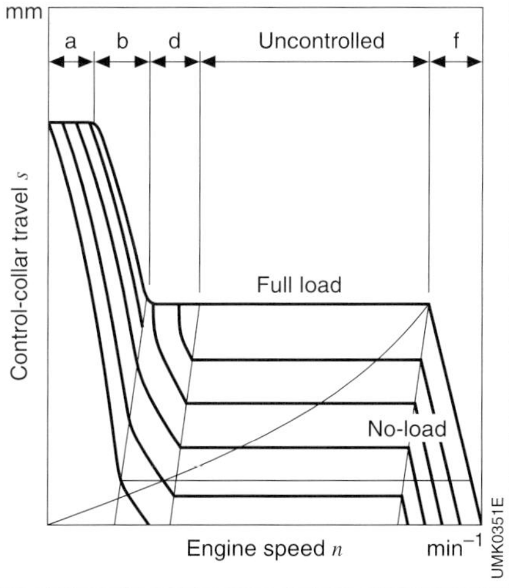

Minimum-maximum-speed governor

The minimum-maximum-speed governor controls (governs) only the idle (minimum) speed and the maximum speed. The speed range between these points is directly controlled by the accelerator pedal (Fig. 6).

Design and construction

The governor assembly with flyweights, and the lever configuration, are comparable with those of the variable-speed governor already dealt with. The main difference lies in the governor spring and its installation. It is in the form of a compression spring and is held in a guide element. Tensioning lever and governor spring are connected by a retaining pin.

Starting

With the engine at standstill, the flyweights are also stationary and the sliding sleeve is in its initial position. This enables the starting spring to push the flyweights to their inner position through the starting lever and the sliding sleeve. On the distributor plunger, the control collar is in the start-quantity position.

Idle control

Once the engine is running and the accelerator pedal has been released, the engine-speed control lever is pulled back to the idle position by its return spring. The centrifugal force generated by the flyweights increases along with engine speed (Fig. 7a) and the inner flyweight legs push the sliding sleeve up against the start lever. The idle spring on the tensioning lever is responsible for the controlling action. The control collar is shifted in the direction of "less delivery" by the pivoting action of the start lever, its position being determined by interaction between centrifugal force and spring force.

Operation under load

If the driver depresses the accelerator pedal, the engine-speed control lever is pivoted through a given angle. The starting and idle springs are no longer effective and the intermediate spring comes into effect. The intermediate spring on the minimum-maximum-speed governor provides a "soft" transition to the uncontrolled range. If the engine-speed control lever is pressed even further in the full-load direction, the intermediate spring is compressed until the tensioning lever abuts against the retaining pin (Fig. 7b). The intermediate spring is now ineffective and the uncontrolled range has been entered. This uncontrolled range is a function of the governor-spring pretension, and in this range the spring can be regarded as a solid element. The accelerator-pedal position (engine-speed control lever) is now transferred directly through the governor lever mechanism to the control collar, which means that the injected fuel quantity is directly determined by the accelerator pedal. To accelerate, or climb a hill, the driver must "give gas", or ease off on the accelerator if less engine power is needed.

If engine load is now reduced, with the engine-speed control lever position unchanged, engine speed increases without an increase in fuel delivery. The flyweights' centrifugal force also increases and pushes the sliding sleeve even harder against the start and tensioning levers. Full-load speed control does not set in, at or near the engine's rated speed, until the governor-spring pre-tension has been overcome by the effect of the sliding-sleeve force.

If the engine is relieved of all load, speed increases to the high idle speed, and the engine is thus protected against over-revving.

Passenger cars are usually equipped with a combination of variable-speed governor and minimum-maximum-speed governor.

Fig. 7

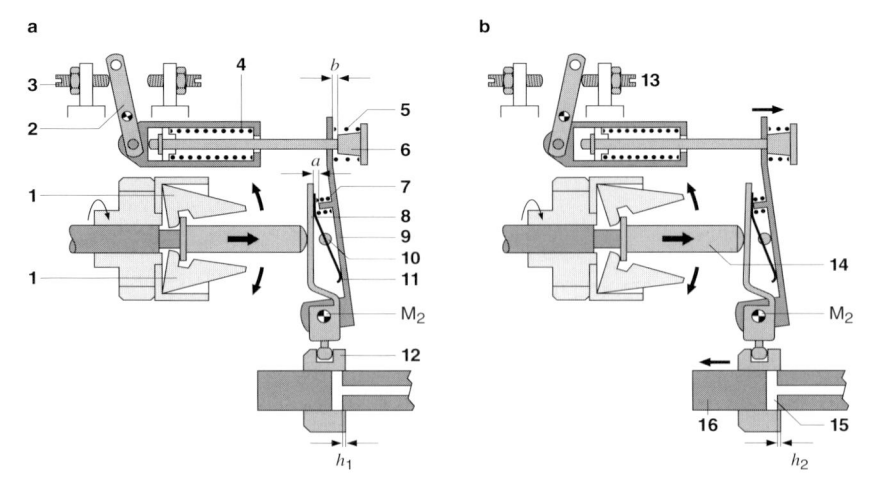

Minimum-maximum-speed governor

a Idle setting, **b** Full-load setting.
1 Flyweights, **2** Engine-speed control lever, **3** Idle-speed adjusting screw, **4** Governor spring,
5 Intermediate spring, **6** Retaining pin, **7** Idle spring, **8** Start lever, **9** Tensioning lever, **10** Tensioning-lever stop, **11** Starting spring, **12** Control collar, **13** Full-load speed control, **14** Sliding sleeve, **15** Distributor plunger cutoff bore, **16** Distributor plunger.
a Start and idle-spring travel, b Intermediate-spring travel, h_1 Idle working stroke, h_2 Full-load working stroke, M_2 fulcrum for 8 and 9.

Injection timing

In order to compensate for the injection lag and the ignition lag, as engine speed increases the timing device advances the distributor pump's start of delivery referred to the engine's crankshaft. Example (Fig. 1):

Start of delivery (FB) takes place after the inlet port is closed. The high pressure then builds up in the pump which, as soon as the nozzle-opening pressure has been reached leads to the start of injection (SB). The period between FB and SB is referred to as the injection lag (SV). The increasing compression of the air-fuel mixture in the combustion chamber then initiates the ignition (VB). The period between SB and VB is the ignition lag (ZV). As soon as the cutoff port is opened again the pump pressure collapses (end of pump delivery), and the nozzle needle closes again (end of injection, SE). This is followed by the end of combustion (VE).

Assignment

During the fuel-delivery process, the injection nozzle is opened by a pressure wave which propagates in the high-pressure line at the speed of sound. Basically speaking, the time required for this process is independent of engine speed, although with increasing engine speed the crankshaft angle between start of delivery and start of injection also increases. This must be compensated for by advancing the start of delivery. The pressure wave's propagation time is determined by the length of the high-pressure line and the speed of sound which is approx. 1,500 m/s in diesel fuel. The interval represented by this propagation time is termed the injection lag. In other words, the start of injection lags behind the start of delivery. This phenomena is the reason for the injector opening later (referred to the engine's piston position) at higher engine speeds than at low engine speeds. Following injection, the injected fuel needs a certain time in

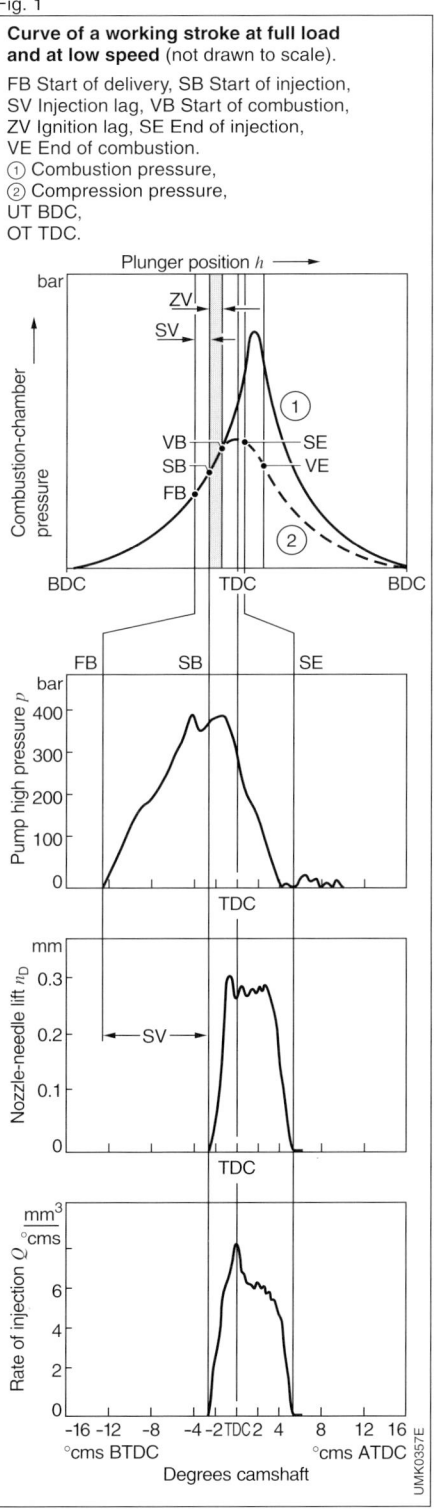

Fig. 1

Curve of a working stroke at full load and at low speed (not drawn to scale).

FB Start of delivery, SB Start of injection,
SV Injection lag, VB Start of combustion,
ZV Ignition lag, SE End of injection,
VE End of combustion.
① Combustion pressure,
② Compression pressure,
UT BDC,
OT TDC.

order to atomize and mix with the air to form an ignitable mixture.

This is termed the air-fuel mixture preparation time and is independent of engine speed. In a diesel engine, the time required between start of injection and start of combustion is termed the ignition lag.

The ignition lag is influenced by the diesel fuel's ignition quality (defined by the Cetane Number), the compression ratio, the intake-air temperature, and the quality of fuel atomization. As a rule, the ignition lag is in the order of 1 millisecond. This means that presuming a constant start of injection, the crankshaft angle between start of injection and start of combustion increases along with increasing engine speed. The result is that combustion can no longer start at the correct point (referred to the engine-piston position).

Being as the diesel engine's most efficient combustion and power can only be developed at a given crankshaft or piston position, this means that the injection pump's start of delivery must be advanced along with increasing engine speed in order to compensate for the overall delay caused by ignition lag and injection lag. This start-of-delivery advance is carried out by the engine-speed-dependent timing device.

Timing device

Design and construction

The hydraulically controlled timing device is located in the bottom of the distributor pump's housing, at right angles to the pump's longitudinal axis (Fig. 2), whereby its piston is free to move in the pump housing. The housing is closed with a cover on each side. There is a passage in one end of the timing device plunger through which the fuel can enter, while at the other end the plunger is held by a compression spring. The piston is connected to the roller ring

Fig. 2

Distributor injection pump with timing device

1 Roller ring, **2** Roller-ring rollers, **3** Sliding block, **4** Pin, **5** Timing-device piston,
6 Cam plate, **7** Distributor plunger.

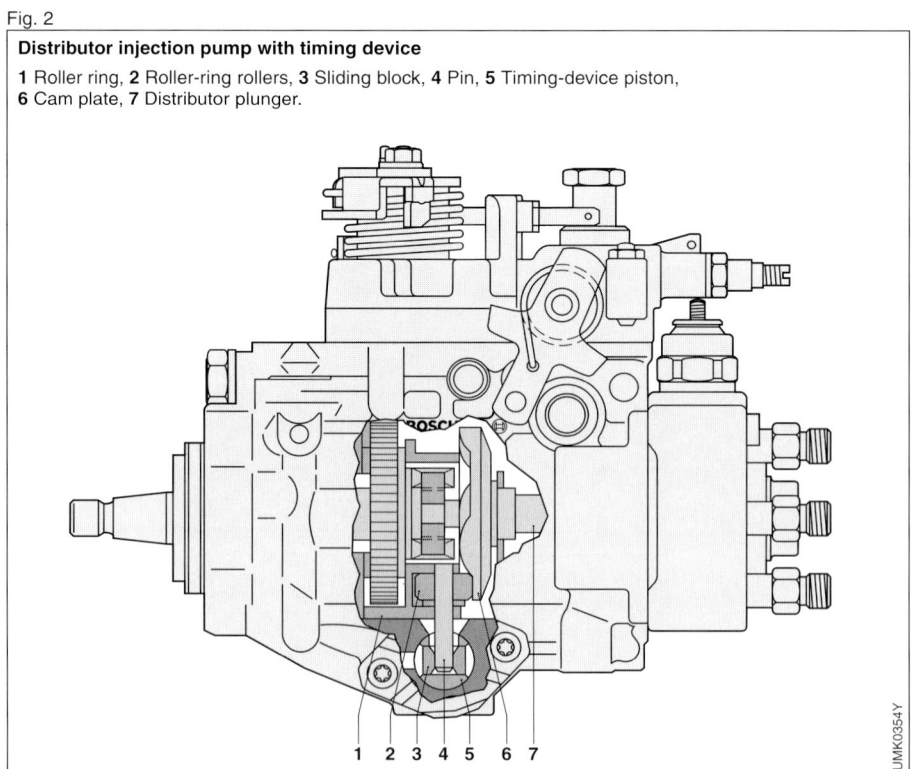

1   2 3 4 5   6 7

UMK0354Y

through a sliding block and a pin so that piston movement can be converted to rotational movement of the roller ring.

Method of operation

The timing-device piston is held in its initial position by the timing-device spring (Fig. 3a). During operation, the pressure-control valve regulates the fuel pressure inside the pump so that it is proportional to engine speed. As a result, the engine-speed-dependent fuel pressure is applied to the end of the timing-device piston opposite to the spring.

As from about 300 min^{-1}, the fuel pressure inside the pump overcomes the spring preload and shifts the timing-device piston to the left and with it the sliding block and the pin which engages in the roller ring (Fig. 3b). The roller ring is rotated by movement of the pin, and the relative position of the roller ring to the cam plate changes with the result that the rollers lift the rotating cam plate at an earlier moment in time. In other words, the roller ring has been rotated through a defined angle with respect to the cam plate and the distributor plunger. Normally, the maximum angle is 12 degrees camshaft (24 degrees crankshaft).

Fig. 3

Timing device, method of operation
a Initial position,
b Operating position.
1 Pump housing, 2 Roller ring,
3 Roller-ring rollers, 4 Pin,
5 Passage in timing-device piston,
6 Cover, 7 Timing-device piston,
8 Sliding block, 9 Timing-device spring.

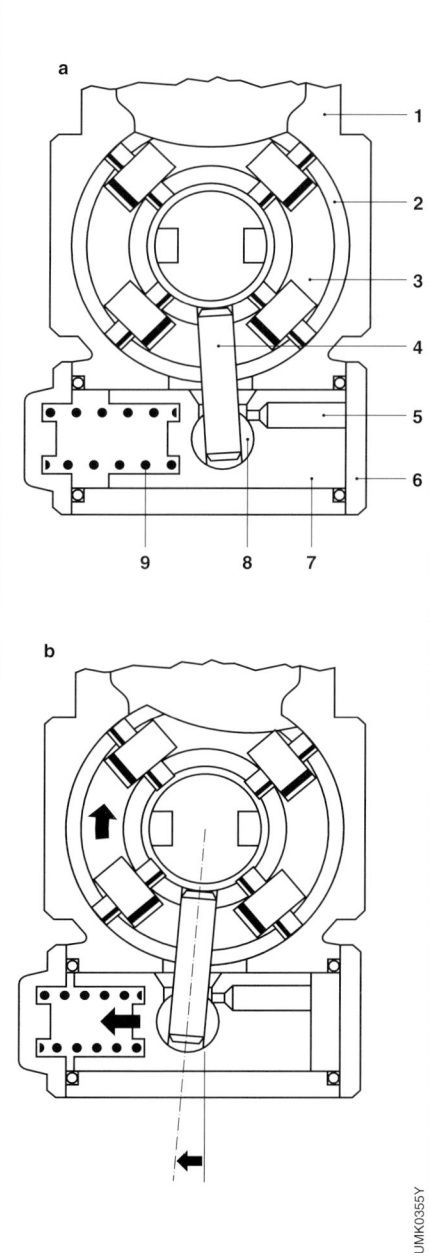

UMK0355Y

Add-on modules and shutoff devices

Application

The distributor injection pump is built according to modular construction principles, and can be equipped with a variety of supplementary (add-on) units (Fig. 1). These enable the implementation of a wide range of adaptation possibilities with regard to optimization of engine torque, power output, fuel economy, and exhaust-gas composition. The overview provides a summary of the add-on modules and their effects upon the diesel engine. The schematic (Fig. 2) shows the interaction of the basic distributor pump and the various add-on modules.

Torque control

Torque control is defined as varying fuel delivery as a function of engine speed in order to match it to the engine's fuel-requirement characteristic. If there are special stipulations with regard to the full-load characteristic (optimization of exhaust-gas composition, of torque characteristic curve, and of fuel economy), it may be necessary

Fig. 1

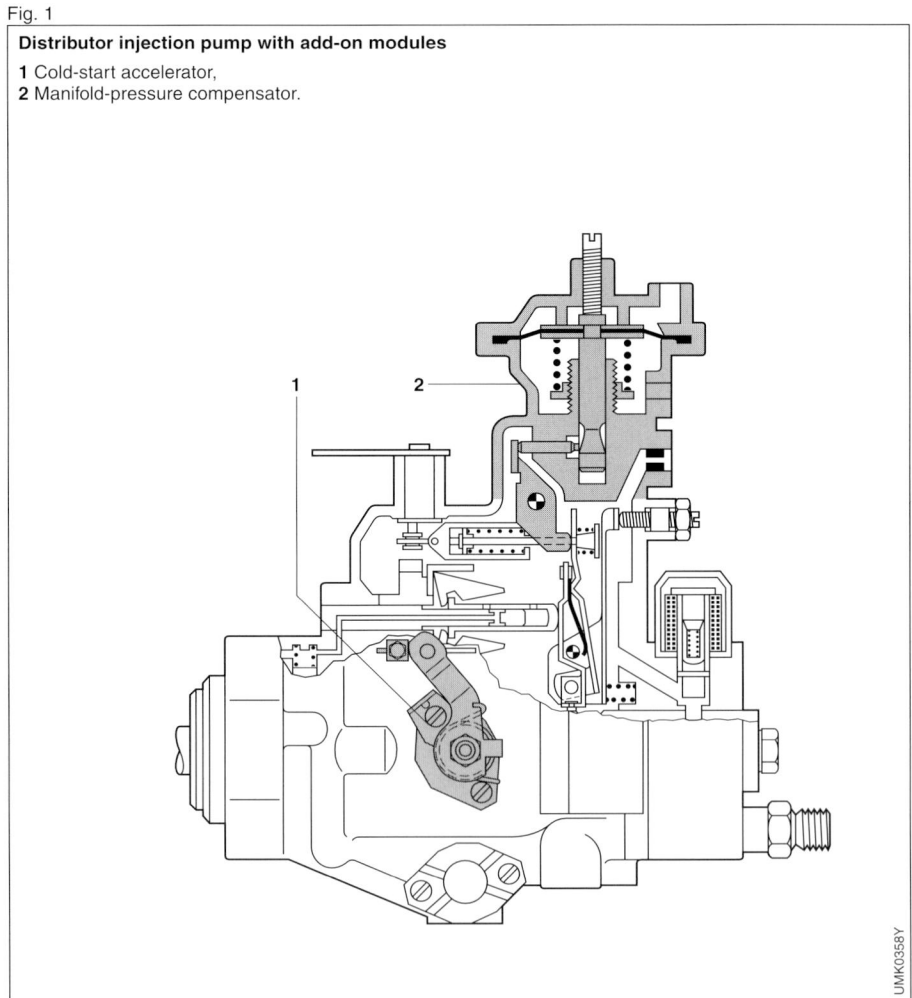

Distributor injection pump with add-on modules

1 Cold-start accelerator,
2 Manifold-pressure compensator.

UMK0358Y

Fig. 2

Schematic of the VE distributor pump with mechanical/hydraulic full-load torque control

LDA Manifold-pressure compensator.
Controls the delivery quantity as a function of the charge-air pressure.

HBA Hydraulically controlled torque control.
Controls the delivery quantity as a function of the engine speed (not for pressure-charged engines with LDA).

LFB Load-dependent start of delivery.
Adaptation of pump delivery to load. For reduction of noise and exhaust-gas emissions.

ADA Altitude-pressure compensator.
Controls the delivery quantity as a function of atmospheric pressure.

KSB Cold-start accelerator.
Improves cold-start behavior by changing the start of delivery.

GST Graded (or variable) start quantity.
Prevents excessive start quantity during warm start.

TLA Temperature-controlled idle-speed increase.
Improves engine warm-up and smooth running when the engine is cold.

ELAB Electrical shutoff device.

A Cutoff port, n_{actual} Actual engine speed (controlled variable), $n_{setpoint}$ Desired engine speed (reference variable), Q_F Delivery quantity, t_M Engine temperature, t_{LU} Ambient-air temperature, p_L Charge-air pressure, p_A Atmospheric pressure, p_i Pump interior pressure.

① Full-load torque control with governor lever assembly, ② Hydraulic full-load torque control.

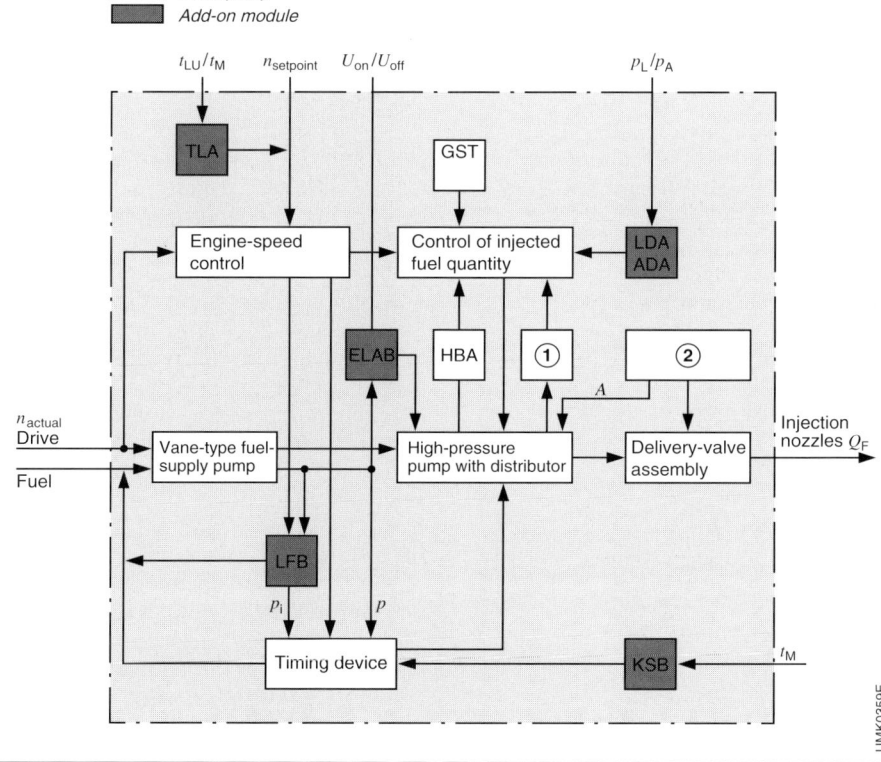

to install torque control. In other words, the engine should receive precisely the amount of fuel it needs. The engine's fuel requirement first of all climbs as a function of engine speed and then levels off somewhat at higher speeds. The fuel-delivery curve of an injection pump without torque control is shown in Fig. 3. As can be seen, with the same setting of the control collar on the distributor plunger, the injection pump delivers slightly more fuel at high speeds than it does at lower speeds. This is due to the throttling effect at the distributor plunger's cutoff port. This means that if the injection pump's delivery quantity is specified so that maximum-possible torque is developed at low engine speeds, this would lead to the engine being unable to completely combust the excess fuel injected at higher speeds and smoke would be the result together with engine overheat. On the other hand, if the maximum delivery quantity is specified so that it corresponds to the engine's requirements at maximum speed and full-load, the engine will not be able to develop full power at low engine speeds due to the delivery quantity dropping along with reductions in engine speed. Performance would be below optimum. The injected fuel quantity must therefore be adjusted to the engine's

Fig. 3

Fuel-delivery characteristics, with and without torque control

a Negative, b Positive torque control.
1 Excess injected fuel,
2 Engine fuel requirement,
3 Full-load delivery with torque control,
Shaded area:
Full-load delivery without torque control.

actual fuel requirements. This is known as "torque control", and in the case of the distributor injection pump can be implemented using the delivery valve, the cutoff port, or an extended governor-lever assembly, or the hydraulically controlled torque control (HBA). Full-load torque control using the governor lever assembly is applied in those cases in which the positive full-load torque control with the delivery valve no longer suffices, or a negative full-load torque control has become necessary.

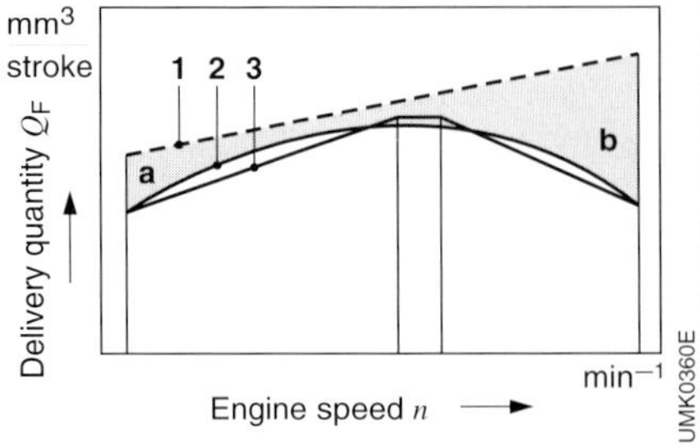

Positive torque control
Positive torque control is required on those injection pumps which deliver too much fuel at higher engine revs. The delivery quantity must be reduced as engine speed increases.

Positive torque control using the delivery valve
Within certain limits, positive torque control can be achieved by means of the delivery valve, for instance by fitting a softer delivery-valve spring.

Positive torque control using the cutoff port
Optimization of the cutoff port's dimensions and shape permit its throttling effect to be utilized for reducing the delivery quantity at higher engine speeds.

Positive torque control using the governor lever assembly (Fig. 4a)
The decisive engine speed for start of torque control is set by preloading the torque-control springs. When this speed is reached, the sliding-sleeve force (F_M) and the spring preload must be in equilibrium, whereby the torque-control lever (6) abuts against the stop lug (5) of the tensioning lever (4). The free end of the torque-control lever (6) abuts against the torque-control pin (7). If engine speed now increases, the sliding-sleeve force acting against the starting lever (1) increases and the common pivot point (M_4) of starting lever and torque-control lever (6) changes its position. At the same time,

the torque-control lever tilts around the stop pin (5) and forces the torque-control pin (7) in the direction of the stop, while the starting lever (1) swivels around the pivot point (M_2) and forces the control collar (8) in the direction of reduced fuel delivery. Torque control ceases as soon as the torque-control-pin collar (10) abuts against the starting lever (1).

Negative torque control

Negative torque control may be necessary in the case of engines which have black-smoke problems in the lower speed range, or which must generate specific torque characteristics. Similarly, turbocharged engines also need negative torque control when the manifold-pressure compensator (LDA) has ceased to be effective. In this case, the fuel delivery is increased along with engine speed (Fig. 3).

Negative torque control using the governor lever assembly (Fig. 4b)
Once the starting spring (9) has been compressed, the torque-control lever (6) applies pressure to the tensioning lever (4) through the stop lug (5). The torque-control pin (7) also abuts against the tensioning lever (4). If the sliding-sleeve force (F_M) increases due to rising engine speed, the torque-control lever presses against the preloaded torque-control spring. As soon as the sliding-sleeve force exceeds the torque-control spring force, the torque-control lever (6) is forced in the direction of the torque-control-pin collar. As a result, the common pivot point (M_4) of the starting lever and torque-control lever changes its position. At the same time the starting lever swivels around its pivot point (M_2) and pushes the control collar (8) in the direction of increased delivery. Torque control ceases as soon as the torque-control lever abuts against the pin collar.

Negative torque control using hydraulically controlled torque control HBA
In the case of naturally aspirated diesel engines, in order to give a special shape to the full-load delivery characteristic as a function of engine speed, a form of torque control can be applied which is similar to the LDA (manifold-pressure compensator).
Here, the shift force developed by the hydraulic piston is generated by the pressure in the pump interior, which in turn depends upon pump speed. In contrast to spring-type torque control, within limits the shape of the full-load characteristic can be determined by a cam on a sliding pin.

Fig. 4

Torque control using the governor-lever assembly

a Positive torque control,
b Negative torque control.
1 Starting lever,
2 Torque-control spring,
3 Governor spring,
4 Tensioning lever,
5 Stop lug,
6 Torque-control lever,
7 Torque-control pin,
8 Control collar,
9 Starting spring,
10 Pin collar,
11 Stop point,
M_2 Pivot point for 1 and 4,
M_4 Pivot point for 1 and 6,
F_M Sliding-sleeve force,
Δs Control-collar travel.

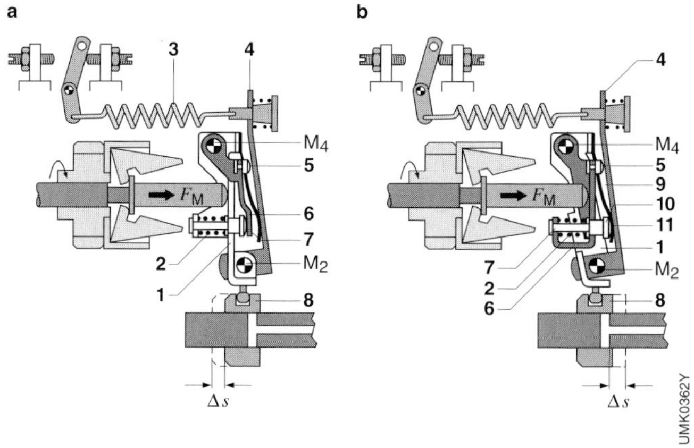

Manifold-pressure compensation

Exhaust-gas turbocharging

Because it increases the mass of air inducted by the engine, exhaust turbo-charging boosts a diesel engine's power output considerably over that of a naturally aspirated diesel engine, with little increase in dimensions and engine speeds. This means that the brake horsepower can be increased corresponding to the increase in air mass (Figure 6). In addition, it is often possible to also reduce the specific fuel consumption. An exhaust-gas turbocharger is used to pressure-charge the diesel engine (Fig. 5).

With an exhaust turbocharger, the engine's exhaust gas, instead of simply being discharged into the atmosphere, is used to drive the turbocharger's turbine at speeds which can exceed 100,000 min⁻¹. Turbine and turbocharger compressor are connected through a shaft. The compressor draws in air, compresses it, and supplies it to the engine's combustion chambers under pressure, whereby not only the air pressure rises but also the air temperature. If temperatures become excessive, some form of air cooling (intercooling) is needed between the turbocharger and the engine intake.

Fig. 5: **Diesel engine with exhaust-gas turbocharger**

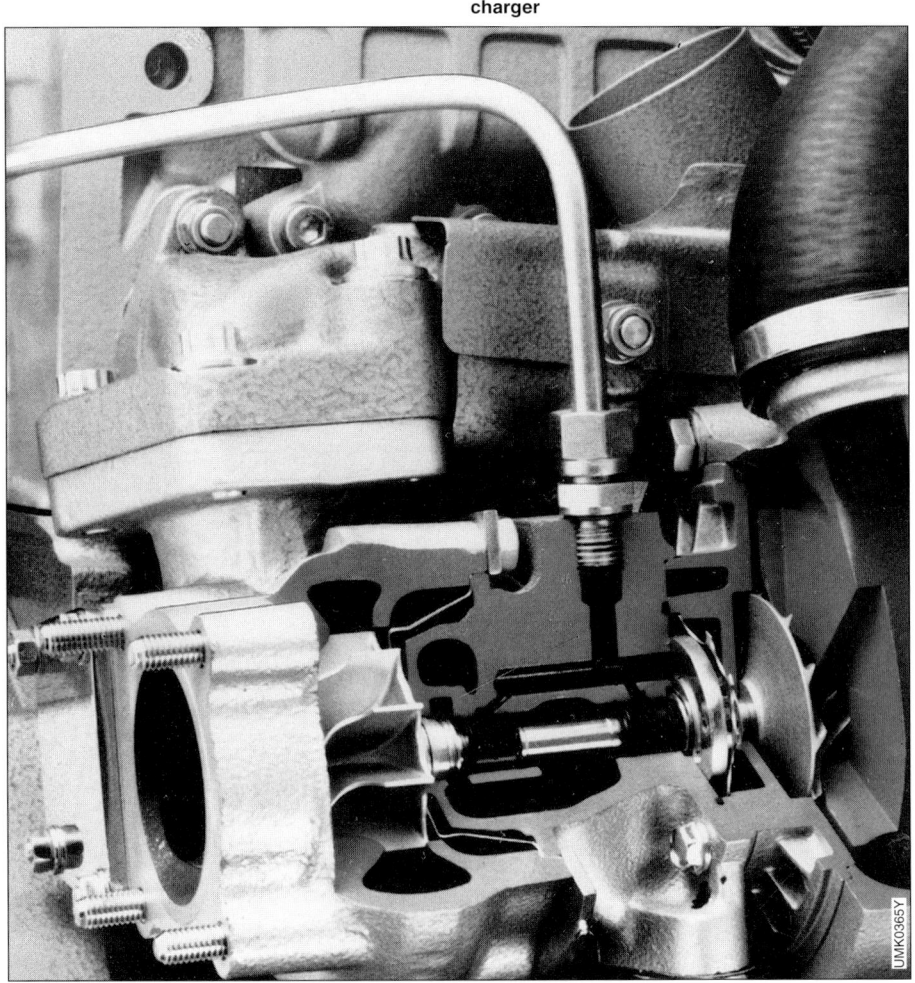

UMK0365Y

Power and torque comparison, naturally aspirated and pressure-charged engines

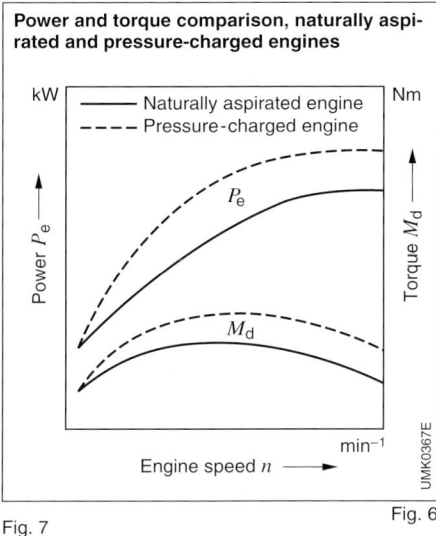

Fig. 7

Fig. 6

Manifold-pressure compensator (LDA)

The manifold-pressure compensator (LDA) reacts to the charge-air pressure generated by the exhaust-gas turbocharger, or the (mechanical) supercharger, and adapts the full-load delivery to the charge-air pressure (Figs. 6 and 7).

Assignment

The manifold-pressure compensator (LDA) is used on pressure-charged diesel engines. On these engines the injected fuel quantity is adapted to the engine's increased air charge (due to pressure-charging). If the pressure-charged diesel engine operates with a reduced cylinder air charge, the in-

Distributor injection pump with manifold-pressure compensator (LDA)

1 Governor spring, **2** Governor cover, **3** Reverse lever, **4** Guide pin, **5** Adjusting nut, **6** Diaphragm, **7** Compression spring, **8** Sliding pin, **9** Control cone, **10** Full-load adjusting screw, **11** Adjusting lever, **12** Tensioning lever, **13** Starting lever, **14** Connection for the charge-air, **15** Vent bore.
M$_1$ pivot for 3.

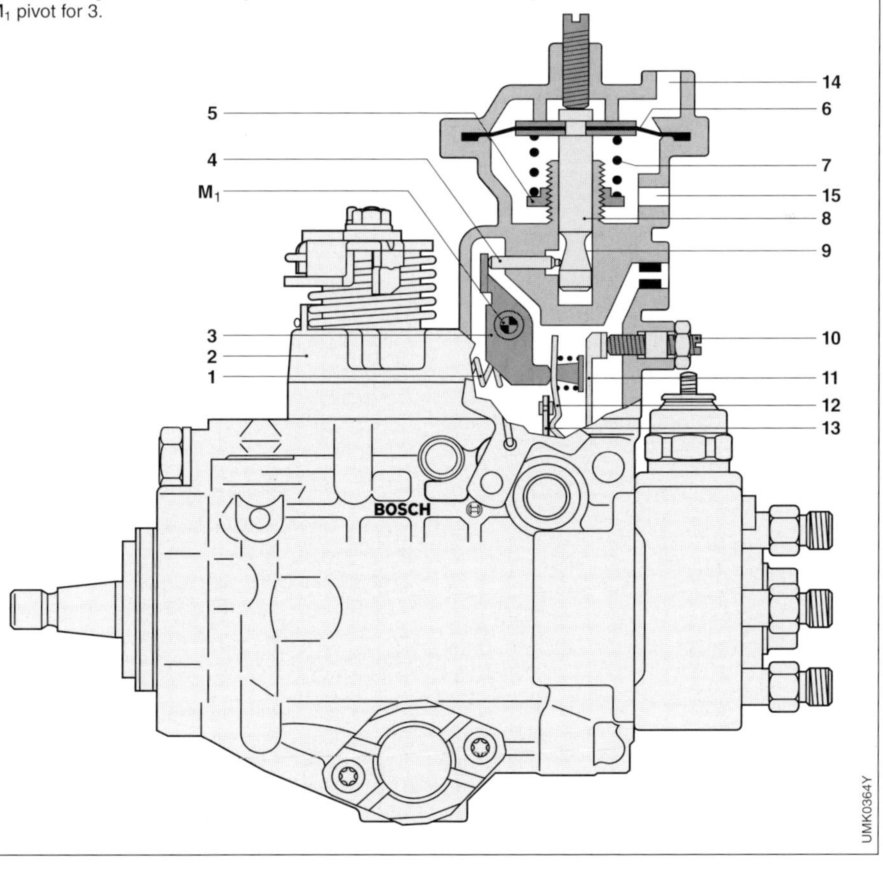

jected fuel quantity must be adapted to the lower air mass. This is performed by the manifold-pressure compensator which, below a given (selectable) charge-air pressure, reduces the full-load quantity.

Design and construction

The LDA is mounted on the top of the distributor pump (Fig. 7). In turn, the top of the LDA incorporates the connection for the charge-air and the vent bore. The interior of the LDA is divided into two separate airtight chambers by a diaphragm to which pressure is applied by a spring. At its opposite end, the spring is held by an adjusting nut with which the spring's preload is set. This serves to match the LDA's response point to the charge pressure of the exhaust turbocharger. The diaphragm is connected to the LDA's sliding pin which has a taper in the form of a control cone. This is contacted by a guide pin which transfers the sliding-pin movements to the reverse lever which in turn changes the setting of the full-load stop. The initial setting of the diaphragm and the sliding pin is set by the adjusting screw in the top of the LDA.

Method of operation

In the lower engine-speed range the charge-air pressure generated by the exhaust turbocharger and applied to the diaphragm is insufficient to overcome the pressure of the spring. The diaphragm remains in its initial position. As soon as the charge-air pressure applied to the diaphragm becomes effective, the diaphragm, and with it the sliding pin and control cone, shift against the force of the spring. The guide pin changes its position as a result of the control cone's vertical movement and causes the reverse lever to swivel around its pivot point M_1 (Fig. 7). Due to the force exerted by the governor spring, there is a non-positive connection between tensioning lever, reverse lever, guide pin, and sliding-pin control cone. As a result, the tensioning lever follows the reverse lever's swivelling movement, causing the

starting lever and tensioning lever to swivel around their common pivot point thus shifting the control collar in the direction of increased fuel delivery. Fuel delivery is adapted in response to the increased air mass in the combustion chamber (Fig. 8). On the other hand, when the charge-air pressure drops, the spring underneath the diaphragm pushes the diaphragm upwards, and with it the sliding pin. The compensation action of the governor lever mechanism now takes place in the reverse direction and the injected fuel quantity is adapted to the change in charge pressure. Should the turbocharger fail, the LDA reverts to its initial position and the engine operates normally without developing smoke. The full-load delivery with charge-air pressure is adjusted by the full-load stop screw fitted in the governor cover.

Fig. 8

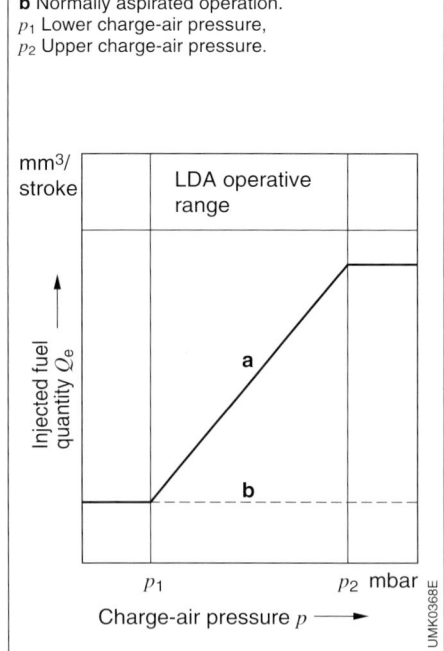

Charge-air pressure: Operative range

a Turbocharger operation,
b Normally aspirated operation.
p_1 Lower charge-air pressure,
p_2 Upper charge-air pressure.

Load-dependent compensation

Depending upon the diesel engine's load, the injection timing (start of delivery) must be adjusted either in the "advance" or "retard" direction.

Load-dependent start of delivery (LFB)

Assignment

Load-dependent start of delivery is designed so that with decreasing load (e.g., change from full-load to part-load), with the control-lever position unchanged, the start of delivery is shifted in the "retard" direction. And when engine load increases, the start of delivery (or start of injection) is shifted in the "advance" direction. These adjustments lead to "softer" engine operation, and cleaner exhaust gas at part- and full-load.

Design and construction

For load-dependent injection timing, modifications must be made to the governor shaft, sliding sleeve, and pump housing. The sliding sleeve is provided with an additional cutoff port, and the governor shaft with a ring-shaped groove, a longitudinal passage and two transverse passages (Fig. 9). The pump housing is provided with a bore so that a connection is established from the interior of the pump to the suction side of the vane-type supply pump.

Method of operation

As a result of the rise in the supply-pump pressure when the engine speed increases, the timing device adjusts the start of delivery in the "advance" direction. On the other hand, with the drop in the pump's interior pressure caused by the LFB it is possible to implement a (relative) shift in the "retard" direction. This is controlled by the ring-shaped groove in the governor shaft and the sliding-sleeve's control port. The control

Fig. 9

Design and construction of the governor assembly with load-dependent start of delivery (LFB)

1 Governor spring, **2** Sliding sleeve, **3** Tensioning lever, **4** Start lever, **5** Control collar,
6 Distributor plunger, **7** Governor shaft, **8** Flyweights.
M_2 Pivot point for 3 and 4.

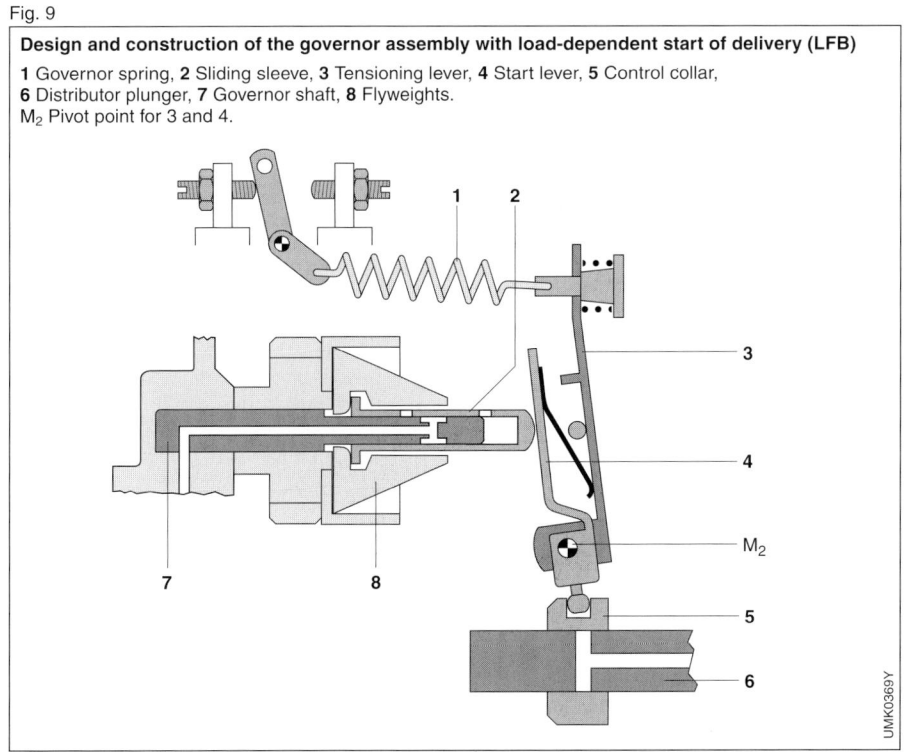

lever is used to input a given full-load speed. If this speed is reached and the load is less than full load, the speed increases even further, because with a rise in speed the flyweights swivel outwards and shift the sliding sleeve. On the one hand, this reduces the delivery quantity in line with the conventional governing process. On the other, the sliding sleeve's control port is opened by the control edge of the governor-shaft groove. The result is that a portion of the fuel now flows to the suction side through the governor shaft's longitudinal and transverse passages and causes a pressure drop in the pump's interior.

This pressure drop results in the timing-device piston moving to a new position. This leads to the roller ring being turned in the direction of pump rotation so that start of delivery is shifted in the "retard" direction. If the position of the control lever remains unchanged and the load increases again, the engine speed drops. The flyweights move inwards and the sliding sleeve is shifted so that its control

port is closed again. The fuel in the pump interior can now no longer flow through the governor shaft to the suction side, and the pump interior pressure increases again. The timing-device piston shifts against the force of the timing-device spring and adjusts the roller ring so that start of delivery is shifted in the "advance" direction (Fig. 10).

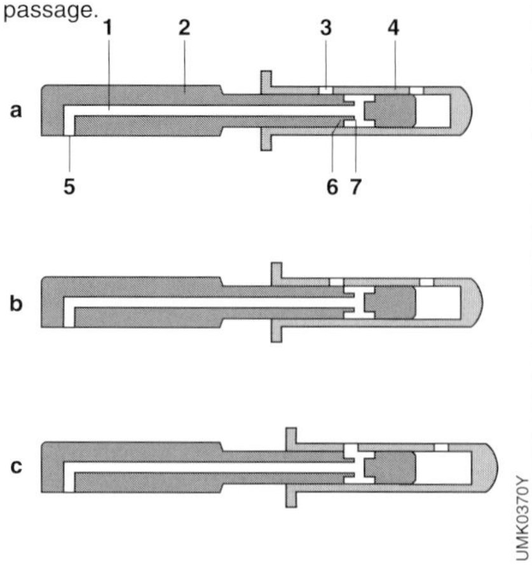

Atmospheric-pressure compensation

At high altitudes, the lower air density reduces the mass of the inducted air, and the injected full-load fuel quantity cannot burn completely. Smoke results and engine temperature rises. To prevent this, an altitude-pressure compensator is used to adjust the full-load quantity as a function of atmospheric pressure.

Altitude-pressure compensator (ADA)

Design and construction
The construction of the ADA is identical to that of the LDA. The only difference being that the ADA is equipped with an aneroid capsule which is connected to a vacuum system somewhere in the vehicle (e.g., the power-assisted brake system). The aneroid provides a constant reference pressure of 700 mbar (absolute).

Method of operation
Atmospheric pressure is applied to the upper side of the ADA diaphragm. The reference pressure (held constant by the aneroid capsule) is applied to the diaphragm's underside. If the atmospheric pressure drops (for instance when the vehicle is driven in the mountains), the sliding bolt shifts vertically away from the lower stop and, similar to the LDA, the reverse lever causes the injected fuel quantity to be reduced.

Fig. 10

Sliding-sleeve positions in the load-dependent injection timing (LFB)

a Start position (initial position),
b Full-load position shortly before the control port is opened,
c Control port opened, pressure reduction in pump interior.
1 Longitudinal bore in the governor shaft,
2 Governor shaft, **3** Sliding-sleeve control port,
4 Sliding-sleeve, **5** Governor-shaft transverse passage, **6** Control edge of the groove in the governor shaft, **7** Governor-shaft transverse passage.

UMK0370Y

Cold-start compensation

The diesel engine's cold-start characteristics are improved by fitting a cold-start compensation module which shifts the start of injection in the "advance" direction. Operation is triggered either by the driver using a bowden cable in the cab, or automatically by means of a temperature-sensitive advance mechanism (Fig. 11).

Mechanical cold-start accelerator (KSB) on the roller ring

Design and construction

The KSB is attached to the pump housing, the stop lever being connected through a shaft to the inner lever on which a ball pin is eccentrically mounted. The ball pin's head extends into the roller ring (a version is available in which the advance mechanism engages in the timing-device piston). The stop lever's initial position is defined by the stop itself and by the helical coiled spring. Attached to the top of the stop lever is a bowden cable which serves as the connection to the manual or to the automatic advance mechanism. The automatic advance mechanism is mounted on the distributor pump, whereas the manual operating mechanism is in the driver's cab (Fig. 12).

Mechanical cold-start accelerator (KSB) engaging in roller ring (cold-start position)

1 Lever, 2 Access window, 3 Ball pin, 4 Longitudinal slot, 5 Pump housing, 6 Roller ring, 7 Roller in the roller ring, 8 Timing-device piston, 9 Torque-control pin, 10 Sliding block. 11 Timing-device spring, 12 Shaft, 13 Coil spring.

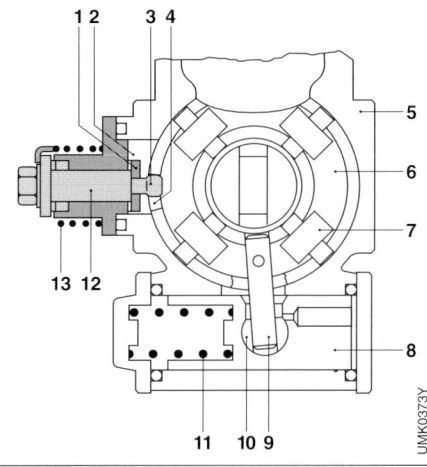

Fig. 12

Method of operation

Automatically and manually operated cold-start accelerators (KSB) differ only with regard to their external advance mechanisms. The method of operation is identical. With the bowden cable not pulled, the coil spring pushes the stop lever up against the stop. Ball pin and roller ring are in their initial position. The force applied by the bowden cable

Fig. 11

Mechanical cold-start accelerator (KSB), advance mechanism with automatic operation (cold-start position)

1 Clamp,
2 Bowden cable,
3 Stop lever,
4 Coil spring,
5 KSB advance lever,
6 Control device sensitive to the temperature of the coolant and the surroundings.

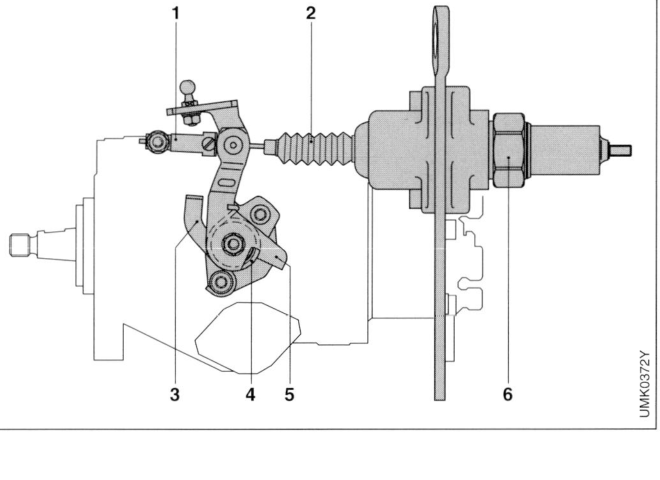

UMK0372Y

causes the stop lever, the shaft, the inner lever and the ball pin, to swivel and change the roller ring's setting so that the start of delivery is advanced. The ball pin engages in a slot in the roller ring, which means that the timing-device piston cannot rotate the roller ring any further in the "advance" direction until a given engine speed has been exceeded.

In those cases in which the KSB is triggered by the driver from the cab (timing-device KSB), independent of the advance defined by the timing device (a), an advance of approx. 2.5° camshaft is maintained (b), as shown in Fig. 13. With the automatically operated KSB, this advance depends upon the engine temperature or ambient temperature.

The automatic advance mechanism uses a control device in which a temperature-sensitive expansion element converts the engine temperature into a stroke movement. The advantage of this method is that for a given temperature, the optimum start of delivery (or start of injection) is always selected.

There are a number of different lever configurations and operating mechanisms in use depending upon the direction of rotation, and on which side the KSB is mounted.

Temperature-controlled idle-speed increase (TLA)

The TLA is also operated by the control device and is combined with the KSB. Here, when the engine is cold, the ball pin at the end of the elongated KSB advance lever presses against the engine-speed control lever and lifts it away from the idle-speed stop screw. The idle speed increases as a result, and rough running is avoided. When the engine has warmed up, the KSB advance lever abuts against its stop and, as a result, the engine-speed control lever is also up against its stop and the TLA is no longer effective (Fig. 14).

Hydraulic cold-start accelerator

Advancing the start of injection by shifting the timing-device piston has only limited applications. In the case of the hydraulic start-of-injection advance, the speed-dependent pump interior pressure is applied to the timing-device piston. In order to implement a start-of-injection advance, referred to the conventional timing-device curve, the pump interior pressure is increased automatically. To do so, the automatic control of pump interior pressure is modified through a bypass in the pressure-holding valve.

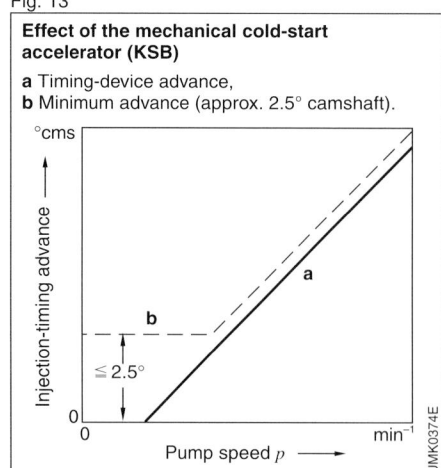

Fig. 13

Effect of the mechanical cold-start accelerator (KSB)

a Timing-device advance,
b Minimum advance (approx. 2.5° camshaft).

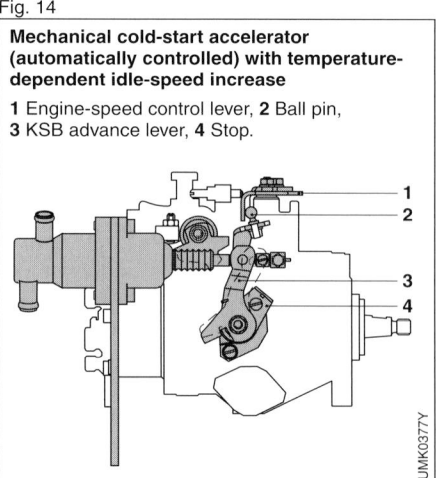

Fig. 14

Mechanical cold-start accelerator (automatically controlled) with temperature-dependent idle-speed increase

1 Engine-speed control lever, **2** Ball pin,
3 KSB advance lever, **4** Stop.

Design and construction

The hydraulic cold-start accelerator comprises a modified pressure-control valve, a KSB ball valve, a KSB control valve, and an electrically heated expansion element.

Method of operation

The fuel delivered by the fuel-supply pump is applied to one of the timing device piston's end faces via the injection pump's interior. In accordance with the injection pump's interior pressure, the piston is shifted against the force of its spring and changes the start-of-injection timing. Pump interior pressure is determined by a pressure-control valve which increases pump interior pressure along with increasing pump speed and the resulting rise in pump delivery (Fig. 15).

There is a restriction passage in the pressure-control valve's plunger in order to achieve the pressure increase needed for the KSB function, and the resulting advance curve shown as a dotted line in Fig. 16. This ensures that the same pressure is effective at the spring side of the pressure-control valve. The KSB ball-type valve has a correspondingly higher pressure level and is used in conjunction with the thermo-element both for switching-on and switching-off the KSB function, as well as for safety switchoff. Using an

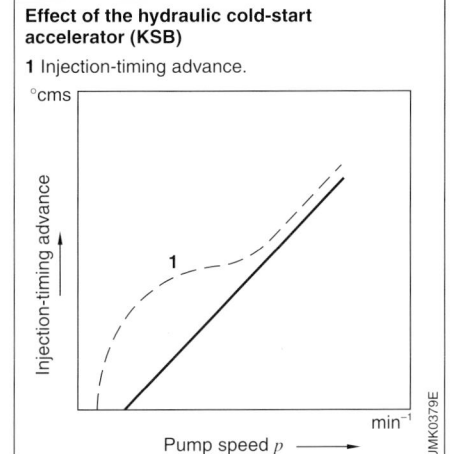

Effect of the hydraulic cold-start accelerator (KSB)

1 Injection-timing advance.

°cms

Injection-timing advance

1

Pump speed p ⟶

min⁻¹

UMK0379E

Fig. 16

adjusting screw in the integrated KSB control valve, the KSB function can be set to a given engine speed. The fuel supply pump pressure shifts the KSB control valve's plunger against the force of a spring. A damping restriction is used to reduce the pressure fluctuations at the control plunger. The KSB pressure characteristic is controlled by its plunger's control edge and the section at the valve holder. The KSB function is adapted by correct selection of the KSB control valve's spring rate and its control section. When the warm engine is started, the expansion element has already opened the ball valve due to the prevailing temperature.

Fig. 15

Hydraulic cold-start accelerator (KSB)

1 Pressure-control valve,
2 Valve plunger,
3 Restriction passage,
4 Internal pressure,
5 Fuel-supply pump,
6 Electrically heated expansion element,
7 KSB ball valve,
8 Pressureless fuel return,
9 KSB control valve, adjustable,
10 Timing device.

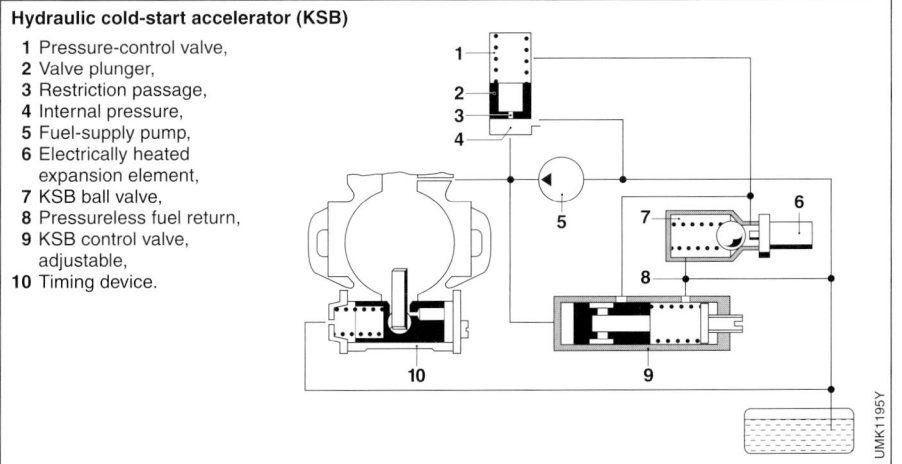

UMK1195Y

Engine shutoff

Assignment

The principle of auto-ignition as applied to the diesel engine means that the engine can only be switched off by interrupting its supply of fuel.

Normally, the mechanically governed distributor pump is switched off by a solenoid-operated shutoff (ELAB). Only in special cases is it equipped with a mechanical shutoff device.

Electrical shutoff device (ELAB)

The electrical shutoff (Fig. 17) using the vehicle's key-operated starting switch is coming more and more to the forefront due to its convenience for the driver.

On the distributor pump, the solenoid valve for interrupting the fuel supply is installed in the top of the distributor head. When the engine is running, the solenoid is energized and the valve keeps the passage into the injection pump's high-pressure chamber open (armature with sealing cone has pulled in). When the driving switch is turned to "OFF", the current to the solenoid winding is also cut, the magnetic field collapses, and the spring forces the armature and sealing cone back onto the valve seat again. This closes the inlet passage to the high-pressure chamber, the distributor-pump plunger ceases to deliver fuel, and the engine stops. From the circuitry point of view, there are a variety of different possibilities for implementing the electrical shutoff (pull or push solenoid).

Mechanical shutoff device

On the injection pump, the mechanical shutoff device is in the form of a lever assembly (Fig. 18). This is located in the governor cover and comprises an outer and an inner stop lever. The outer lever is operated by the driver from inside the vehicle (for instance by means of bowden cable). When the cable is pulled, both levers swivel around their common pivot point, whereby the inner

stop lever pushes against the start lever of the governor-lever mechanism. This swivels around its pivot point M_2 and shifts the control collar to the shutoff position. The distributor plunger's cutoff port remains open and the plunger delivers no fuel.

Fig. 17

Electrical shutoff device
(pull solenoid)

1 Inlet passage, **2** Distributor plunger,
3 Distributor head, **4** Push or pull solenoid,
5 High-pressure chamber.

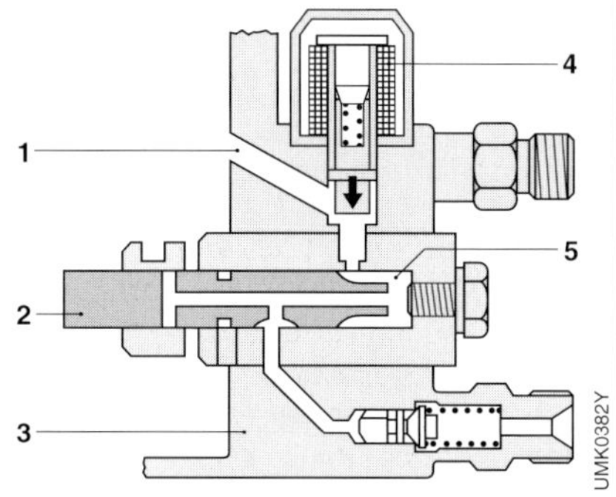

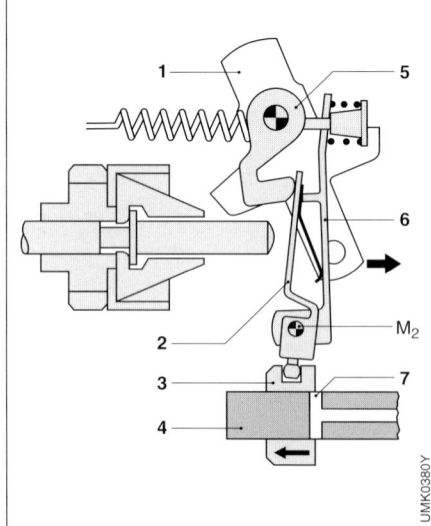

Fig. 18

Mechanical shutoff device

1 Outer stop lever, **2** Start lever,
3 Control collar, **4** Distributor plunger,
5 Inner stop lever, **6** Tensioning lever,
7 Cutoff port.
M_2 Pivot point for 2 and 6.

Testing and calibration

Injection-pump test benches

Precisely tested and calibrated injection pumps and governors are the prerequisite for achieving the optimum fuel-consumption/performance ratio and compliance with the increasingly stringent exhaust-gas legislation. And it is at this point that the injection-pump test bench becomes imperative. The most important framework conditions for the test bench and for the testing itself are defined in ISO-Standards which, in particular, place very high demands upon the rigidity and uniformity of the pump drive.

The injection pump under test is clamped to the test-bench bed and connected at its drive end to the test-bench coupling. Drive is through an electric motor (via hydrostatic or manually-switched transmission to flywheel and

Fig. 18

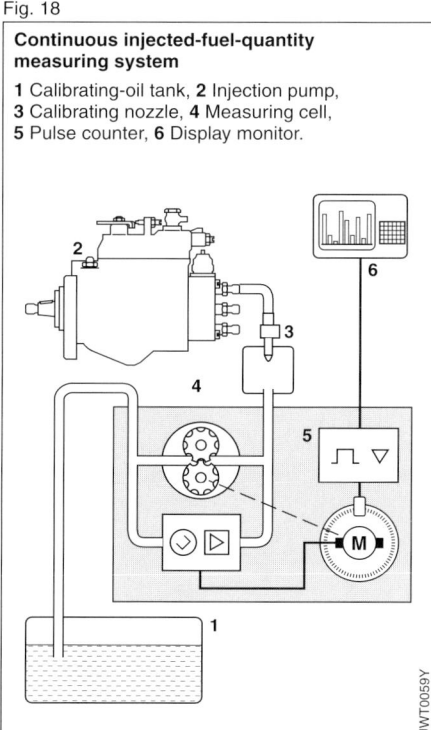

Continuous injected-fuel-quantity measuring system

1 Calibrating-oil tank, **2** Injection pump,
3 Calibrating nozzle, **4** Measuring cell,
5 Pulse counter, **6** Display monitor.

UWT0059Y

coupling, or with direct frequency control). The pump is connected to the bench's calibrating-oil supply via oil inlet and outlet, and to its delivery measuring device via high-pressure lines. The measuring device comprises calibrating nozzles with precisely set opening pressures which inject into the bench's measuring system via spray dampers. Oil temperature and pressure is adjusted in accordance with test specifications. There are two methods for fuel-delivery measurement. One is the so-called continuous method. Here, a precision gear pump delivers per cylinder and unit of time, the same quantity of calibrating-oil as the quantity of injected fuel. The gear pump's delivery is therefore a measure of delivery quantity per unit of time. A computer then evaluates the measurement results and displays them as a bar chart on the screen. This measuring method is very accurate, and features good reproducibility (Fig. 1).

The other method for fuel-delivery measurement uses glass measuring graduates. The fuel to be measured is at first directed past the graduates and back to the tank with a slide. When the specified number of strokes has been set on the stroke-counting mechanism the measurement starts, and the slide opens and the graduates fill with oil. When the set number of strokes has been completed, the slider cuts off the flow of oil again. The injected quantity can be read off directly from the graduates.

Engine tester for diesel engines

The diesel-engine tester is necessary for the precise timing of the injection pump to the engine. Without opening the high-pressure lines, this tester measures the start of pump delivery, injection timing, and engine speeds. A sensor is clamped over the high-pressure line to cylinder 1, and with the stroboscopic timing light or the TDC sensor for detecting crankshaft position, the tester calculates start of delivery and injection timing.

Nozzles and nozzle holders

The injection nozzles and their respective nozzle holders are vitally important components situated between the in-line injection pump and the diesel engine. Their assignments are as follows:
– Metering the injection of fuel,
– Management of the fuel,
– Defining the rate-of-discharge curve,
– Sealing-off against the combustion chamber.

Considering the wide variety of combustion processes and the different forms of combustion chamber, it is necessary that the shape, "penetration force", and atomization of the fuel spray injected by the nozzle are adapted to the prevailing conditions. This also applies to the injection time, and the injected fuel quantity per degree camshaft.

Since the design of the nozzle-holder combination makes maximum use of standardized components and assemblies, this means that the required flexibility can be achieved with a minimum of components. The following nozzles and nozzle holders are used with in-line injection pumps:
– Pintle nozzles (DN..) for indirect-injection (IDI) engines, and
– Hole-type nozzles (DLL../DLSA..) for direct-injection (DI) engines,
– Standard nozzle holders (single-spring nozzle holders), with and without needle-motion sensor, and
– Two-spring nozzle holders, with and without needle-motion sensor.

Pintle nozzles

Application
Pintle nozzles are used with in-line injection pumps on indirect-injection engines (pre-chamber and whirl-chamber engines).
In this type of diesel engine, the air/fuel mixture is for the most part formed by the air's vortex work. The injected fuel spray serves to support this mixture-formation process.

The following types of pintle nozzle are available:
– Standard pintle nozzles (Fig. 1),
– Throttling pintle nozzles, and
– Flat-cut pintle nozzles (Fig. 2).

Design and construction
All pintle nozzles are of practically identical design, the only difference being in the pintle's geometry:

Standard pintle nozzle
On the standard pintle nozzle, the nozzle needle is provided with a pintle which extends into the injection orifice of the nozzle body in which it is free to move with a minimum of play. The injection spray can be matched to the engine's requirements by appropriate choice of dimensions and pintle designs.

Fig. 1

Standard pintle nozzle

1 Lift stop surface, **2** Ring groove, **3** Needle guide, **4** Nozzle-body shaft, **5** Pressure chamber, **6** Pressure shoulder, **7** Seat lead-in, **8** Inlet port, **9** Nozzle-body shoulder, **10** Nozzle-body collar, **11** Sealing surface, **12** Pressure shaft, **13** Pressure-pin contact surface.

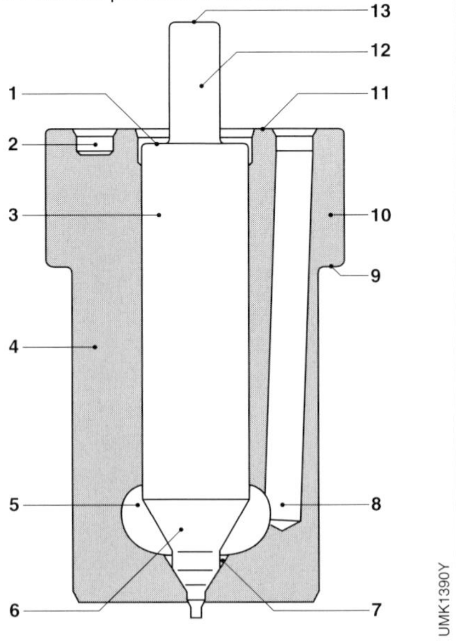

UMK1390Y

Throttling pintle nozzle

The throttling pintle nozzle is a pintle nozzle with special pintle dimensions. The special pintle design serves to define the shape of the rate-of-discharge curve. When the nozzle needle lifts it first of all opens a small annular gap so that only a small amount of fuel is injected (throttling effect).

As needle lift increases (due to pressure rise), the spray orifice is opened increasingly until the major portion of the injection (main injection) takes place towards the end of needle lift. Since the pressure in the combustion chamber rises less sharply, this shaping of the rate-of-injection curve leads to "softer" combustion. This results in quieter combustion in the part-load range. In other words, it is possible to shape the required rate-of-discharge curve by means of the pintle shape, the characteristic of the nozzle needle's spring, and the throttling gap.

Flat-cut pintle nozzle

This nozzle's pintle has a ground surface which opens a flow cross-section in addition to the annular gap when the pintle opens (only slight needle lift). The resulting increased flow volume prevents deposits forming in this flow channel. This is the reason why flat-cut pintle nozzles coke-up far less, and any coking which does take place is more uniform. The annular gap between spray orifice and throttling pintle is very small (less than 10 µm). Very often, the flat-cut pintle surface is parallel to the nozzle-needle axis. Referring to Fig. 3, with an additional inclined cut on the pintle, the gradient of the injected-fuel-quantity curve's flat portion can be increased so that the transition to full nozzle opening is less abrupt. Specially shaped pintles, such as the "radius" or "profile surface" types, can be applied to match the flow curve to engine-specific requirements. Part-load noise and vehicle driveability are both improved as a result.

Fig. 2

Flat-cut pintle nozzle

a Side view, b Front view.
1 Needle seat, 2 Nozzle-body floor,
3 Throttling pintle, 4 Flat cut, 5 Injection orifice,
6 Profiled pintle, 7 Total overlap,
8 Cylindrical overlap, 9 Nozzle-body seat.

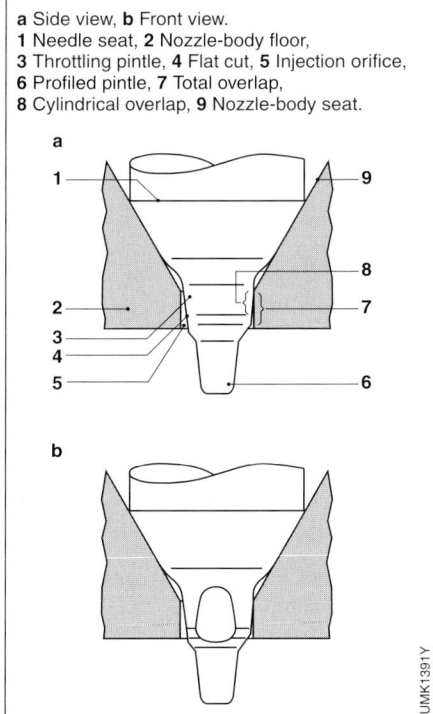

Fig. 3

Flow quantity as a function of needle lift and nozzle version

1 Throttling pintle nozzle,
2 Throttling pintle nozzle with inclined cut on pintle (flat-cut pintle nozzle)

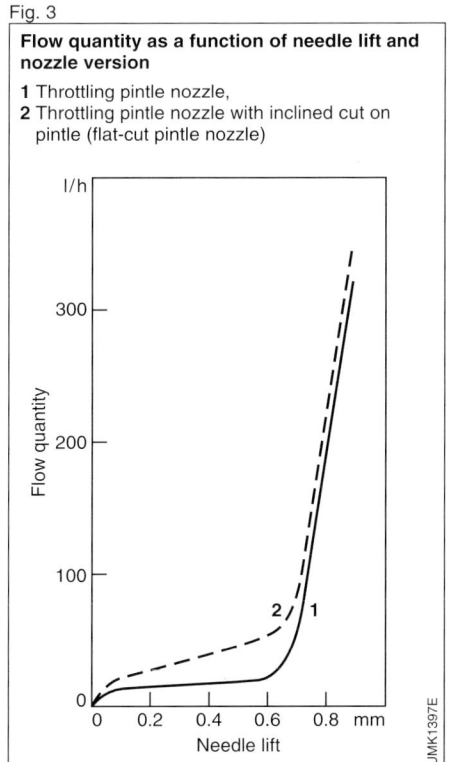

Hole-type nozzles

Application

Hole-type nozzles are used with in-line injection pumps on direct-injection engines.

One differentiates between:
- Sac-hole, and
- Seat-hole nozzles.

The hole-type nozzles also vary according to their size:
- Type P with 4 mm needle diameter, and
- Type S with 5 and 6 mm needle diameters.

Design and construction

The spray holes are located on the envelope of a spray cone (Fig. 4). The number of spray holes and their diameter depend upon:
- The injected fuel quantity,
- The combustion-chamber shape, and
- The air swirl in the combustion chamber.

The input edges of the spray holes can be rounded by hydro-erosive (HE) machining.

Fig. 4

Spray cone

γ Spray-cone offset angle, d Spray cone.

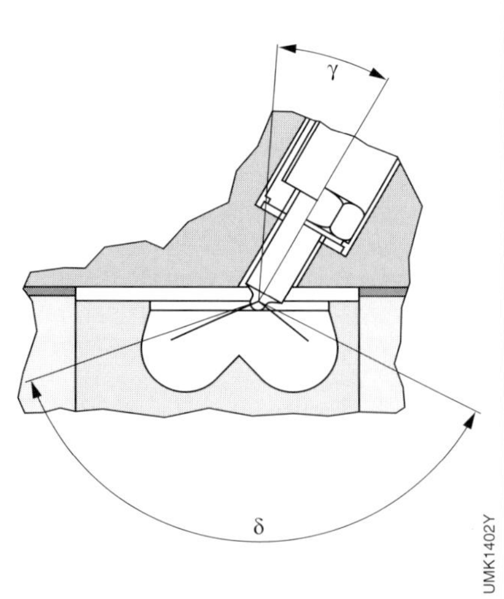

At those points where high flow rates occur (spray-hole entrance), the abrasive particles in the hydro-erosive (HE) medium cause material loss.

This so-called HE-rounding process can be applied to both sac-hole and seat-hole nozzles, whereby the target is:
- Prevent in advance the edge wear caused by abrasive particles in the fuel and/or
- Reduce the flow tolerance.

For low hydrocarbon emissions, it is highly important that the volume filled with fuel (residual volume) below the edge of the nozzle-needle seat is kept to a minimum. Seat-hole nozzles are therefore used.

Designs

Sac-hole nozzle

The spray holes of the sac-hole nozzle (Fig. 5) are arranged in the sac hole.

In the case of a round nozzle tip (Fig. 6a), depending upon design the spray holes are drilled mechanically or by means of electrochemical machining (e.c.m.).

Sac-hole nozzles with conical tip (Figs. 6b and 6c) are always drilled using e.c.m.

Sac-hole nozzles are available
- With cylindrical, and
- Conical sac holes

in a variety of different dimensions.

Sac-hole nozzle with cylindrical sac hole and round tip (Fig. 6a):

This nozzle's sac hole has a cylindrical and a semispherical portion, and permits a high level of design freedom with respect to
- Number of spray holes,
- Spray-hole length, and
- Injection angle.

The nozzle tip is semispherical, and together with the shape of the sac hole, ensures that the spray holes are of identical length.

Sac-hole nozzle with cylindrical sac hole and conical tip (6b):
This type of nozzle is used exclusively with spray-hole lengths of 0.6 mm. The tip's conical shape enables the wall thickness to be increased between the throat radius and the nozzle-body seat with an attending improvement of nozzle-tip strength.

Sac-hole nozzle with conical sac hole and conical tip (Fig. 6c):
Due to the conical shape of this nozzle's sac hole, its volume is less than that of a nozzle with cylindrical sac hole. The volume is between that for a seat-hole nozzle and a sac-hole nozzle with cylindrical sac hole. In order to achieve uniform tip-wall thickness, the tip's conical design corresponds to that of the sac hole.

Fig. 5

Sac-hole nozzle

1 Pressure shaft, **2** Needle-lift stop face,
3 Inlet passage, **4** Pressure shoulder,
5 Needle shaft, **6** Nozzle tip,
7 Nozzle-body shaft, **8** Nozzle-body shoulder,
9 Pressure chamber, **10** Needle guide,
11 Nozzle-body collar, **12** Locating hole,
13 Sealing surface,
14 Pressure-pin contact surface.

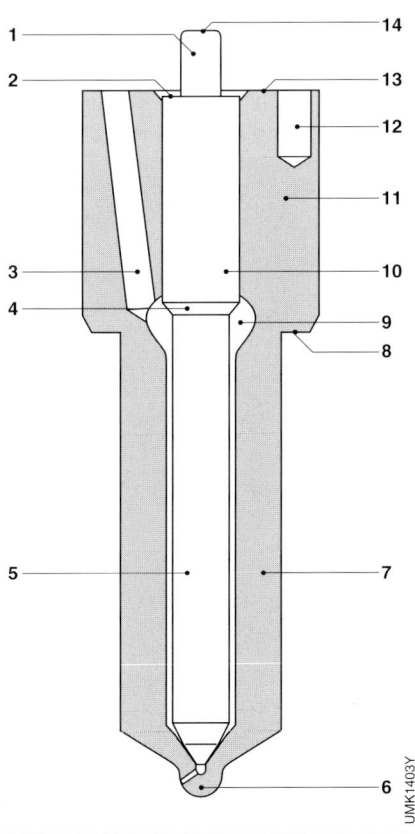

Fig. 6

Sac-hole shapes

a Cylindrical sac hole with round tip,
b Cylindrical sac hole with conical tip,
c Conical sac hole with conical tip.
1 Shoulder, **2** Seat entrance, **3** Needle seat,
4 Needle tip, **5** Injection orifice,
6 Injection-orifice entrance, **7** Sac hole,
8 Throat radius, **9** Nozzle-tip cone,
10 Nozzle-body seat, **11** Damping cone.

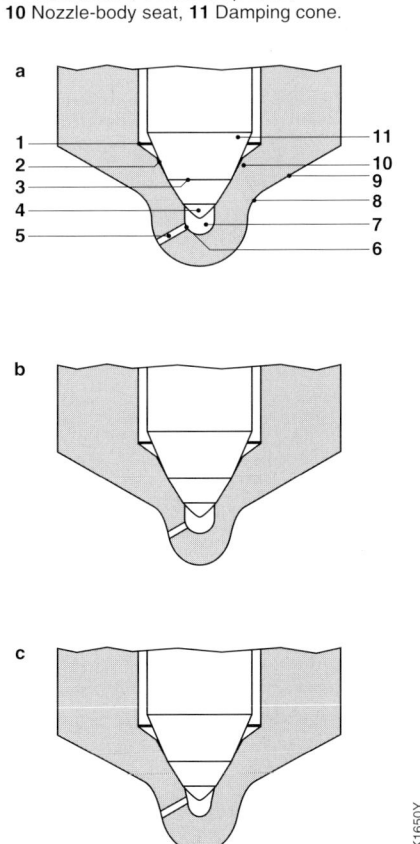

Seat-hole nozzle

In order to minimise the residual volume – and therefore the HC emissions – the start of the spray hole is located in the seat taper, and with the nozzle closed it is covered almost completely by the nozzle needle. This means that there is no direct connection between the sac hole and the combustion chamber (Figs. 7 and 8). The sac-hole volume here is much lower than that of the sac-hole nozzle. Compared to sac-hole nozzles, seat-hole nozzles have a much lower loading limit and are therefore only manufactured as Size P with a spray-hole length of 1 mm.

For reasons of strength, the nozzle tip is conically shaped. The spray holes are always formed using e.c.m. methods.

Fig. 7

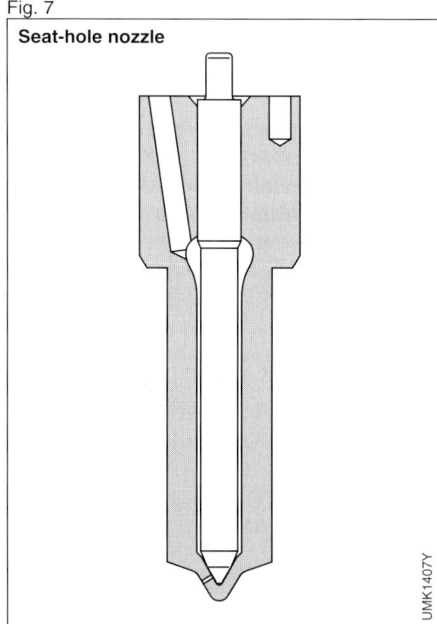

Seat-hole nozzle

UMK1407Y

Fig. 8

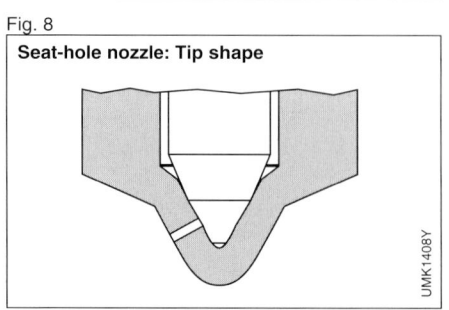

Seat-hole nozzle: Tip shape

UMK1408Y

Standard nozzle holders

Assignments and designs

Nozzle holders with hole-type nozzles in combination with a radial-piston distributor injection pump are used on DI engines.

With regard to the nozzle holders, one differentiates between
– Standard nozzle holders (single-spring nozzle holders) with and without needle-motion sensor, and
– Two-spring nozzle holders, with and without needle-motion sensor.

Application

The nozzle holders described here have the following characteristics:
– Cylindrical external shape with diameters between 17 and 21 mm,
– Bottom-mounted springs (leads to low moving masses),
– Pin-located nozzles for direct-injection engines, and
– Standardised components (springs, pressure pin, nozzle-retaining nut) make combinations an easy matter.

Design

The nozzle-and-holder assembly is composed of the injection nozzle and the nozzle holder.

The nozzle holder comprises the following components (Fig. 9):
– Nozzle-holder body,
– Intermediate element,
– Nozzle-retaining nut,
– Pressure pin,
– Spring,
– Shim, and
– Locating pins.

The nozzle is centered in the nozzle body and fastened using the nozzle-retaining nut. When nozzle body and retaining nut are screwed together, the intermediate element is forced up against the sealing surfaces of nozzle body and retaining nut. The intermediate element serves as the needle-lift stop and with its locating pins centers the nozzle in the nozzle-holder body.

The nozzle-holder body contains the
- Pressure pin,
- Spring, and
- Shim.

The spring is centered in position by the pressure pin, whereby the pressure pin is guided by the nozzle-needle's pressure shaft.

The nozzle is connected to the injection pump's high-pressure line via the nozzle-holder feed passage, the intermediate element, and the nozzle-body feed passage. If required, an edge-type filter can be installed in the nozzle holder.

Method of operation

The nozzle-holder spring applies pressure to the nozzle needle through the pressure pin. The spring's initial tension defines the nozzle's opening pressure which can be adjusted using a shim.

On its way to the nozzle seat, the fuel passes through the nozzle-holder inlet passage, the intermediate element, and the nozzle nody. When injection takes place, the nozzle needle is lifted by the injection pressure and fuel is injected through the injection orifices into the combustion chamber. Injection terminates as soon as the injection pressure drops far enough for the nozzle spring to force the nozzle needle back onto its seat.

Two-spring nozzle holders

Application

The two-spring nozzle holder is a further development of the standard nozzle holder, and serves to reduce combustion noise particularly in the idle and part-load ranges.

Design

The two-spring nozzle holder features two springs located one behind the other. At first, only one of these springs has an influence on the nozzle needle and as such defines the initial opening pressure. The second spring is in contact with a stop sleeve which limits the needle's initial stroke.

Fig. 9

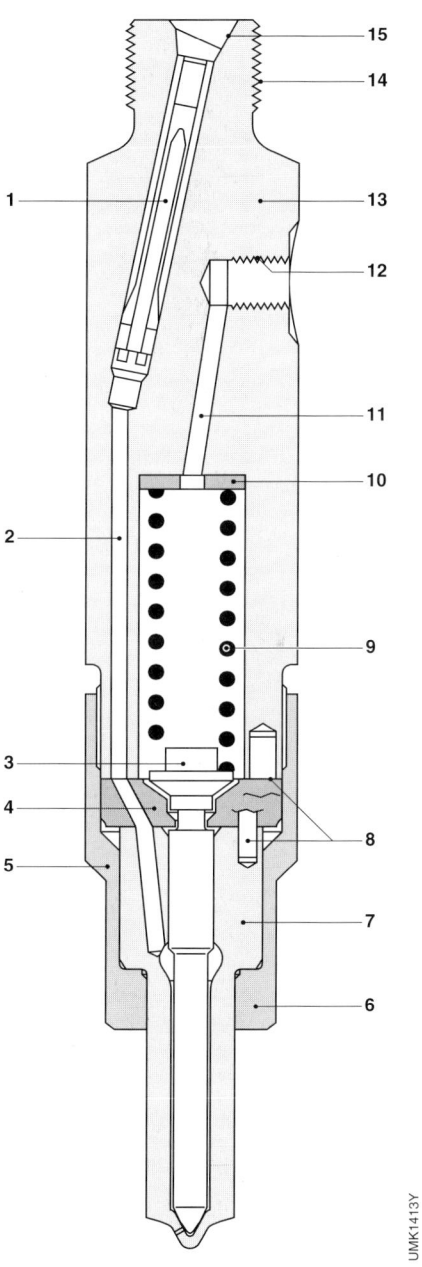

Standard nozzle holder

1 Edge-type filter, 2 Inlet passage,
3 Pressure pin, 4 Intermediate element,
5 Nozzle-retaining nut, 6 Wall thickness,
7 Nozzle, 8 Locating pins, 9 Spring,
10 Shim, 11 Leak-fuel passage,
12 Leak-fuel connection thread,
13 Nozzle-holder body, 14 Connection thread,
15 Sealing cone.

UMK1413Y

Fig. 10

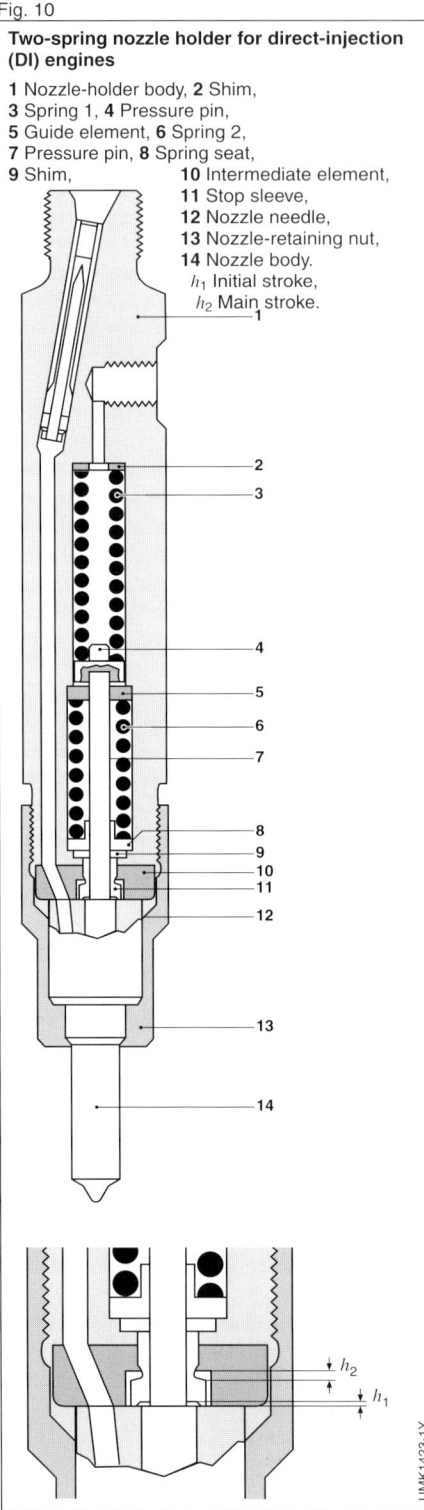

Two-spring nozzle holder for direct-injection (DI) engines

1 Nozzle-holder body, **2** Shim,
3 Spring 1, **4** Pressure pin,
5 Guide element, **6** Spring 2,
7 Pressure pin, **8** Spring seat,
9 Shim, **10** Intermediate element,
11 Stop sleeve,
12 Nozzle needle,
13 Nozzle-retaining nut,
14 Nozzle body.
h_1 Initial stroke,
h_2 Main stroke.

When strokes take place in excess of the initial stroke, the stop sleeve lifts and both springs have an effect upon the nozzle needle (Fig. 10).

Method of operation
During the actual injection process, the nozzle needle first of all opens an initial amount so that only a small volume of fuel is injected into the combustion chamber.

Along with increasing injection pressure in the nozzle holder though, the nozzle needle opens completely and the main quantity is injected (Fig. 11). This 2-stage rate-of-discharge curve leads to "softer" combustion and to a reduction in noise.

Nozzle holders with needle-motion sensor

Application
The start-of-injection point is an important parameter for optimum diesel-engine operation. For instance, its evaluation permits load and speed-dependent injection timing, and/or control of the exhaust-gas recirculation (EGR) rate.

Fig. 11

Comparison of needle-lift curves

a Standard nozzle holder (single-spring nozzle holder),
b Two-spring nozzle holder.
h_1 Initial stroke, h_2 Main stroke.

This necessitates a nozzle holder with needle-motion sensor (Fig. 13) which outputs a signal as soon as the nozzle needle opens.

Design
When it moves, the extended pressure pin enters the current coil.
The degree to which it enters the coil (overlap length "X" in Fig. 14) determines the strength of the magnetic flux.

Method of operation
The magnetic flux in the coil changes as a result of nozzle-needle movement and induces a signal voltage which is proportional to the needle's speed of movement but not to the distance it has travelled. This signal is processed directly in an evaluation circuit (Fig. 12).
When a given threshold voltage is exceeded, this serves as the signal to the evaluation circuit for the start of injection.

Two-spring nozzle holder with needle-motion sensor for direct-injection (DI) engines
1 Nozzle-holder body, **2** Needle-motion sensor, **3** Spring 1, **4** Guide element, **5** Spring 2, **6** Pressure pin, **7** Nozzle-retaining nut.

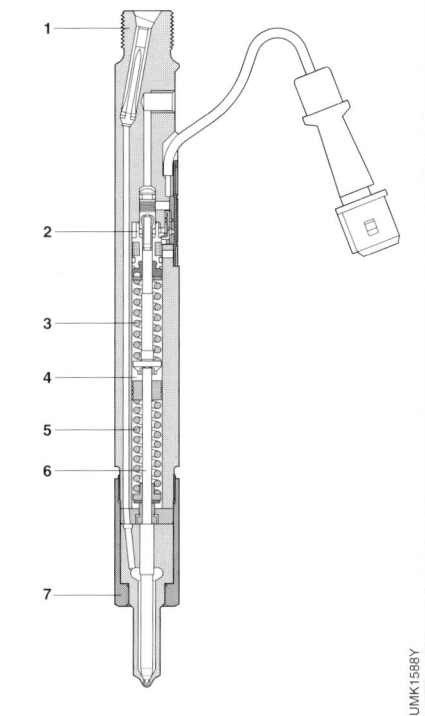

Fig. 13

Fig. 12

Comparison between a needle-lift curve and the corresponding signal-voltage curve of the needle-motion sensor

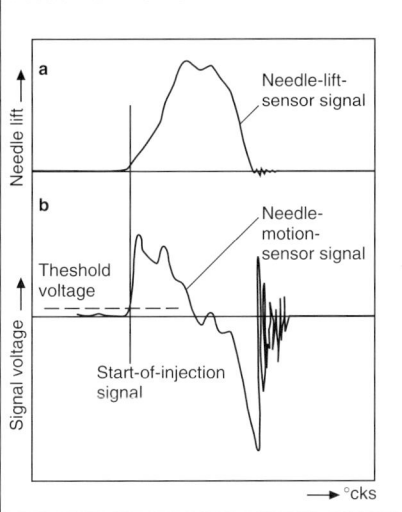

Fig. 14

Needle-motion sensor in a two-spring nozzle holder for direct-injection (DI) engines
1 Adjusting pin, **2** Terminal, **3** Current coil, **4** Pressure pin, **5** Spring seat.
X Overlap length.

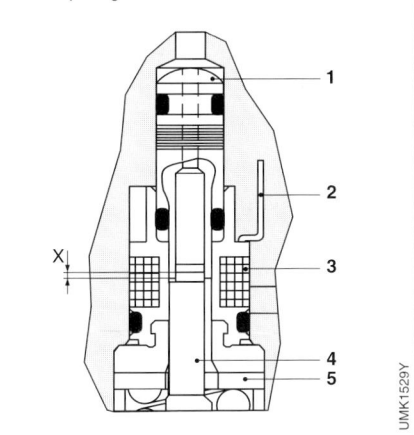

Electronic diesel control EDC

Technical requirements

The reduction of fuel consumption along with an increase in power output or torque, are the decisive factors behind present-day developments in the diesel fuel-injection field. In the past years this has led to an increase in the use of direct-injection (DI) diesel engines. Compared to prechamber or whirl-chamber engines, the so-called indirect-injection (IDI) engines, the DI engine operates with far higher injection pressures. This leads to improved mixture formation, and fuel combustion is more complete. In the DI engine, the improved mixture formation and the fact that there are no overflow losses between pre-chamber/whirl chamber and the main combustion chamber results in a fuel-consumption reduction of 10...15% compared to the IDI engine.

In addition, modern-day engines are subject to more severe requirements with regard to exhaust-gas and noise emissions.

This has led to higher demands being made on the injection system and its control:
– High injection pressures,
– Structured rate-of-discharge curve,
– Variable start of injection,
– Pilot injection,
– Adaptation of injected fuel quantity, boost pressure, and injected fuel quantity to the given operating state,
– Temperature-dependent start quantity,
– Load-independent idle-speed control,
– Cruise control,
– Closed-loop-controlled exhaust-gas recirculation (EGR), and
– Reduced tolerances and higher accuracy throughout the vehicle's useful life.

Conventional mechanical (flyweight) governors use a number of add-on devices to register the various operating conditions, and ensure that mixture formation is of high quality. Such governors, though, are restricted to simple open-loop control operations at the engine, and there are many important actuating variables which they cannot register at all or not quickly enough.

System overview

In the past years, the marked increase in the computing power of the microcontrollers available on the market has made it possible for the EDC (Electronic Diesel Control) to comply with the above-named stipulations.

In contrast to diesel-engined vehicles with conventional in-line or distributor injection pumps, the driver of an EDC-controlled vehicle has no direct influence, for instance through the accelerator pedal and Bowden cable, upon the injected fuel quantity. On the contrary, the injected fuel quantity is defined by a variety of actuating variables, e.g. operating state, driver input, pollutants emission, etc. This of course means that an extensive safety concept must be implemented that detects errors and malfunctions and, depending upon their severity, initiates appropriate countermeasures (e.g. limitation of torque, or emergency (limp-home) running in the idle-speed range) EDC also permits the exchange of data with other electronic systems in the vehicle (e.g. with the traction control system (TCS), and with the electronic transmission-shift control). This means that it can be integrated in the overall vehicle system.

EDC data processing

Input signals

Together with the actuators, the sensors represent the interface between the vehicle and its data-processing unit the ECU.

The signals from the sensors are passed to the ECU (or to several ECU's) via protective circuitry and, where necessary, via signal transducers and amplifiers (Fig. 1):

– Analog input signals (e.g. information from analog sensors on the quantity of air drawn in by the engine, engine and intake-air temperatures, battery voltage, etc.) are converted to digital values by an A/D converter in the ECU microprocessor.

– Digital input signals (e.g. On/Off switching signals, or digital sensor signals such as the rotational-speed pulses from a Hall sensor) can be processed directly by the microprocessor.

– In order to suppress interference pulses, the pulse-shaped input signals from inductive sensors which carry information on engine speed and reference mark are conditioned by a special circuit in the ECU and converted to square-wave form.

Depending upon the level of integration, signal conditioning can take place completely or partially in the sensor. The operating conditions encountered at its installation point determine the sensor's loading.

Fig. 1

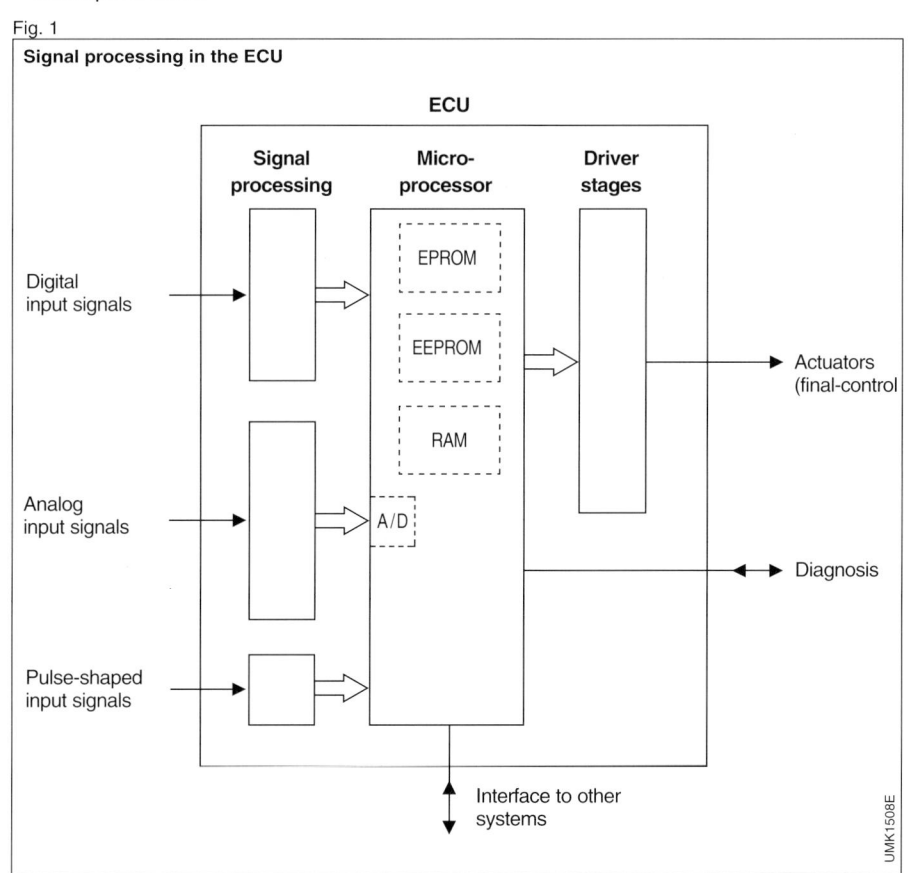

Signal processing in the ECU

ECU

Signal processing | Micro-processor | Driver stages

Digital input signals

EPROM

EEPROM

RAM

Analog input signals

A/D

Pulse-shaped input signals

Actuators (final-control

Diagnosis

Interface to other systems

UMK1508E

Signal conditioning

Protective circuitry is used to limit the incoming signals to a maximum voltage level. The effective signal is freed almost completely of superimposed interference signals by means of filtering, and is then amplified to match it to the ECU input voltage.

Signal processing in the ECU

The ECU microprocessors (Fig. 1) mostly process the input signals digitally, and therefore need a special program. This is stored in a Read Only Memory (ROM or Flash-EPROM).

In addition, engine-specific curves and engine-management maps are stored in a Flash-EPROM. Immobilizer data, calibration and manufacturing data, as well as data on errors/malfunctions which may have occurred during operation are stored in a <u>non-volatile read/write memory</u> (EEPROM).

Due to the large number of engine and equipment variants, the ECU's are provided with a so-called variant code. Using this code, a selection of the maps stored in the Flash-EPROM takes place at the manufacturer or in the workshop, in order to provide the specific functions required for the vehicle variant in question. This selection is also stored in the EEPROM.

Other ECU variants are designed so that complete data sets can be programmed into the Flash-EPROM at the end of vehicle production. This reduces the number of different ECU types required by the vehicle manufacturer.

A <u>volatile random access memory</u> (RAM) is needed to store variable data such as calculations data and signal values. In order to function correctly, the RAM requires a permanent power supply. In other words, it loses its complete data stock when the ECU is switched off via the ignition switch or when the vehicle battery is disconnected. In such cases, the adaptation values (values which have been learnt regarding engine and operating conditions) would have to be re-established when the ECU is switched on again. To prevent this, the adaptation values are stored in an EEPROM and not in a RAM.

Output signals

With their output signals, the microprocessors trigger output stages which usually are powerful enough for direct connection to the actuators. The triggering of the individual actuators is dealt with in the particular system description. These output stages are proof against short-circuit to ground or to battery voltage, as well as against destruction due to electrical overload. Such faults are recognized by the output stages and reported to the microprocessor. This also applies to conductor open-circuits.

In addition, a number of the output signals are transmitted through interfaces to other systems in the vehicle.

Data transmission to other systems

System overview

The increasing use of electronic open and closed-loop controls in the vehicle necessitates the individual ECU's being networked with each other. These controls include:
- Transmission-shift control,
- Electronic engine management and/ or fuel-injection-pump control,
- Antilock braking system (ABS),
- Traction control system (TCS),
- Electronic stability program (ESP),
- Engine drag-torque control (MSR),
- Electronic immobilizer (EWS),
- On-board computer etc.

The exchange of information between the systems reduces the total number of sensors required, and improves the utilisation of the individual systems. The interfaces of the communications systems specifically designed for automotive applications can be sub-divided into two categories:
– Conventional interfaces, and
– Serial interfaces, e.g. Controller Area Network (CAN).

Conventional data transmission

Conventional data transmission in the vehicle is characterized by the fact that every signal is allocated an individual conductor of its own (Fig. 2). Binary signals can only be transmitted using the two states "1" or "0" (binary code), for instance air-conditioner compressor "On" or "Off".

On/Off ratios can be used to transmit continually changing parameters such as the status of the (accelerator) pedal-travel sensor for instance.

The increase in the exchange of data betwen the electrical components in the vehicle, has today reached such dimensions that it is no longer a sensible proposition to attempt to handle it via conventional wiring and plug-in connec-

tors. Today, considerable costs are already involved in keeping wiring-harness complexity down to a controllable level. And the demands made upon the exchange of data between the ECU's continues to climb steadily.

Serial data transmission (CAN)

The problems with data transfer via conventional interfaces can be solved by using bus systems (data highways). One example is CAN, a bus system specifically developed for automotive applications. The above-named signals can be transmitted via CAN provided the electronic control units have a serial "CAN" interface.

There are three major areas for CAN application in the vehicle:
– ECU networking,
– Bodywork, and comfort and convenience electronics,
– Mobile communications.

Fig. 2

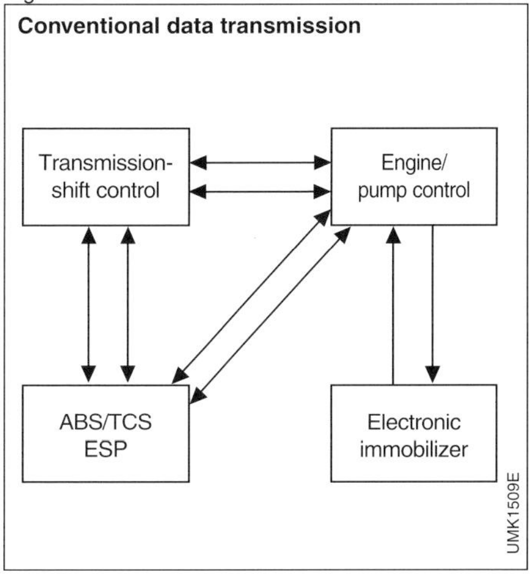

Conventional data transmission

Transmission-shift control ⟷ Engine/pump control

ABS/TCS ESP

Electronic immobilizer

UMK1509E

Fig. 3

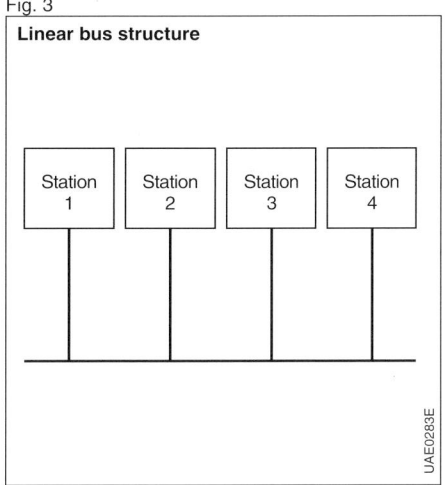

Linear bus structure

| Station 1 | Station 2 | Station 3 | Station 4 |

UAE0283E

The following description is limited to ECU networking.

ECU networking

Here, electronic systems such as engine or injection-pump management, ABS, TCS, electronic transmission-shift control, and electronic stability program (ESP) etc. are networked with each other. The ECU's are assigned equal priority and connected together using a linear bus structure (Fig. 3). One advantage of this structure is the fact that should one of the stations (subscribers) fail, all remaining stations continue to have full access to the network. The probability of total failure is therefore much lower than with other logical arrangements (such as loop or star structures). With loop or star structures namely, the failure of one of the stations or the central ECU leads to total system failure.

Typical CAN transfer rates are between approx. 125 kBit/s and 1 MBit/s (for instance, the engine-management ECU and the pump ECU for the EDC on the radial-piston pump, communicate with each other using 500 kBit/s). The transfer rates must be so high in order to guarantee the required real-time response.

Content-based addressing

Instead of addressing the individual stations, the addressing scheme used by CAN assigns a label to every "message". Each message thus has its own unique 11 or 29-bit "identifier" which identifies the contents of the message (e.g. engine speed).

A given station processes only those messages whose identifiers are stored in its acceptance list (message filtering, Fig. 4). All other messages are simply ignored.

Content-based addressing means that a signal can be sent to a number of stations. The sensor only needs to send its signal directly (or via an ECU) to the bus network where it is then distributed accordingly. In addition, since it is an easy matter to add further stations to an existing CAN bus system, a large number of equipment variations can be implemented.

Priority assignment

The identifier labels both the data content and the priority of the message being sent. A signal which changes quickly, for instance the signal for engine speed, must be transmitted immediately and is therefore allocated a higher priority than a signal which changes relatively slowly (e.g. for the engine-temperature).

Fig. 4

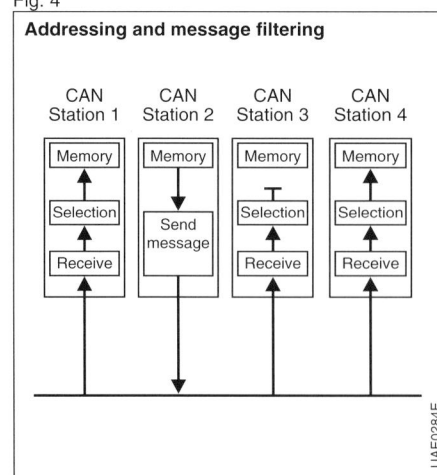

Addressing and message filtering

Fig. 5

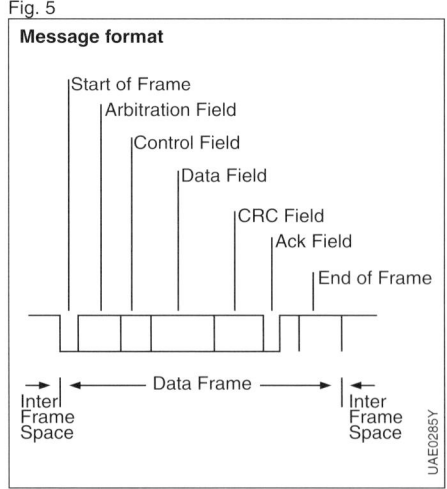

Message format

Bus arbitration

As soon as the bus is free, each station can begin transmitting its message. If several stations start to transmit simultaneously, arbitration awards first access to the message with the highest priority, with no loss of either time or data bits. The stations with lower-priority messages automatically switch to receive and repeat their transmission attempt as soon as the bus is free again.

Message format

A data frame of maximum 130 bits length (standard format), or 150 bits (extended format), is generated for transmissions to the bus. This ensures that the queue time until the next, possibly extremely urgent, data transmission is always kept to a minimum. The data frame is comprised of seven consecutive fields (Fig. 5):

– "Start of frame" indicates the start of the message and synchronises all stations.

– "Arbitration field" comprises the message's identifier and an additional control bit. While this field is being transmitted, the transmitter accom-panies the transmission of each bit with a check to ensure that no higher-priority station is also transmitting. The control bit decides whether the message is classified under "data frame" or "remote frame".

– "Control field" contains the code indicating the number of data bytes in the data field.

– "Data field" has an information content of between 0 and 8 bytes. A message with data length 0 can be used for synchronizing distributed processes.

– "CRC field" (CRC = Cyclic Redundancy Check) contains the frame check word for the identification of possible transmission interference.

– "Ack field" contains the acknowledgement signals with which all receivers indicate reception of non-corrupted messages.

– "End of Frame" indicates the end of the message.

Integrated diagnostics

The CAN bus system is provided with a number of monitoring functions for detecting errors. These include the check signal in the "data frame", and "monitoring" in which each transmitter receives its own transmitted message again, and is thereby able to detect any deviations.

If a station detects an error, it sends an "error flag" which stops the current transmission. This prevents other stations possibly receiving the faulty message.

In case a station is defective, it could occur that all messages, including the faultless ones, would be terminated with an error flag. To prevent this happening, the CAN bus system incorporates a function which can distinguish between intermittent and permanent errors, and which can localise station failures. This process is based upon statistical evaluation of error situations.

Standardization

ISO (International Organization for Standardization) has defined standards for CAN data transfer in automotive applications:
– ISO 11 519-2 for applications up to 125 kBit/s, and
– ISO 11 898 for applications above 125 kBit/s.

Other committees (for instance, from the commercial and utility-vehicle markets in the USA), and vehicle manufacturers have also selected CAN.

Electronically controlled PE-EDC in-line fuel-injection pumps

Standard PE in-line fuel-injection pumps

Thanks to electrical measuring techniques, flexible electronic data processing, and closed control loops with electric actuators, the EDC is able to process actuating variables which could not be taken into account with the previous all-mechanical governor systems.

EDC also permits data interchange with other electronic systems in the vehicle (e.g. traction control, electronic transmission-shift control); this means that it can be integrated into the overall vehicle system.

System blocks

The EDC system comprises three system blocks (Fig. 2):

1. Sensors and desired-value generators for the detection of operating conditions and the generation of desired values. These convert the various physical quantities into electric signals.

2. Electronic control unit (ECU) which uses specific control algorithms to process the above information into suitable electric output signals.

3. Solenoid actuator to convert the ECU's electrical output signals into mechanical control-rack movement.
The actuator is fastened to the injection pump and shifts the control rack by means of a linear solenoid. It takes the place of the mechanical governor.

Components

Pump-speed sensor
An inductive-type sensor in the in-line injection pump's actuator monitors the pump's rotational speed.

Rack-travel sensor
The rack-travel sensor is also incorporated in the pump actuator and registers the pump's rack position.

Charge-air-pressure sensor
The charge-air pressure at the pressure side of the turbocharger is measured by a piezo-resistive sensor.

Temperature sensors
Temperature sensors are used to measure the temperature of the intake or charge air, coolant, and diesel fuel.

Vehicle-speed sensor
The trip-recorder signal (always available on commercial vehicles) or the signal from a separate vehicle-speed sensor is used to determine the vehicle speed.

Accelerator-pedal sensor
The accelerator-pedal setting, and therefore the driver's torque input or speed input to the engine, is registered by a potentiometer which takes the place of the mechanical accelerator-pedal linkage.

Operator panel
The driver can enter or cancel the desired values for vehicle speed and intermediate speed. It is also possible to carry out minor changes to the idle speed.

Switches for brakes, exhaust brake and clutch
Each time the brakes, exhaust brake, or clutch are actuated, switches transmit the corresponding signal to the ECU.

Fuel-injection system with electronically controlled in-line fuel-injection pump

1 Fuel tank, **2** Supply pump, **3** Fuel filter, **4** In-line fuel-injection pump, **5** Electrical shutoff device (ELAB), **6** Fuel-temperature sensor, **7** Rack-travel sensor, **8** Actuator with linear-motion solenoid, **9** Pump-speed sensor, **10** Nozzle, **11** Coolant-temperature sensor, **12** Accelerator-pedal sensor, **13** Switches for brakes, exhaust brake, clutch, **14** Operator panel, **15** Warning lamp and diagnosis connection, **16** Tachograph or vehicle-speed sensor, **17** ECU, **18** Air-temperature sensor, **19** Charge-air-pressure sensor, **20** Exhaust-gas turbocharger, **21** Battery, **22** Glow-plug and starter switch.

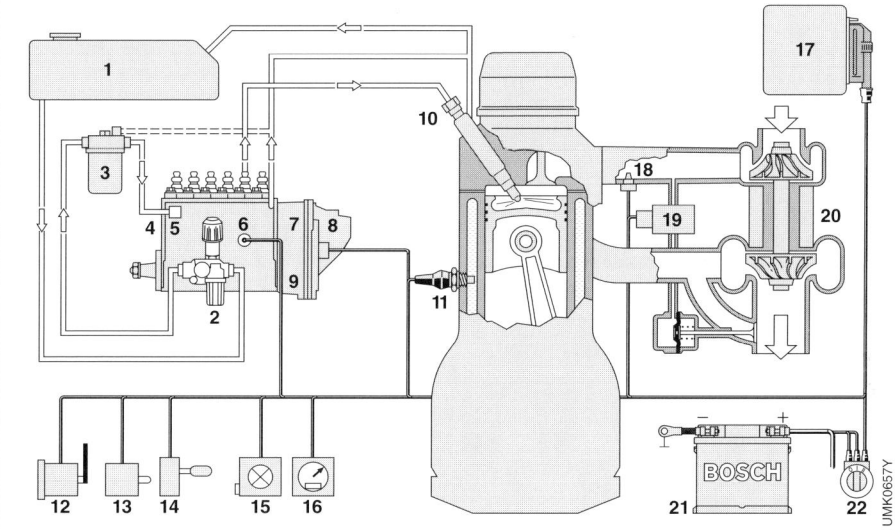

Fig. 1

Fig. 2

System blocks for electronic diesel control (EDC)

Sensors	ECU	Actuators
Charge-air-pressure sensor	Injected fuel quantity	Fuel-injection pump
Temperature sensors (coolant, air, fuel)	Stop	
Injection-pump speed sensor		
Substitute speed signal (alternator)	Micro-processor	OBD
Rack-travel sensor (injection pump)		Diagnosis display / Diagnosis call-up
Vehicle-speed sensor		

Desired-value generators

- Accelerator-pedal sensor
- Operator panel (vehicle speed, intermediate speed)
- Switches for brakes, exhaust brake, clutch

Vehicle ECU (option)
- Data receive
- Fuel-quantity override

Data Maps Characteristic curves

Variant programming
- Programming device (only for data program)

UMK0655E

Electronic control unit (ECU)

The ECU uses digital technology. It registers and processes the signals from the various sensors and desired-value generators.

The ECU's circuitry incorporates microprocessors with integrated input and output interface circuits, as well as memory units and devices for converting the input signals into computer-compatible form.

Depending upon the parameters concerned, a number of different maps can be stored in the ECU (For instance: Load, rotational speed, coolant temperature, fuel temperature, air temperature, and charge-air pressure).

Load and rotational speed are the basic parameters which are influenced by the driver through the accelerator-pedal setting. The other parameters are auxiliary variables.

This means that the ECU can be adapted to the engine-specific and vehicle-specific requirements of the particular application. The engine-specific data are stored in the ECU's program immediately after manufacture, or on the engine or vehicle manufacturer's premises.

The fact that the ECU can be adapted in this manner means that it can be used for a wide variety of different engine and vehicle variants without it being necessary to modify the hardware. The ECU is designed to operate in temperatures typical of automotive applications.

It can therefore be installed in the cab or in a suitable position in the engine compartment. High demands are made on the ECU's interference immunity, and to comply with them it is equipped with short-circuit-proof inputs and outputs; these are also protected against spurious pulses from the vehicle electrical system. A high level of EMC (electromagnetic compatibility) against external interference is provided by the filters and shielding fitted to the ECU.

Fig. 3

EDC solenoid actuator

1 Control rack,
2 Return spring,
3 Short-circuiting ring for rack-travel sensor,
4 Linear solenoid,
5 Pump-speed sensor,
6 Toothed wheel for pump-speed sensor,
7 Injection-pump camshaft.

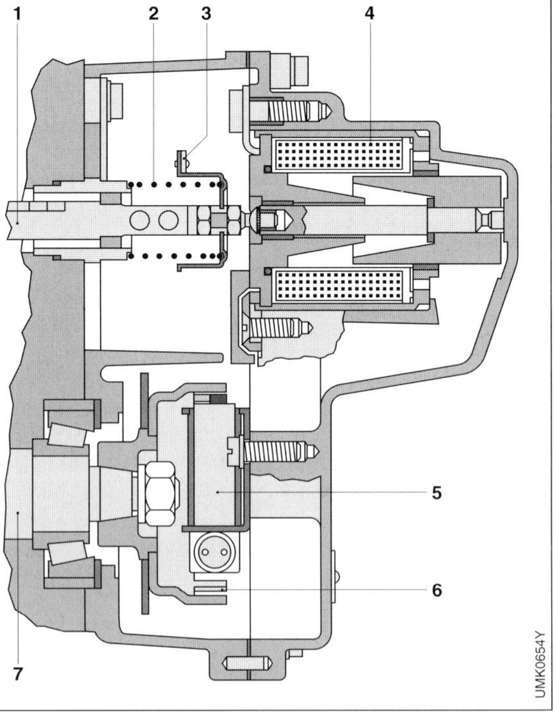

UMK0654Y

Solenoid actuator

As is the case with the in-line pump equipped with a mechanical governor, the injected fuel quantity is a function of control-rack position and pump speed. The solenoid actuator is attached directly to the pump and its linear-motion solenoid shifts the pump's control rack. When the solenoid is de-energized, a spring forces the control rack to the Stop (shutoff) position and interrupts the supply of fuel to the engine.

The solenoid, when energized, exerts force on the control rack in opposition to the spring. This force increases along with rising current and shifts the control rack in the direction of greater injected fuel quantity.

This means that the control-rack travel is continuously adjusted as a function of current level, and varies the injected fuel quantity between zero and maximum (Fig. 3) as required.

Control loops (Figure 4)

Injected fuel quantity

The injected fuel quantity has a decisive influence upon the engine's starting characteristics, as well as upon its idle speed, power, driveability, and particulate emissions. Accordingly, the ECU is programmed with maps for starting, idle, full-load, accelerator-pedal characteristic, smoke limitation, and pump characteristic.

The rack position serves as the equivalent signal for the injected fuel quantity. Standard control characteristics adopted from the RQ and RQV mechanical governors can be used for driveability. The driver stipulates the torque or engine-speed requirements through the potentiometer which is used to register the accelerator-pedal position. Taking the stored map values and the actual-value inputs from the sensors, the ECU calculates the required injected fuel quantity, in other words, the required rack position.

Fig. 4

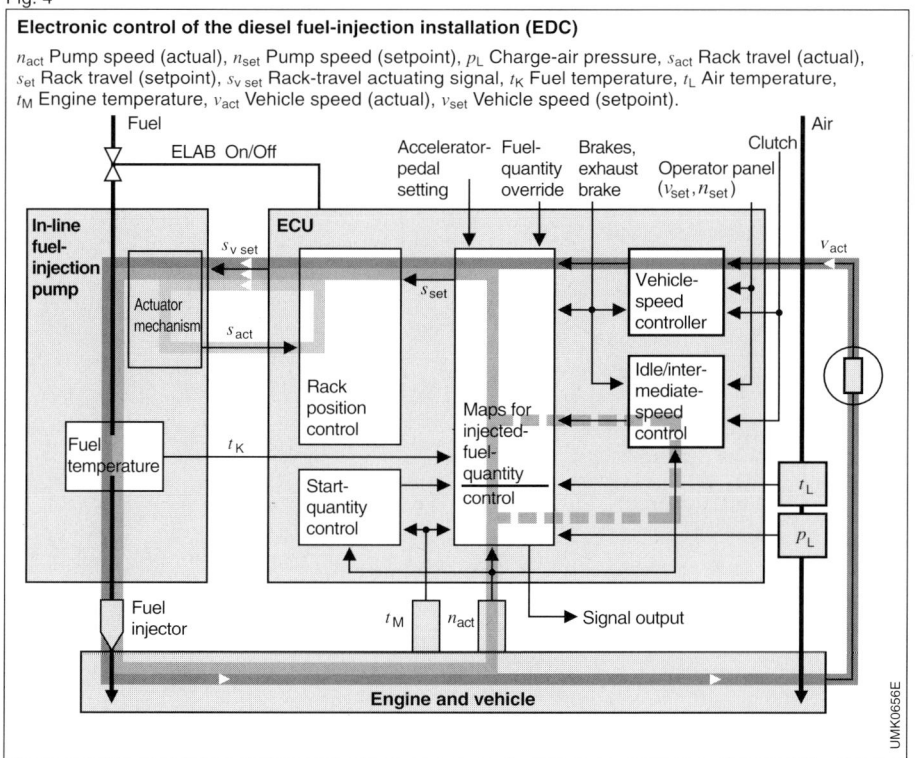

Electronic control of the diesel fuel-injection installation (EDC)

n_{act} Pump speed (actual), n_{set} Pump speed (setpoint), p_L Charge-air pressure, s_{act} Rack travel (actual), s_{set} Rack travel (setpoint), $s_{v\,set}$ Rack-travel actuating signal, t_K Fuel temperature, t_L Air temperature, t_M Engine temperature, v_{act} Vehicle speed (actual), v_{set} Vehicle speed (setpoint).

UMK0656E

This setpoint value is the reference variable for the closed control loop. The ECU incorporates a position controller which registers the actual rack position and therefore the control-system deviation. This position controller ensures that the control rack quickly assumes its correct setting.

Idle speed

Independent of load, the idle speed is controlled to a given setpoint. If necessary, this can be adjusted on the vehicle-speed control device (Cruise Control) panel.

Intermediate speed

It is possible to activate an intermediate-speed control facility for auxiliary power take-offs (e.g. for crane operation). The control device then maintains the engine at the specified speed, regardless of load. It is activated through the vehicle-speed controller panel with the vehicle at a standstill. By pressing a key, a fixed engine speed can be called up from the data storage. In addition, arbitrary speeds can be preselected through the vehicle-speed controller panel.

Vehicle speed

In order to determine and limit the vehicle speed, the vehicle-speed controller (also known as Cruise Control) evaluates the speed signal from the trip recorder, or from the vehicle-speed sensor. This signal is compared with the specified setpoint value and is applied for engine-speed limitation purposes.

A 4-function operator panel is used to enter and to delete the road-speed setpoints:

1. Accelerate and select (store).
When the appropriate key is pressed, the vehicle accelerates. The speed at the moment the key is released (speed select) is stored as the control system's reference setpoint for vehicle-speed.

2. Decelerate and select (store).
When the appropriate key is pressed, the vehicle decelerates. The speed at the moment the key is released (speed select) is again stored as the vehicle-speed setpoint to which the controller now adjusts.

3. Reactivation.
When this key is pressed, speed adjustment takes place to the last vehicle-speed setpoint stored.

4. Switch-off.
The vehicle-speed control facility is switched-off by pressing this key.

Further functions

Engine (exhaust) brake functions

When the engine brake (or exhaust brake) is operated, the injected fuel quantity is set to either zero delivery or idle delivery. To do so the ECU evaluates the setting of the engine-brake switch.

Protection against overheating

The maximum torque is reduced as soon as a specified coolant temperature is exceeded.

Roll-start block

With the EDC switched off, a return spring forces the rack into the stop position. This prevents the engine being "roll-started" should the vehicle inadvertently move off when parked on a slope.

Key-operated stop

The stop function using the starting key supersedes the conventional mechanical shutoff device. It interrupts the current to the electrical shutoff device (ELAB) as well as to the control rack's solenoid actuator and in doing so shuts off the fuel supply to the engine.

Interface

By means of a signal line, it is possible to transmit EDC-relevant quantities (e.g. injected fuel quantity, accelerator-pedal setting) to other systems in the vehicle such as the transmission-shift control. These systems can stipulate the desired injected fuel quantity between idle and full load (fuel-quantity override) through a separate line.

Compatibility with TCS (Traction Control) is possible.

Safety concept

Self-monitoring

This system's safety concept features extensive function redundancy, with the ECU assuming "self-monitoring" duties for the sensors, the solenoid actuator, and the microprocessors. When the switch is activated, the diagnosis system indicates faulty components by means of a warning lamp in the instrument panel.

Limp-home (substitute) functions

Extensive limp-home functions are integrated in the system. For instance, if the pump-speed sensor should fail, the signal from Terminal W of the alternator is used as a substitute. If an important sensor fails, this fact is indicated by the warning lamp.

Shutoff function

In addition to the fuel-blocking effect of the control rack in the Stop (shutoff) position, the solenoid valve in the fuel circuit also interrupts the fuel supply when it is de-energized. This separate electrical shutoff device (ELAB) also switches off the engine should the injected-fuel-quantity actuator fail.

Advantages

- The scanning of engine-specific maps results in optimum engine behavior for each and every operating point.
- Clear-cut delineation of the individual functions. The governor characteristics and the curve for injected fuel quantity are no longer interdependent. During applications engineering, this increases the adaptation versatility.
- Extensive processing of parameters which previously could not be processed mechanically (e.g. fuel-temperature compensation, load-independent idle-speed control).
- The reduction in the effects of tolerances improves control precision and control consistency throughout the entire life of the engine.
- Performance and driveability are improved: Map storage makes it possible to select parameters within very wide limits, thus permitting the optimization of the overall engine/vehicle system.
- Extended functional scope. Such functions as Cruise Control and intermediate-speed control are realizable with minimum outlay.
- Coupling with other electronic vehicle systems will enable vehicles to be more comfortable and more ecconomical in future, as well as improving their environmental compatibility and making them safer (for instance through transmission-shift control and TCS).
- Since mechanical add-on modules are no longer required on the injection pump, this results in significant reduction of the under-hood space required by the pump.
- Variations are possible in accordance with customer wishes. Individual maps and/or parmeters are entered by Bosch when the ECU comes off of the production line (EOL), or by the vehicle or engine manufacturers themselves. This means that the ECU can be applied for a number of different engine and vehicle variants.

Table 1. **ECU reactions**

Failure	Monitoring of	Reaction	Warning lamp	Diagnosis output
Correction sensors	Signal range	Reduced injected fuel quantity		●
System sensors	Signal range	Limp-home or emergency function (graded)	●	●
Computer	Program runtime (self-test)	Limp-home or emergency function		●
Injected-fuel-quantity actuator	Permanent control deviation	Engine switchoff	●	●

Control-sleeve in-line fuel-injection pumps

Start-of-injection control

With the control-sleeve in-line fuel-injection pump, an additional electrical actuator in combination with the start-of-injection control can be used to adjust at will not only the injected fuel quantity but also the start of injection. This enables the following functions to be implemented:
– Minimization of pollutant emissions,
– Optimization of fuel consumption in all operating states, and
– Improvement of the start phase, and in particular of the warm-up phase.

The components for the start-of-injection control are as follows:
– Needle-motion sensor,
– Rotational-speed sensor,
– ECU, and
– Electrical actuator mechanism.

The fuel-delivery control is identical for control-sleeve and conventional in-line fuel-injection pumps fitted with EDC. The extended functional scope in the ECU for the control-sleeve pump is mainly the result of the additional expanded programs. The circuitry scope has only been increased slightly.

Components

Needle-motion sensor

An inductive pulse generator is installed in one of the injection nozzles (reference nozzle). When the nozzle needle opens or closes, this pulse generator outputs a pulse. The signal outputted by this "needle-motion sensor" when the nozzle opens serves to inform the ECU of the start of injection. This means that in a closed-loop control circuit the start of injection can be precisely corrected to the setpoint value for the particular operating point.

Rotational-speed sensor

An inductive sensor scans the engine's ring gear and ascertains the crankshaft's rotation and position. The sensor's output signal is used for determining engine speed and, together with the signal from the needle-motion sensor, serves to calculate the start of injection.

ECU

In addition to the signals required for fuel-delivery control, the ECU processes the signal from the needle-motion sensor. Following interference suppression and amplification, the "raw" signal from the sensor is processed in the respective evaluation circuit to form a square-wave signal which is ideal for precise evaluation and which indicates start of injection for the engine's reference cylinder. This reference signal is generated once per complete revolution of the pump camshaft.

The desired values for the start of injection as a function of injected fuel quantity and engine speed are stored in a special map in the non-volatile data storage. Further correction maps take engine temperature and charge-air pressure into account. The closed control loop for start of injection is in the form of an arithmetic program implemented by the microprocessor. Using a switching signal (pulse-width modulation), the processor triggers a driver stage which energises the start-of-injection controlling element in the actuator mechanism.

Actuator mechanism

In addition to the fuel-quantity solenoid and the rack-travel sensor, the actuator mechanism for the control-sleeve in-line injection pump is also equipped with a start-of-delivery solenoid which actuates the control-sleeve adjusting shaft via a lever. The adjusting force of this solenoid is a function of the current and operates against the return force of an integrated spring. The solenoid stroke results from the balance of these two forces. This system does not need a position-feedback signal for the start of delivery.

The lines of force for solenoid and spring always have a clearly defined intersection, which means that the distance travelled by the start-of-delivery solenoid is proportional to the applied current. A low-level current results in a small shift of the start-of-delivery solenoid together with a retarded start of delivery and start of injection. A higher current results in a shift of the start of injection in the advance direction.

Closed control loop

Using the rotational-speed pulses and the start-of-injection signals from the needle-motion sensor in the reference nozzle, the computer calculates the start-of-injection's actual value in degrees crankshaft (°cks) referred to reference-cylinder TDC. The digital start-of-injection controller adjusts the current through the start-of-delivery actuator mechanism so that the actual start-of-injection always corresponds to the current setpoint value (Fig. 1). The closed-loop control's accuracy and dynamic response are further improved by an additional digital current controller which, practically without delay, corrects the actual value of the current to the setpoint value from the start-of-injection controller.

The start-of-injection signal can only be evaluated as long as fuel is being injected and rotational speed is stable. Before the engine is cranked (and during cranking), the needle-motion-sensor signal is unsuitable for evaluation. This lack of check-back signal for start of injection means that the control loop for start of injection cannot be closed. The controller is then switched off and the start of injection must be open-loop controlled. In order to also ensure the start-of-injection accuracy in open-loop control, the start-of-injection solenoid is calibrated in order to reduce the effects of tolerances. The current controller compensates for the influence of the solenoid coil's temperature-dependent resistance. This ensures that the current setpoint calculated from the start map always results in the appropriate stroke of the start-of-delivery solenoid, and therefore in the required start of injection.

Fig. 1

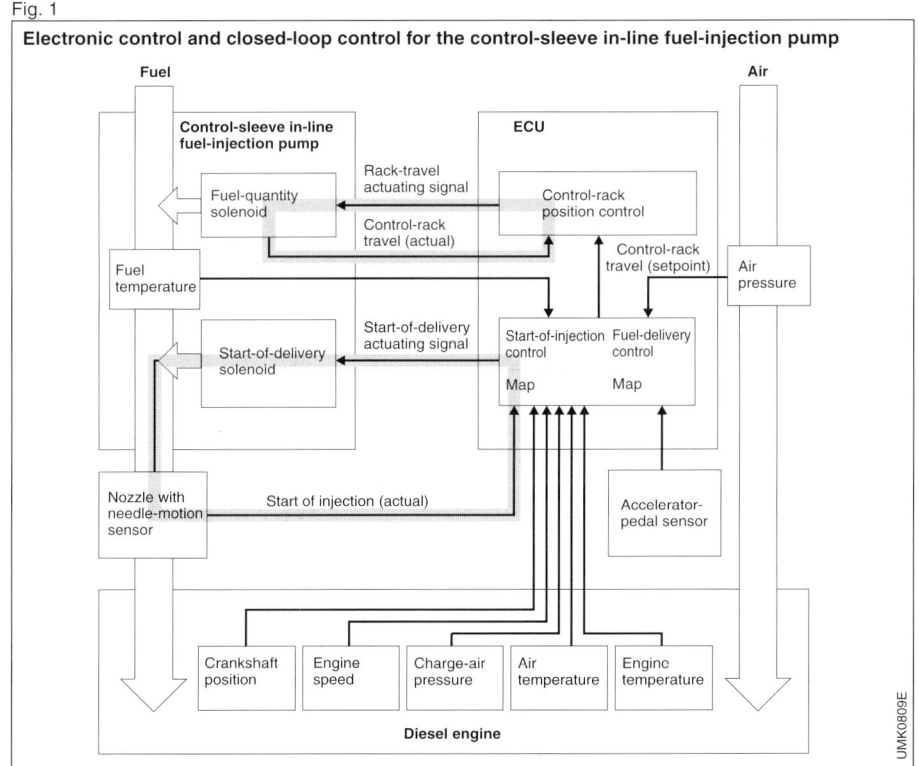

Electronic control and closed-loop control for the control-sleeve in-line fuel-injection pump

Electronically-controlled VE-EDC axial-piston distributor fuel-injection pumps

Mechanical diesel-engine speed control (mechanical governing) registers a wide variety of different operating statuses and permits high-quality A/F mixture formation.

The Electronic Diesel Control (EDC) takes additional requirements into account. By applying electronic measurement, highly-flexible electronic data processing, and closed control loops with electric actuators, it is able to process mechanical influencing variables which it was impossible to take into account with the previous purely mechanical control (governing) system.

The EDC permits data to be exchanged with other electronic systems in the vehicle (for instance, traction control system (TCS), and electronic transmission-shift control). In other words, it can be integrated completely into the overall vehicle system.

System blocks

The electronic control is divided into three system blocks (Fig. 1):

1. Sensors for registering operating conditions. A wide variety of physical quantities are converted into electrical signals.

2. Electronic control unit (ECU) with microprocessors which processes the in-

Fig. 1

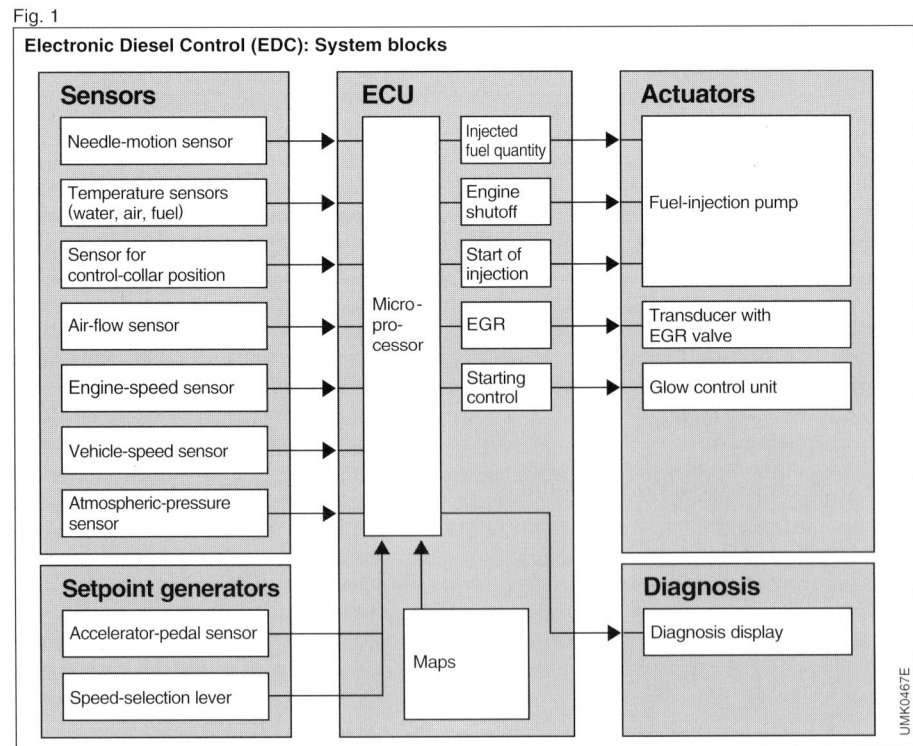

Electronic Diesel Control (EDC): System blocks

Sensors	ECU	Actuators
Needle-motion sensor		Injected fuel quantity →
Temperature sensors (water, air, fuel)		Engine shutoff → Fuel-injection pump
Sensor for control-collar position		Start of injection →
Air-flow sensor	Micro-processor	EGR → Transducer with EGR valve
Engine-speed sensor		Starting control → Glow control unit
Vehicle-speed sensor		
Atmospheric-pressure sensor		
Setpoint generators		**Diagnosis**
Accelerator-pedal sensor	Maps	Diagnosis display
Speed-selection lever		

UMK0467E

formation in accordance with specific control algorithms, and outputs corresponding electrical signals.

3. Actuators which convert the ECU's electrical output signals into mechanical quantities.

Components

Sensors

The positions of the accelerator and the control collar in the injection pump are registered by the angle sensors. These use contacting and non-contacting methods respectively. Engine speed and TDC are registered by inductive sensors. Sensors with high measuring accuracy and long-term stability are used for pressure and temperature measurements. The start of injection is registered by a sensor which is directly integrated in the nozzle holder and which detects the start of injection by sensing the needle movement (Figs. 2 and 3).

Electronic control unit (ECU)

The ECU employs digital technology. The microprocessors with their input and output interface circuits form the heart of the ECU. The circuitry is completed by the memory units and devices for the conversion of the sensor signals into computer-compatible quantities. The ECU is installed in the passenger compartment to protect it from external influences.

There are a number of different maps stored in the ECU, and these come into effect as a function of such parameters as: Load, engine speed, coolant temperature, air quantity etc. Exacting demands are made upon interference immunity. Inputs and outputs are short-circuit-proof and protected against spurious pulses from the vehicle electrical system. Protective circuitry and mechanical shielding provide a high level of EMC (Electro-Magnetic Compatibility) against outside interference.

Fig. 2

Sensor signals

1 Untreated signal from the needle-motion sensor (NBF),
2 Signal derived from the NBF signal,
3 Untreated signal from the engine-speed signal,
4 Signal derived from untreated engine-speed signal,
5 Evaluated start-of-injection signal.

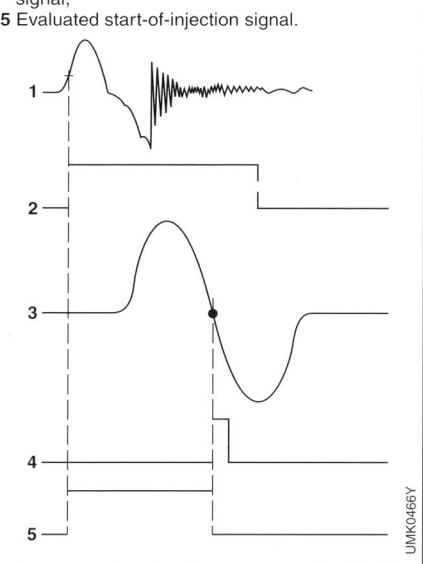

UMK0466Y

Fig. 3

Nozzle-and-holder assembly with needle-motion sensor (NBF)

1 Setting pin, **2** Sensor winding, **3** Pressure pin, **4** Cable, **5** Plug.

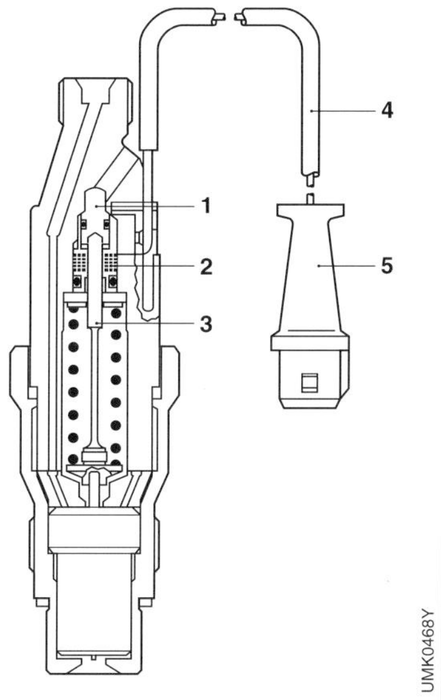

UMK0468Y

Solenoid actuator for injected-fuel quantity control

The solenoid actuator (rotary actuator) engages with the control collar through a shaft (Fig. 4). Similar to the mechanically governed fuel-injection pump, the cutoff ports are opened or closed depending upon the control collar's position. The injected fuel quantity can be infinitely varied between zero and maximum (e.g., for cold starting). Using an angle sensor (e.g., potentiometer), the rotary actuator's angle of rotation, and thus the position of the control collar, are reported back to the ECU and used to determine the injected fuel quantity as a function of engine speed. When no voltage is applied to the actuator, its return springs reduce the injected fuel quantity to zero.

Solenoid valve for start-of-injection control

The pump interior pressure is dependent upon pump speed. Similar to the mechanical timing device, this pressure is applied to the timing-device piston (Fig. 4). This pressure on the timing-device pressure side is modulated by a clocked solenoid valve.

With the solenoid valve permanently opened (pressure reduction), start of injection is retarded, and with it fully closed (pressure increase), start of injection is advanced. In the intermediate range, the on/off ratio (the ratio of solenoid valve open to solenoid valve closed) can be infinitely varied by the ECU.

Fig. 4

Distributor injection pump for electronic diesel control

1 Control-collar position sensor, **2** Solenoid actuator for the injected fuel quantity, **3** Electromagnetic shutoff valve, **4** Delivery plunger, **5** Solenoid valve for start-of-injection timing, **6** Control collar.

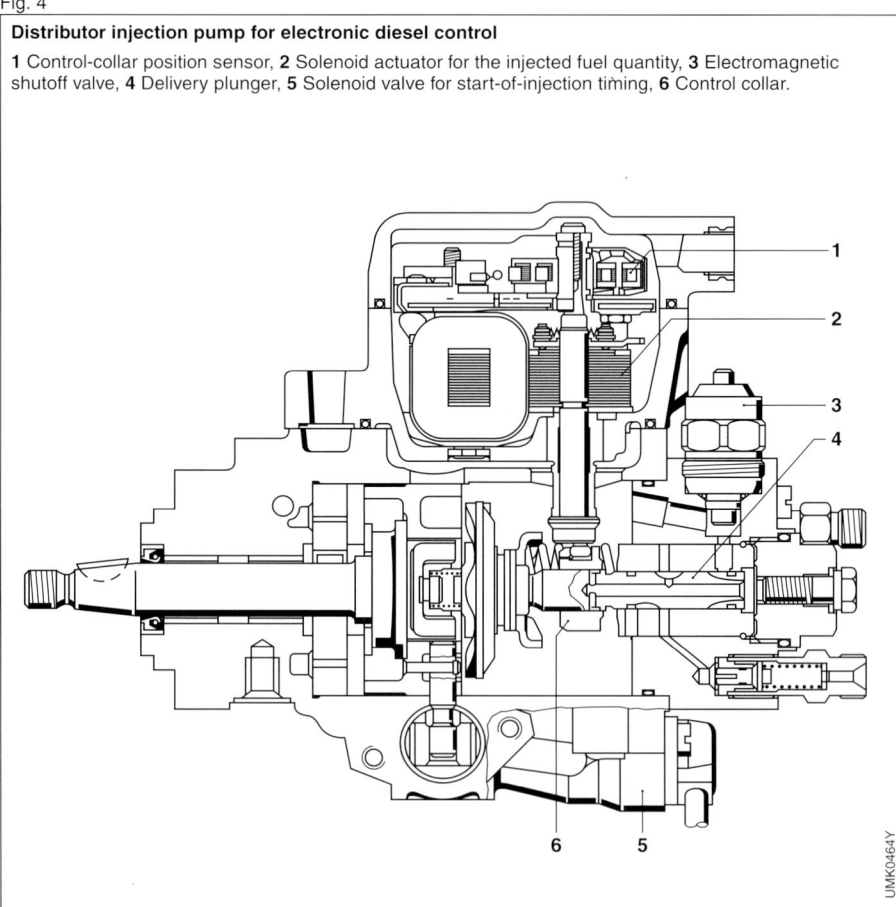

UMK0464-Y

Closed control loops (Fig. 5)

Injected fuel quantity

The injected fuel quantity has a decisive influence upon the vehicle's starting, idling, power output and driveability characteristics, as well as upon its particulate emissions. For this reason, the corresponding maps for start quantity, idle, full load, accelerator-pedal characteristic, smoke limitation, and pump characteristic, are programmed into the ECU. The driver inputs his or her requirements regarding torque or engine speed through the accelerator sensor.

Taking into account the stored map data, and the actual input values from the sensors, a setpoint is calculated for the setting of the rotary actuator in the pump. This rotary actuator is equipped with a check-back signalling unit and ensures that the control collar is correctly set.

Start of injection

The start of injection has a decisive influence upon starting, noise, fuel consumption, and exhaust emissions. Start-of-injection maps programmed into the ECU take these interdependencies into account. A closed control loop is used to guarantee the high accuracy of the start-of-injection point. A needle-motion sensor (NBF) registers the actual start of injection directly at the nozzle and compares it with the programmed start of injection (Figs. 2 and 3). Deviations result in a change to the on/off ratio of the timing-device solenoid valve, which continues until deviation reaches zero.

Fig. 5

Closed control loop of the electronic diesel control (EDC)

Q Air-flow quantity, n_{act} Engine speed (actual), p_A Atmospheric pressure, s_{set} Control-collar signal (setpoint), s_{act} Control-collar position (actual), $s_{v\,set}$ Timing-device signal (setpoint), t_K Fuel temperature, t_L Intake-air temperature, t_M Engine temperature, $t_{i\,act}$ Start of injection (actual).

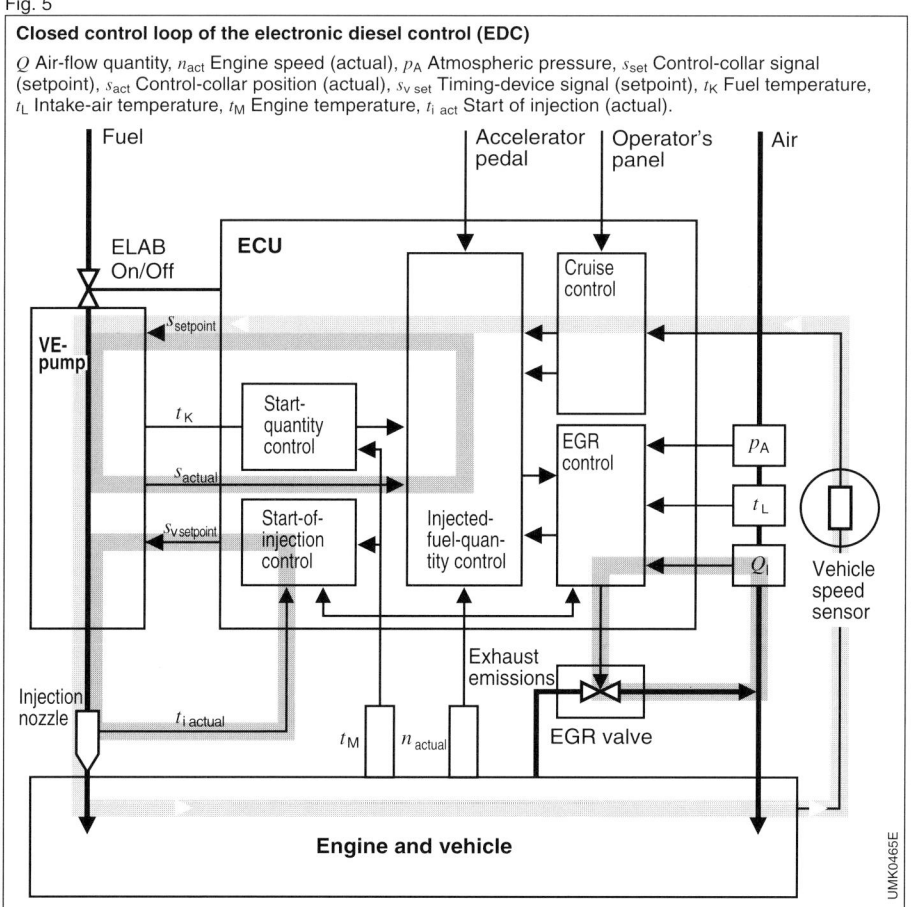

This clocked solenoid valve is used to modulate the positioning pressure at the timing-device piston, and this results in the dynamic behavior being comparable to that obtained with the mechanical start-of-injection timing.

Because during engine overrun (with injection suppressed) and engine starting there are either no start-of-injection signals available, or they are inadequate, the controller is switched off and an open-loop-control mode is selected. The on/off ratio for controlling the solenoid valve is then taken from a control map in the ECU.

Exhaust-gas recirculation (EGR)

EGR is applied to reduce the engine's toxic emissions. A defined portion of the exhaust gas is tapped-off and mixed with the fresh intake air. The engine's intake-air quantity (which is proportional to the EGR rate) is measured by an airflow sensor and compared in the ECU with the programmed value for the EGR map, whereby additional engine and injection data for every operating point are taken into account.

In case of deviation, the ECU modifies the triggering signal applied to an electropneumatic transducer. This then adjusts the EGR valve to the correct EGR rate.

Cruise control

An evaluated vehicle-speed signal is compared with the setpoint signal inputted by the driver at the cruise-control panel. The injected fuel quantity is then adjusted to maintain the speed selected by the driver.

Supplementary functions

The electronic diesel control (EDC) provides for supplementary functions which considerably improve the vehicle's driveability compared to the mechanically governed injection pump.

Active anti-buck damping

With the active anti-buck damping (ARD) facility, the vehicle's unpleasant longitudinal oscillations can be avoided.

Idle-speed control

The idle-speed control avoids engine "shake" at idle by metering the appropriate amount of fuel to each individual cylinder.

Safety measures

Self-monitoring

The safety concept comprises the ECU's monitoring of sensors, actuators, and microprocessors, as well as of the limp-home and emergency functions provided in case a component fails. If malfunctions occur on important components, the diagnostic system not only warns the driver by means of a lamp in the instrument panel but also provides a facility for detailed trouble-shooting in the workshop.

Limp-home and emergency functions

There are a large number of sophisticated limp-home and emergency functions integrated in the system. For instance if the engine-speed sensor fails, a substitute engine-speed signal is generated using the interval between the start-of-injection signals from the needle-motion sensor (NBF). And if the injected-fuel quantity actuator fails, a separate electrical shutoff device (ELAB) switches off the engine. The warning lamp only lights up if important sensors fail. The Table below shows the ECU's reaction should certain faults occur.

Diagnostic output

A diagnostic output can be made by means of diagnostic equipment, which can be used on all Bosch electronic automotive systems. By applying a special test sequence, it is possible to systematically check all the sensors and their connectors, as well as the correct functioning of the ECU's.

Table 1. ECU reactions

Failure	Monitoring of	Reaction	Warning lamp	Diagnostic output
Correction sensors	Signal range	Reduce injected fuel quantity		●
System-sensors	Signal range	Limp-home or emergency function (graded)	●	●
Computer	Program runtime (self-test)	Limp-home or emergency function	●	●
Fuel-quantity actuator	Permanent deviation	Engine shutoff	●	●

Advantages

– Flexible adaptation enables optimization of engine behavior and emission control.
– Clear-cut delineation of individual functions: The curve of full-load injected fuel quantity is independent of governor characteristic and hydraulic configuration.
– Processing of parameters which previously could not be performed mechanically (e.g., temperature-correction of the injected fuel quantity characteristic, load-independent idle control).
– High degree of accuracy throughout complete service life due to closed control loops which reduce the effects of tolerances.
– Improved driveability: Map storage enables ideal control characteristics and control parameters to be established independent of hydraulic effects. These are then precisely adjusted during the optimisation of the complete engine/vehicle system. Bucking and idle shake no longer occur.
– Interlinking with other electronic systems in the vehicle leads the way towards making the vehicle safer, more comfortable, and more economical, as well as increasing its level of environmental compatibility (e.g., glow systems or electronic transmission-shift control).
The fact that mechanical add-on units no longer need to be accomodated, leads to marked reductions in the amount of space required for the fuel-injection pump.

Engine shutoff

As already stated on Page 40, the principle of auto-ignition as applied to the diesel engine means that the engine can only be switched off by interrupting its supply of fuel.
When equipped with Electronic Diesel Control (EDC), the engine is switched off by the injected-fuel quantity actuator (Input from the ECU: Injected fuel quantity = Zero). As already dealt with, the separate electrical engine shutoff device serves as a standby shutoff in case the actuator should fail.

Electrical shutoff device

The electrical shutoff device is operated with the "ignition key" and is above all used to provide the driver with a higher level of sophistication and comfort.
On the distributor fuel-injection pump, the solenoid valve for interrupting the supply of fuel is fitted in the top of the distributor head. With the diesel engine running, the inlet opening to the high-pressure chamber is held open by the energized solenoid valve (the armature with sealing cone is pulled in). When the "ignition switch" is turned to "Off", the power supply to the solenoid is interrupted and the solenoid de-energized. The spring can now push the armature with sealing cone onto the valve seat and close off the inlet opening to the high-pressure chamber so that the distributor plunger can no longer deliver fuel.

Solenoid-valve-controlled axial-piston distributor fuel-injection pumps VE-MV

Prospects

On the electronically-controlled distributor pumps of the future, the electrical actuator mechanism with control collar for fuel metering will be superseded by a high-pressure solenoid valve. This will permit an even higher degree of flexibility in the fuel metering and in the variability of the start of injection.

Design and construction

This pump is of modular design. The field-proven distributor injection pump can thus be combined with a new electronically controlled fuel-metering system (Fig. 1). Basically speaking, the solenoid-valve-controlled distributor pump's dimensions, installation conditions, and drivetrain including the pump's cam drive, are identical to those of the conventional distributor pump. The most important new components are:

- Angle-of-rotation sensor (in the form of an incremental angle/time system [IWZ]) which is located in the injection pump on the driveshaft between the vane-type supply pump and the roller ring,
- Electronic pump ECU, which is mounted as a compact unit on the top side of the pump and connected to the engine ECU,
- High-pressure solenoid valve, installed in the center of the distributor head.

With regard to its installation and hydraulic control, the timing device with pulse valve is identical to the one in the previous electronically-controlled distributor pump.

Components

Angle-of-rotation sensor

Angle-of-rotation detection uses the following components: Sensor, sensor retaining ring on the driveshaft, and the trigger wheel with a given tooth pitch. Detection is based upon the signals generated by the sensor.

The pulses generated by the sensor are inputted to the ECU where they are processed by an evaluation circuit. The fact that the sensor is coupled to the pump's roller ring ensures the correct assignment of the angular increment to the position of the cam when the roller ring is rotated by the timing device.

Pump ECU

The pump ECU is mounted on the upper side of the pump and uses hybrid techniques. In addition to the mechanical loading with which it is confronted in the vehicle's under-hood environment, the pump must also fulfill the following assignments:

- Data exchange with the separately mounted engine ECU via the serial bus system,
- Evaluation of the signal from the angle-of-rotation sensor (IWZ),
- Triggering of the high-pressure solenoid valve,
- Triggering of the timing device.

Maps are stored in the pump ECU which not only take into account the setting points for the particular vehicle application and certain engine characteristics, but also permit the plausibility of the received signals to be checked. In addition, they form the basis for defining a number of different computational values.

High-pressure solenoid valve

The high-pressure solenoid valve must fulfill the following assignments:
- Large valve cross-section for efficient filling of the high-pressure chamber, even at very high rotational speeds,
- Low weight (low moving masses), to keep the loading of the parts to a minimum,
- Short switching times to guarantee high-precision fuel metering, and
- Magnetic forces which are powerful enough to cope with the high pressures.

The high-pressure solenoid valve is comprised of:
- The valve body,
- The valve needle, and
- The electromagnet with electrical connection to the pump ECU.

The magnetic circuit is concentric to the valve. This fact permits a compact assembly comprising high-pressure solenoid valve and distributor head.

Method of operation

Principle

Pressure generation in the solenoid-valve-controlled distributor injection pump is based on the same principle as that in the conventional electronically-controlled VE pump.

Fuel supply and delivery

Via the distributor head and the opened high-pressure solenoid valve, the vane-type supply pump delivers fuel to the high-pressure chamber at a pressure of approx. 12 bar.

No fuel is delivered when the high-pressure solenoid valve is de-energized (open). The valve's instant of closing defines the injection pump's start of delivery. This can be located at the bottom dead center (BDC) of the cam or on the rise portion of the cam slope. Similarly, the valve's instant of opening defines the pump's end of delivery. The length of time the valve is closed determines the injected fuel quantity.

The high pressure generated in the high-pressure chamber (the fuel from the supply pump is compressed by the axial piston when this is forced up by the cam plate riding over the rollers of the roller ring) opens the delivery valve and the fuel is forced through the pressure line to the injection nozzle in the nozzle holder. Injection pressure at the nozzle is 1400 bar. Excess fuel is directed back to the tank through return lines.

Since there are no additional intake ports available, if the high-pressure solenoid valve should fail, fuel injection stops. This prevents uncontrolled "racing" of the engine.

Fig. 1

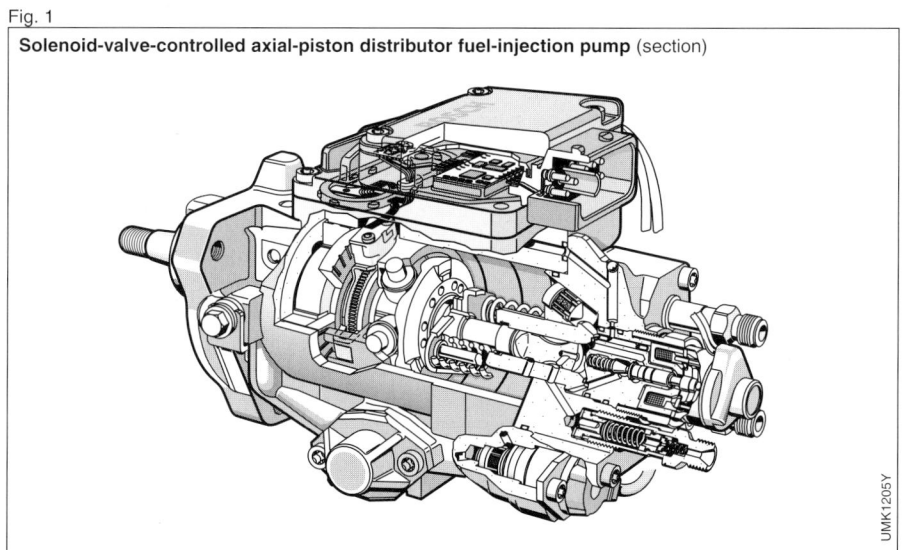

Solenoid-valve-controlled axial-piston distributor fuel-injection pump (section)

UMK1205Y

VR radial-piston distributor fuel-injection pumps

System overview

Field of application

The small, high-speed diesel engines used for passenger cars and light commercial vehicles, as well as for tractors and stationary engines, require a high-performance fuel-injection system, featuring rapid injection sequences, low weight and small installation volume. Regulations on fuel-consumption reduction and emission limits have also made it necessary to increase the injection pressure. The distributor injection pump is ideally suited to this field of application. In each case, the pump's power and its design are determined by the rated speed, power output, and design of the diesel engine concerned. The radial-piston distributor pump was introduced in 1964, and following years of further development it became the most commonly installed passenger-car fuel-injection pump. The traditional, mechanically governed, VE distributor pumps for indirect injection (IDI) engines (prechamber and swirl-chamber engines) generate injection pressures of up to 350 bar at the injection nozzle. On the other hand, electronically controlled VE pumps with electrical actuator mechanism, or with high-pressure solenoid valve, are also suitable for direct-injection (DI) engines. They generate injection pressures of up to 800 bar for low-speed, and 1,400 bar for high-speed diesel engines with up to 25 kW per cylinder. Bosch developed the VR radial-piston distributor pump specifically for high-speed DI diesel engines with output powers up to 37 kW per cylinder. It is characterized by highly dynamic fuel-delivery control and start-of-injection control, as well as by injection pressures of up to 1,600 bar at the injection nozzle.

Functions

The diesel fuel-injection system with VR radial-piston distributor pump is equipped with two ECU's for the electronic diesel control (EDC): An engine ECU and an injection-pump ECU. On the one hand, the provision of two ECU's is necessary in order to prevent the overheating of certain electronic components, and on the other to prevent interference signals which can be caused by the in some cases very high currents (up to 20 A) required for controlling the solenoid valve.

The pump ECU registers the signals from the sensors inside the pump. These provide information on the angle of rotation and upon fuel temperature which is then processed by the ECU to define the injection parameters. At the same time, the engine ECU processsses all the data as registered by the external sensors concerning the engine and surrounding conditions, and uses this to calculate the actuator-mechanism adjustments necessary at the engine.

The sensors register all the required operating data, such as

– Intake-air, coolant, and fuel temperature,
– Engine speed,
– Charge-air pressure,
– Accelerator-pedal setting,
– Vehicle speed, etc.

These signals are conditioned by the ECU input circuits and used by the microprocessors for calculating the command signals needed for optimal driving operations.

A number of different system components have been networked so that

– Signals can be used for a number of different purposes,

– Actuator adjustments can be precisely coordinated,
– Fuel can be saved, and
– All the operating components can be operated for maximum efficiency and service life.

The interchange of data between the engine and the pump ECU takes place via the CAN bus system.

Fig. 1 shows an example of a diesel fuel-injection system with a VR radial-piston distributor pump fitted to a 4-cylinder diesel engine, together with a variety of different components.

Basic functions

The basic functions serve to control the injection of diesel fuel at the correct instant in time, in the correct quantities, and at as high a pressure as possible. This ensures that the diesel engine runs quietly, as economically as possible, and with low levels of exhaust emissions.

Auxiliary functions

The auxiliary closed and open-loop control functions serve to reduce the exhaust-gas emissions and the fuel consumption, or they increase safety, security, comfort and convenience. Examples are:
– Exhaust-gas recirculation (EGR),
– Boost-pressure control,
– Vehicle-speed control (Cruise Control),
– Electronic immobilizer etc.

The CAN bus system enables an exchange of data to take place with the other ECU's in the vehicle (e.g. ABS, electronic transmission-shift control). A diagnosis interface permits the evaluation of stored system data during vehicle inspections. The chapter "System control using EDC" describes the processes of operating-data collection and processing, as well as the method of functioning of the individual sensors and actuators.

Fig. 1

Diesel fuel-injection system with VR radial-piston distributor pump

1 Engine ECU, **2** Glow-control unit, **3** Fuel filter, **4** Air-mass meter, **5** Injection nozzles, **6** Glow plugs, **7** Radial-piston distributor pump with pump ECU, **8** Alternator, **9** Coolant-temperature sensor, **10** Crankshaft sensor, **11** Accelerator-pedal sensor.

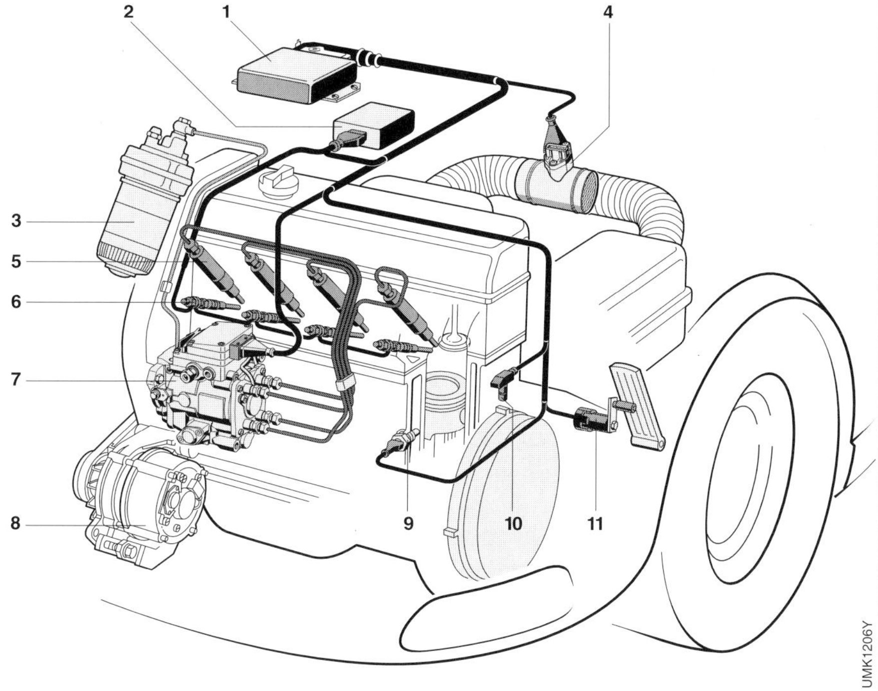

UMK1206Y

Fuel system

The fuel system in a fuel-injection system with radial-piston distributor pump (Fig. 1) is comprised of a low-pressure stage for the low-pressure delivery of fuel, a high-pressure stage for the high-pressure delivery, and an ECU.

Low-pressure delivery

The low-pressure stage of the fuel system incorporates:
– Fuel tank,
– Low-pressure fuel lines,
– Fuel filter, and
– Injection-pump components

Fuel tank
(Excerpt from §45 StVZO (FMVSS/CUR))
The fuel tank must be of non-corroding material, and must remain free from leaks at double the operating pressure, and in any case at 0.3 bar. Suitable openings or safety valves must be provided, or appropriate measures taken, to permit excess pressure to escape of its own accord. Fuel must not leak past the filler cap or through pressure-compensation devices when the vehicle is subjected to minor shocks, as well as when cornering and when standing or driving on an incline.

Fuel tank and engine must be so far apart from each other that in case of accident there is no danger of fire.

Special regulations apply concerning the height of the fuel tank and its protective shielding in vehicles with open cabs, tractor vehicles, and buses.

Fuel lines for the low-pressure stage
(§ 46 StVZO (FMVSS/CUR))
As an alternative to steel pipes, flame-inhibiting steel-braid-armoured flexible fuel lines can be used for the low-pressure stage. They must be routed so that they cannot be damaged mechanically, and fuel which has dripped or evaporated must not be able to accumulate, nor must it be able to ignite. When the vehicle twists, or the engine moves etc., this must have no derogatory effects upon fuel-line function. All parts which carry fuel must be protected

Fig. 1

Fig. 1: Fuel system for a fuel-injection system with radial-piston distributor pump

1 Fuel tank, **2** Presupply pump, **3** Fuel filter, **4** Radial-piston distributor pump, **5** High-pressure fuel lines, **6** Nozzle-and-holder assembly, **7** ECU.

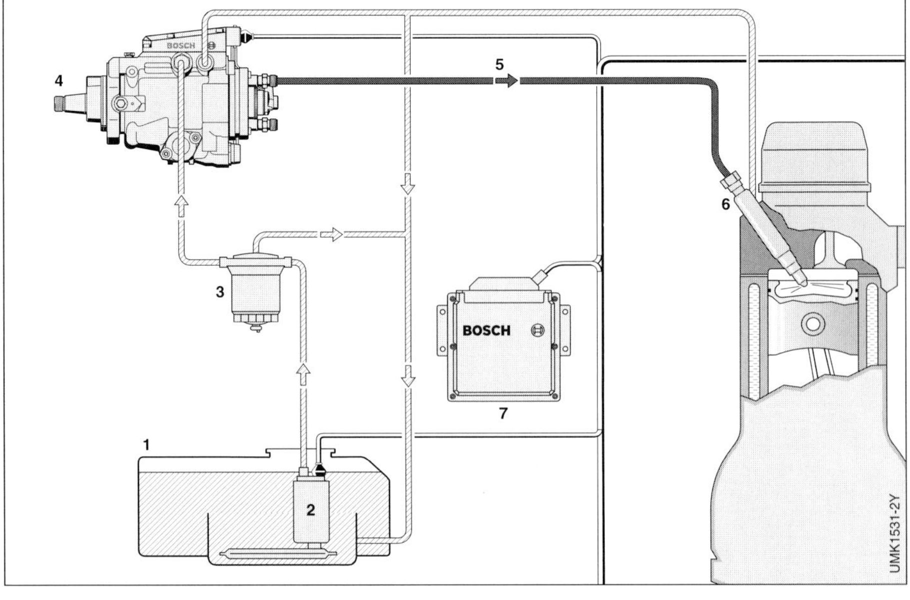

against the effects of heat. In the case of buses, fuel lines must not be located in the passenger compartment or in the driver's cab, nor may fuel be delivered by force of gravity.

Fuel filter

Inadequate filtering can lead to damage at the pump components, delivery valves, and injection nozzles. The fuel filter cleans the fuel before it reaches the radial-piston distributor pump, and thereby prevents premature wear at the pump's sensitive components.

Low-pressure injection-pump components

Vane-type fuel-supply pump
The vane-type fuel-supply pump draws the fuel from the fuel tank and with each revolution delivers a practically constant quantity of fuel to the radial-piston high-pressure pump.

Pressure-control valve
The pressure-control valve regulates the fuel pump's delivery pressure. It opens when the fuel pressure climbs excessively, and closes when it drops too far.

Overflow throttle valve
Upon reaching a pre-selected opening pressure, the overflow throttle valve opens and permits a defined quantity of fuel to flow back to the fuel tank. The overflow throttle valve also facilitates automatic bleeding of the pump.

High-pressure delivery

The high-pressure stage of the fuel system utilises a radial-piston high-pressure pump to generate the pressure required for the injection process. For each injection process, the following components are used to deliver and inject the fuel:

– Injection-pump components,
– High-pressure line,
– Nozzle holder, and
– Injection nozzle.

High-pressure components of the injection pump

Radial-piston high-pressure pump
With the high-pressure solenoid valve opened, the fuel moves from the low-pressure stage to the delivery plunger in the high-pressure stage. The cam ring, with lobes on its inner wall, forces the delivery plunger inwards radially and with each stroke compresses the fuel for injection into the cylinder in question.

High-pressure solenoid valve
The high-pressure solenoid valve is controlled by the pump ECU. It regulates the entry of fuel to the radial-piston high-pressure pump, and defines the injected-fuel quantity and the instant of injection for each injection process.

Distributor shaft with distributor head
Every time it rotates, the distributor shaft distributes fuel to each cylinder through a line fitting on the distributor head and the associated high-pressure delivery line.

Return-flow throttle valves
These are installed in the distributor-head line fittings. By dampening the reflected fuel pressure waves which are generated when the injectors close, they serve to prevent wear in the high-pressure stage and uncontrolled opening of the nozzles.

Fuel lines in the high-pressure section
The high-pressure fuel lines (usually seamless steel tubing) lead from the fuel-injection pump to the nozzles. The injection lines between the rail and the injectors are matched to the rate-of-discharge curve and must therefore all be of the same length. The differences in length between the injection pump and the nozzles are compensated for by using slight or pronounced bends in the individual lengths of tubing.

Nozzles and nozzle-holders
The nozzles are installed in the nozzle holders and inject the fuel into the engine cylinders in precisely metered quantities. The nozzles define the rate-of-discharge curve. Excess fuel returns to the fuel tank under very low pressure.

Design and function

Subassemblies

The following subassemblies are combined in and around the pump housing of the VR radial-piston distributor pump (Fig. 1):
– Vane-type fuel-supply pump (1) with pressure-control valve and overflow throttle,
– Radial-piston high-pressure pump (4) with distributor shaft and delivery valve,
– High-pressure solenoid valve (6),
– Timing device (5) with solenoid valve,
– Angle-of-rotation sensor (DWS system [2]), and
– Pump ECU (3).

The combination of these subassemblies to form a compact unit, permits the interaction of the various functional units to be matched to each other with a very high degree of precision. This enables the very tight requirements, and the stipulated performance specs to be complied with in full.

Vane-type supply pump with pressure-control valve and overflow throttle valve

A rugged distributor shaft is mounted in the distributor housing of the radial-piston pump. It is held at one end in a plain bearing and at the other in a ball or rolling bearing. The vane-type supply pump is located inside the pump and attached to the distributor shaft. It draws in the fuel, generates an accumulator-chamber pressure, and supplies the radial-piston high-pressure pump with fuel.

Radial-piston high-pressure pump with distributor shaft and delivery valve

The radial-piston high-pressure pump is driven directly from the drive shaft. It generates the high pressure needed for the

Fig. 1

Components of the radial-piston distributor injection pump

1 Vane-type fuel-supply pump with pressure-control valve, **2** Angle-of-rotation sensor, **3** Pump ECU, **4** Radial-piston high-pressure pump with distributor shaft and delivery valve, **5** Timing device and solenoid valve (pulse valve), **6** High-pressure solenoid valve.

UMK1533Y

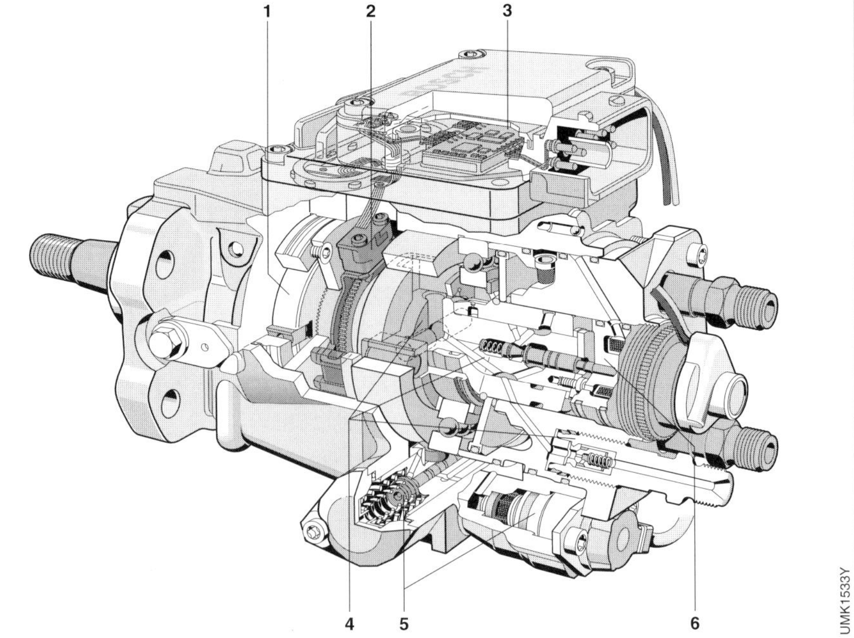

fuel injection, and distributes the fuel to the individual engine cylinders. The distributor shaft is driven by the drive shaft through a driver disc.

High-pressure solenoid valve
The high-pressure solenoid valve is located in the center of the hydraulic head. Its needle projects into the drive shaft and turns in synchronism with it. The valve opens and closes in accordance with a variable pulse ratio depending upon the signals from the pump ECU. Its closed period defines the high-pressure pump's delivery period. This enables precision metering of the injected fuel quantity.

Timing device
The hydraulic timing device together with the pulse valve and the working piston at right angles to the pump axis are attached to the underside of the injection pump. By rotating the cam ring as a function of the load and rotational speed, the timing device adjusts the start of delivery, and with it the instant of injection. This variable form of control is also designated "Electronic injection timing".

Angle-of-rotation sensor (DWS system)
The increment wheel and the mount for the angle-of-rotation sensor are attached to the driveshaft. These serve to measure the angle which results between driveshaft and cam ring during driveshaft rotation. The results are used to calculate the current pump speed, the timing-device setting, and the angular setting of the camshaft.

Pump ECU
The pump ECU, which is provided with cooling ribs, is screwed to the upper side of the injection pump. From the information received from the DWS system and the engine ECU, it calculates the triggering signal for the high-pressure solenoid valve and for the timing-device solenoid valve.

Pump installation and pump drive

Installation
The radial-piston distributor pump is flanged directly to the diesel engine. To avoid confusion with the engine-cylinder numbering when the fuel-injection lines are connected, the distributor pump's outlet ports are identified by the letters A, B, ..., F in accordance with the cylinder in question. Radial-piston distributor pumps are particularly suitable for diesel engines with up to 6 cylinders.

Drive
The radial-piston distributor pump's driveshaft is driven by a mechanism which is matched to the design of the engine in question. In the case of 4-stroke engines, the pump's driveshaft turns at half the speed of the engine crankshaft. In other words at the same speed as the engine's camshaft. The injection pump's drive is aligned to the engine's piston movements. Chains, gearwheels, toothed belts, or splined shafts are used to ensure synchronism between engine and pump.

Low-pressure stage

It is the job of the low-pressure stage to provide the high-pressure stage with enough fuel. Its most important components are the vane-type supply pump, the pressure-control valve, and the overflow throttle valve (Fig. 2).

Vane-type supply pump
The vane-type supply pump is located around the radial-piston distributor pump's driveshaft (Fig. 3). The eccentric locating ring (3) with its shaped inner running surface, is held between the pump housing's inside wall and the support ring serving as a cover. The inside wall of the housing is provided with two recesses which permit the fuel to enter the pump (4) and to leave it (7). Thanks to their special shape, they are designated the "suction kidney" and the "pressure kidney".

The vane wheel (2), which is driven by splines on the driveshaft (1), rotates inside the locating ring. The spring-loaded vanes (5) are free to move inside the vane-wheel guide slots, and are pressed against the locating ring by centrifugal force when the vane wheel rotates. The space, designated "cell" (6), is defined by the following elements (Fig. 3):

– The pump-housing inside wall,
– The support ring,
– The shaped inner running surface of The locating ring,
– The outside surface of the vane wheel, and
– Two adjacent vanes.

The fuel reaches the suction kidney in the "cell" through the inlet passage in the housing and via other internal passages. In the cell, the rotation of the vane wheel transports the fuel in the direction of the pressure kidney. During rotation, the shaped inner wall of the eccentric support ring causes the cell volume to decrease so that the fuel is compressed. The vol- ume decrease causes the fuel pressure to increase considerably by the time the fuel reaches the outlet leading to the pressure "kidney". The various sub-assemblies are supplied with pressurized fuel through internal passages in the pump housing. The pressure-control valve is also supplied through one of these connections.

The radial-piston distributor pump operates at a pressure which is relatively high compared to that of other distributor pumps. This high pressure is the reason for the vanes (5) having a hole in the center of their end face. This leads to only one of the face-end edges sliding along the shaped surface of the locating ring. As a result, the complete end face of the vane is not subjected to pressure, and unwanted radial movement is avoided. When changing over from one edge to the other (for instance when changeover takes place from inlet to outlet), the pressure applied at the end face of the vane can propagate through the passage to the other end of the vane. To a great extent, the opposing pressure forces cancel

Fig. 2

Radial-piston distributor pump: Low-pressure stage

To facilitate ease of understanding, a number of components have been turned with respect to their real position.**1** Vane-type supply pump (turned through 90°), **2** Pressure-control valve, **3** Overflow throttle valve.

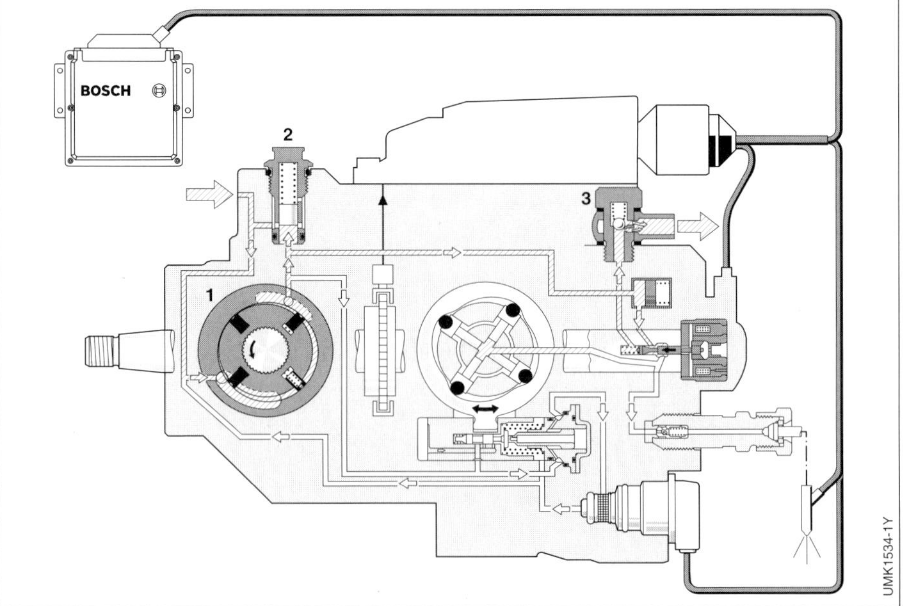

UMK1534-1Y

each other out and, as described above, the vanes contact the locating ring's inner running surface due to centrifugal force or spring force.

Pressure-control valve

The fuel pressure generated in the pressure kidney by the vane-type supply pump is a function of the pump speed. To prevent excessive pressure at very high speeds, there is a pressure-control valve (spring-loaded slide valve, Fig. 4) in the immediate vicinity of the supply pump and connected to the pressure kidney (5) by a passage. It changes the supply pump's delivery pressure as a function of the delivered fuel quantity. If the fuel pressure climbs beyond a given level, the front edge of the valve plunger (3) opens radially arranged passages (4) through which the fuel can return to the supply pump's suction kidney (6). If the fuel pressure is not high enough, the radially arranged passages remain closed due to the spring pressure. The opening pressure is determined by the adjustable pretension of the spring.

Overflow throttle valve

In order to cool and vent the radial-piston distributor pump, fuel flows through the overflow throttle valve (Fig. 5), which is screwed to the pump's housing, and back to the tank.

The overflow throttle valve is connected to the overflow (5) of the distributor head. There is a spring-loaded ball valve (3) inside the valve body which does not open to permit fuel to flow back to the tank until a given pressure has been exceeded.

In the valve body there is a secondary passage to the ball valve. This is connected to the pump overflow through a small-diameter throttle bore (4). This throttled connection facilitates automatic pump venting. The overall low-pressure stage is so designed that a defined quantity of fuel can flow back to the fuel tank via the pump overflow (Fig.5).

Fig. 3

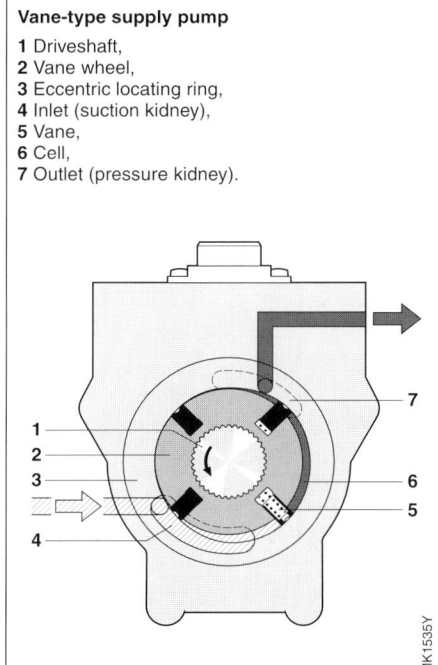

Vane-type supply pump

1 Driveshaft,
2 Vane wheel,
3 Eccentric locating ring,
4 Inlet (suction kidney),
5 Vane,
6 Cell,
7 Outlet (pressure kidney).

UMK1535Y

Fig. 4

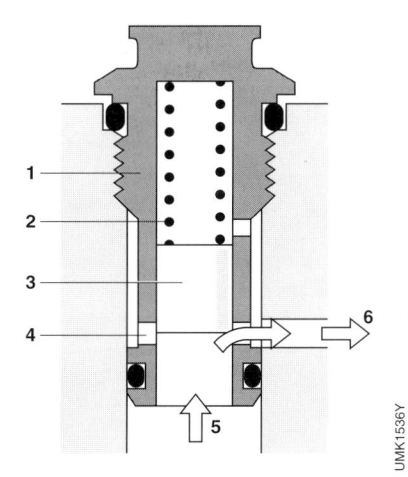

Pressure-control valve

1 Valve body,
2 Spring,
3 Valve plunger,
4 Passage (radially arranged),
5 From the pressure kidney,
6 To the pressure kidney.

UMK1536Y

Fuel filter

For trouble-free operation, it is imperative that a fuel filter is used which is specifically aligned to the requirements of the particular fuel-injection system. Otherwise, contaminants in the fuel can lead to damage at the pump components, delivery valve, and nozzles.

Diesel fuel can contain water either in bound form (emulsion) or in free form (e.g. condensation of water due to temperature change). If this water enters the injection system, it can lead to damage as a result of corrosion.

Similar to other injection systems, the radial-piston distributor pump also needs a fuel filter with paper filter element and water reservoir (Fig. 6), from which the water must be drained by opening the water drain screw at regular intervals.

High-pressure stage

In addition to high-pressure generation, fuel distribution, fuel-metering, and start-of-delivery control also take place in the high-pressure stage (Fig. 7). A single actuator (high-pressure solenoid valve) is all that is required.

Generation of high pressure using the radial-piston high-pressure pump

The pressure required for injection (approx. 1000 bar at the pump) is generated by the radial-piston high-pressure pump. It is driven by the driveshaft and comprises (Fig. 8):
– The driver disc,
– The roller supports (4) with rollers (2),
– The cam ring (1),
– The delivery plunger (5), and
– The front section (head) of the distributor shaft (6).

The rotation of the driveshaft is transferred directly to the distributor shaft through a driver disc which engages in the radial guide slots (3) at the end of the

Fig. 5

Overflow throttle valve

1 Valve body,
2 Spring,
3 Ball valve,
4 Throttle bore,
5 To overflow.

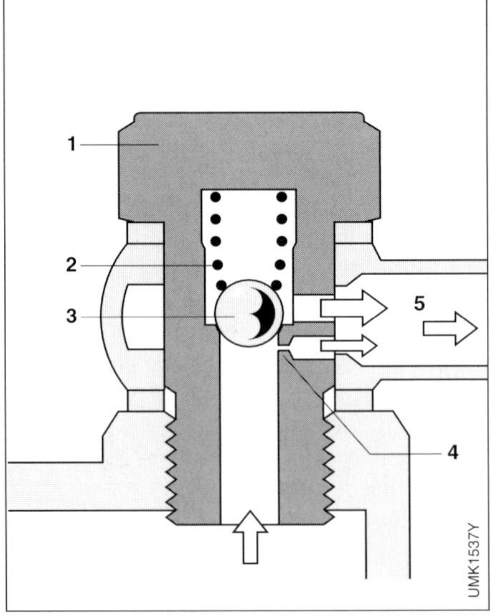

UMK1537Y

Fig. 6

Fuel filter

1 Filter cover, **2** Fuel-inlet connections, **3** Paper filter element, **4** Filter case, **5** Water reservoir, **6** Water drain screw, **7** Fuel outlet.

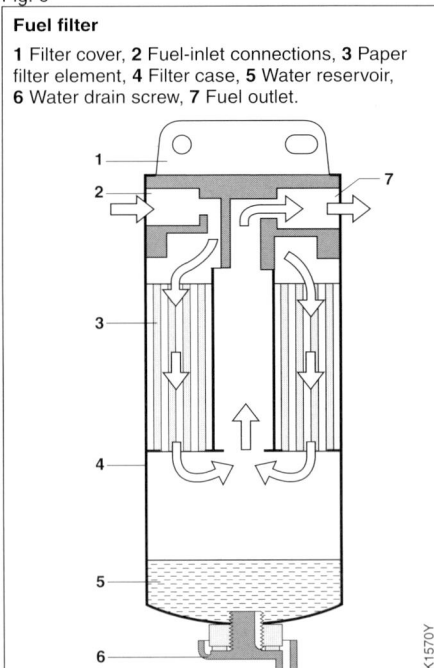

UMK1570Y

driveshaft. The guide slots also serve to hold the roller supports (4) which, together with their rollers (2) travel around the inner cam track of the cam ring (1) located around the driveshaft. There are lobes in the inner cam track, the number of which correspond to the number of cylinders in the engine. The delivery pistons are located radially in the distributor head (hence the term "radial-piston high-pressure pump"). The rollers are held by their supports, and when the distributor shaft rotates they travel around the track on the inside of the cam ring. In the process, the lobes on the track force the pistons inwards so that they compress the fuel in the central high-pressure area (7). Depending upon the number of engine cylinders and the particular application, there are versions available with 2, 3, or 4 delivery pistons. (Fig. 8a, b, c).

Fig. 7

High-pressure stage of the radial-piston distributor pump

To facilitate ease of understanding, a number of components have been turned with respect to their real position. **1** ECU, **2** Radial-piston high-pressure pump (turned through 90°), **3** Distributor head, **4** High-pressure solenoid valve, **5** Injection-line fitting.

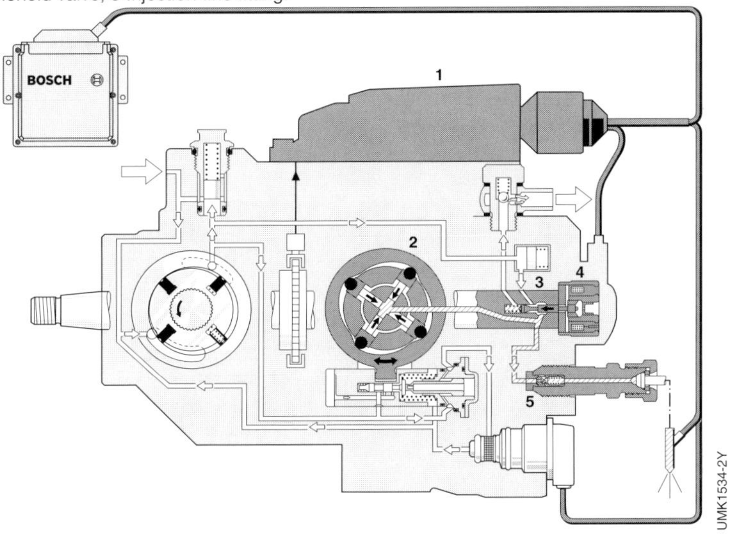

UMK1534-2Y

Fig. 8

Delivery-piston configuration in the radial-piston high-pressure pump (examples)

a 4 and 6-cylinder version, **b** 6-cylinder version, **c** 4-cylinder version. **1** Cam ring, **2** Roller, **3** Driveshaft guide slot, **4** Roller support, **5** Delivery piston, **6** Distributor shaft, **7** High-pressure area.

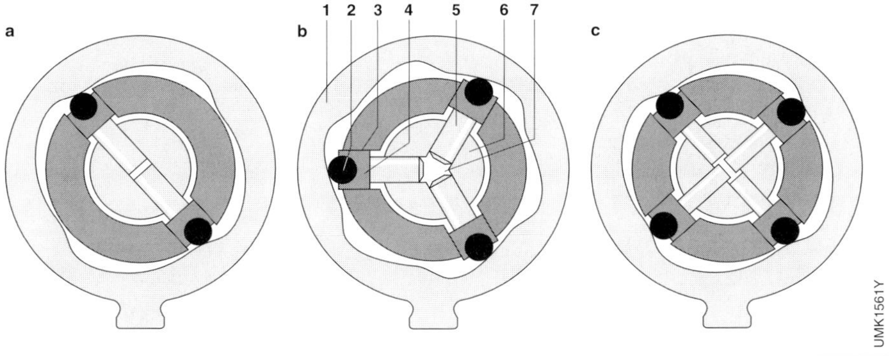

UMK1561Y

Fuel distribution with the distributor head

The distributor head (Fig. 9) comprises:
- The flange (6),
- The control bushing (3) shrunk into the flange,
- The rear section of the distributor shaft (2) held in the control bushing,
- The needle (4) of the high-pressure solenoid valve (7),
- The accumulator diaphragm (10), and
- The injection-line fitting (16) with the return-flow throttle (15).

During the filling phase (Fig. 9a), the delivery pistons (1) are forced outward by the cam lobes. The valve needle (4) is open. Fuel can therefore enter the distributor head and fill the high-pressure volume (8) via the low-pressure inlet (12), the annular passage (9), and the valve needle (4). Excess fuel flows off through the fuel return (5).

During the delivery phase (Fig. 9b), the delivery pistons (1) are forced inwards by the cam lobes. The valve needle (4) is

Fig. 9

Distributor head

a Filling phase, b Delivery phase. 1 Delivery piston, 2 Distributor shaft, 3 Control bushing, 4 Valve needle, 5 Fuel return, 6 Flange, 7 High-pressure solenoid valve, 8 High-pressure volume, 9 Annular passage, 10 Accumulator diaphragm, 11 Diaphragm chamber, 12 Low-pressure inlet, 13 Distributor slot, 14 High-pressure outlet, 15 Return-flow throttle valve, 16 Injection-line fitting.

UMK1540Y

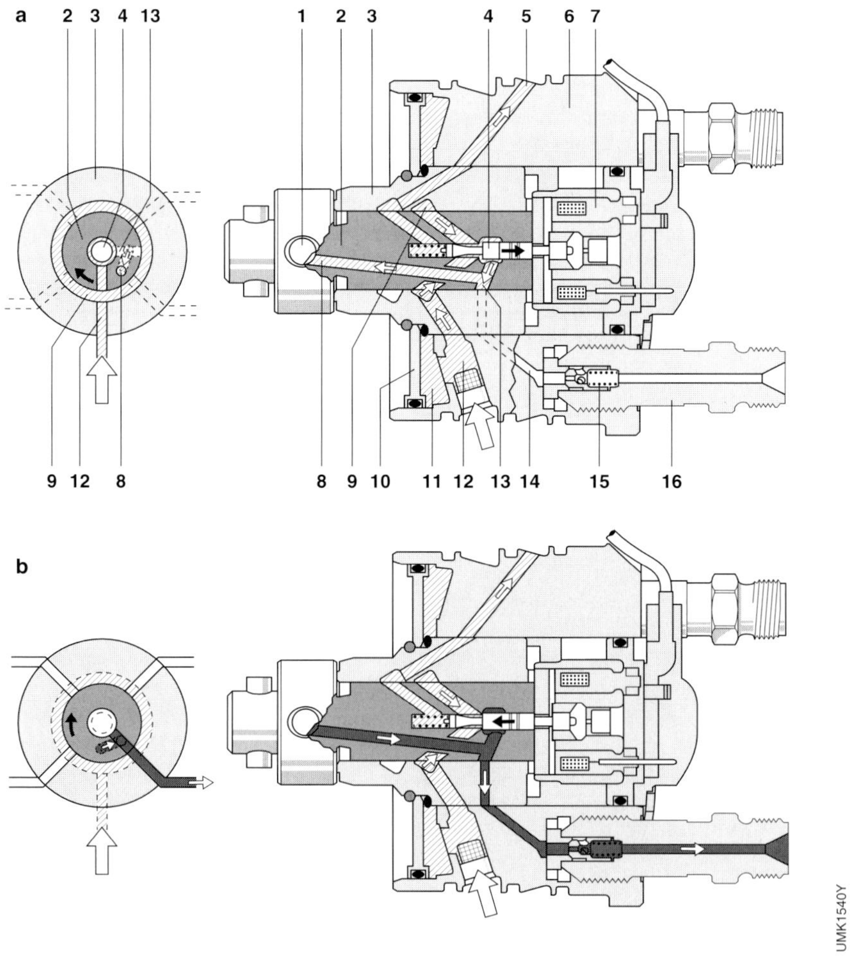

closed. The pistons compress the fuel inside the closed-off high-pressure volume (8). The distributor slot (13) is connected to the high-pressure outlet (14) by the rotation of the distributor shaft (2). The pressurized fuel is now free to reach the nozzle via the distributor slot (13), injection-line fitting (16) with return-flow throttle valve (15), high-pressure line and nozzle holder. The nozzle injects the fuel into the engine's combustion chamber.

Fuel metering with the high-pressure solenoid valve

At the BDC setting of the cam lobe, a control pulse from the pump ECU closes the high-pressure solenoid valve (Fig. 9, item 7) with valve needle (4). The solenoid valve's instant of closing defines the pump's start of delivery. The pump ECU is provided with precise information on the valve's instant of closing through the electronic closing-point detection facility (BIP = Beginning of Injection Period).

Fuel metering takes place between the start of delivery and the high-pressure solenoid valve's end-of-triggering point. The fuel-metering period is known as the delivery or injection period. In other words, the injected fuel quantity is a function of the length of time the high-pressure solenoid valve remains closed. The high-pressure fuel injection ceases immediately the high-pressure solenoid valve opens. The excess fuel which is delivered until the cam reaches TDC is diverted away to the diaphragm cham-

ber. The high pressure peaks which are generated on the low-pressure side as a result, are damped by the accumulator diaphragm. The fuel stored in the diaphragm chamber also supports the filling process for the next injection.

To shut down the engine, the high-pressure delivery is stopped completely by the high-pressure solenoid valve. An additional shutoff valve as used with the VE distributor pump is therefore not needed.

Damping the pressure waves with the return-flow throttle valve

The return-flow throttle valve (Fig. 10) prevents the pressure waves which are generated at the end of the injection process being reflected back and opening the nozzle needle again (secondary injection). Secondary injection has a negative effect upon exhaust-gas emissions.

As soon as delivery starts, the fuel pressure causes the valve-cone (3) to lift from its seat. The fuel is now directed via the injection-line fitting (5), and the high-pressure injection tubing to the injection nozzle. As soon as fuel delivery stops, the fuel pressure drops abruptly and the valve spring (4) forces the valve cone back onto its seat again (1).

The reflected pressure waves which are generated when the nozzle closes are now reduced to such an extent by the throttle (2) that no damaging pressure-wave reflections are generated at all.

Fig. 10

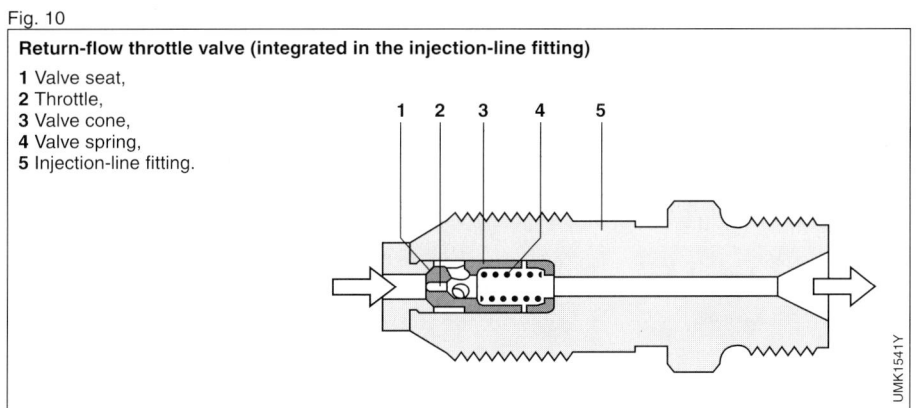

Return-flow throttle valve (integrated in the injection-line fitting)
1 Valve seat,
2 Throttle,
3 Valve cone,
4 Valve spring,
5 Injection-line fitting.

UMK1541Y

Injection timing

Assignment

If the start of injection is left constant, the faster the engine turns the greater is the crank angle between start of injection and start of combustion. This means that the start of combustion can no longer take place at the correct instant in time (referred to the engine-piston setting).

The diesel engine features its most efficient combustion and delivers its maximum power at one specific position of the crankshaft (or pistons). As engine speed increases, it is the job of the injection timing to advance the pump's start of delivery with reference to the crankshaft. The injection timing comprises an angle-of-rotation sensor, timing device, and timing-device solenoid valve. By compensating for the time lag resulting from the injection and combustion delays, it optimally adapts the instant of injection to the engine's operating status (Fig. 1). Examples of working-stroke characteristics are shown in Figs. 2 to 4:

The start of delivery (FB) starts after the high-pressure solenoid valve has closed. High fuel pressure builds up in the high-pressure delivery lines. This nozzle-side line pressure p_D (Fig. 3) opens the nozzle needle upon reaching the nozzle-opening pressure, and injection starts (SB).

The time between start of delivery and start of injection is referred to as the injection lag (SV). Combustion (VB) starts when the pressure in the combustion chamber (Fig. 2) climbs further. This time interval between start of injection and combustion is known as the ignition lag (ZV).

As soon as the high-pressure solenoid valve opens again, the fuel high pressure collapses (end of delivery) and the nozzle needle closes (end of injection, SE).

Fig. 1

Injection timing in the radial-piston distributor pump

To facilitate ease of understanding, a number of components have been turned with respect to their real position. **1** Engine ECU, **2** Pump ECU, **3** Vane-type supply pump (turned through 90°), **4** Angle-of-rotation sensor, **5** Cam ring (turned through 90°), **6** High-pressure solenoid valve, **7** Timing device, **8** Timing-device solenoid valve.

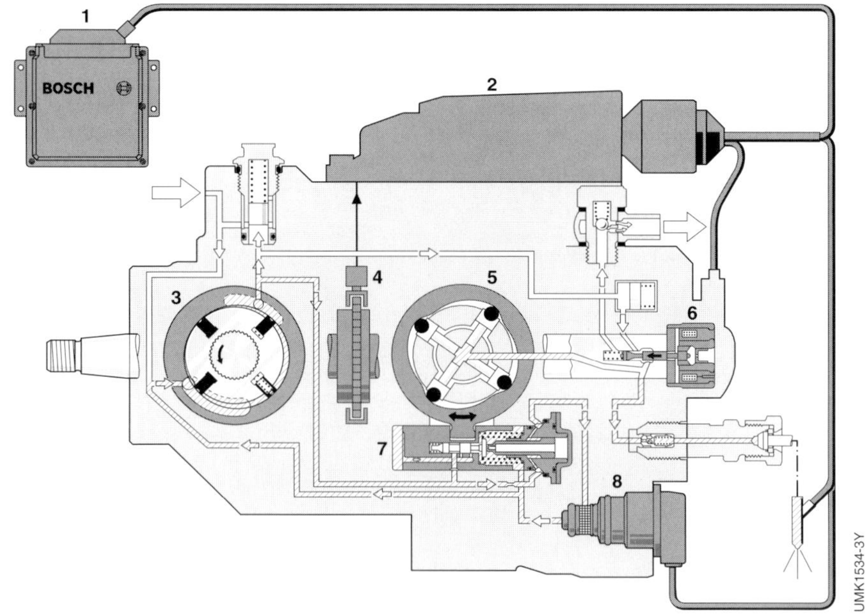

This is followed by the end of combustion (VE). During the injection pump's delivery process, the injection nozzle is opened by a pressure wave travelling at the speed of sound in the high-pressure delivery line. The pressure wave's propagation time is defined by the length of the line and the speed of sound in diesel fuel which is approx. 1500 m/s. The propagation time is the time between the start of delivery and the start of injection, and is therefore also known as the injection lag (SV).

Basically, the injection lag is independent of the engine speed. The crank angle between start of delivery and start of injection, though, increases along with rising engine speed. This leads to the nozzle opening later and later (referred to the engine-piston position).

Following the injection into the cylinder, the diesel fuel needs a certain time to convert to the gaseous state and form an ignitable mixture with the air. The required period of time between start of injection and start of combustion is independent of engine speed, and on the diesel engine is referred to as the ignition lag.

The ignition lag is defined by the following parameters:
– The diesel fuel's ignition quality (given by means of the Cetane number),
– The compression ratio,
– The air temperature, and
– The fuel atomization.

As a rule, the ignition lag lasts about 1 millisecond.

Fig. 3

Characteristic of the nozzle-end line pressure p_D at WOT and high engine speeds

FB Start of delivery,
SB Start of injection,
SE End of injection,
OT Top Dead Center (TDC),
p_0 Nozzle opening pressure.

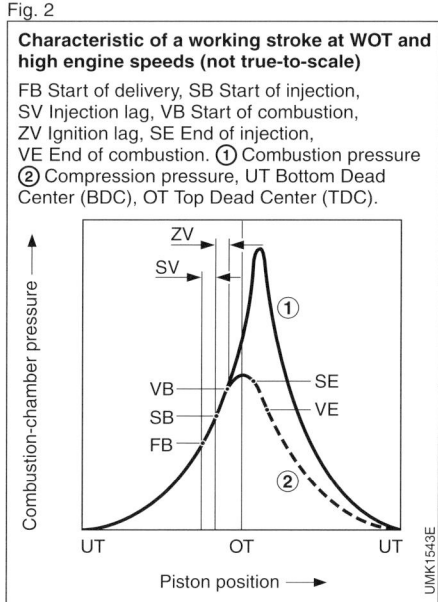

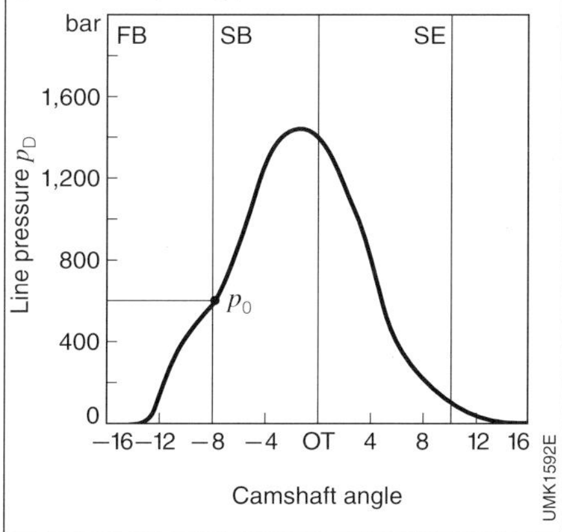

Fig. 2

Characteristic of a working stroke at WOT and high engine speeds (not true-to-scale)

FB Start of delivery, SB Start of injection,
SV Injection lag, VB Start of combustion,
ZV Ignition lag, SE End of injection,
VE End of combustion. ① Combustion pressure
② Compression pressure, UT Bottom Dead
Center (BDC), OT Top Dead Center (TDC).

Fig. 4

Characteristic of the nozzle-needle lift at WOT and high engine speeds

FB Start of delivery,
SB Start of injection,
SV Injection lag,
SE End of injection,
OT Top Dead Center (TDC).

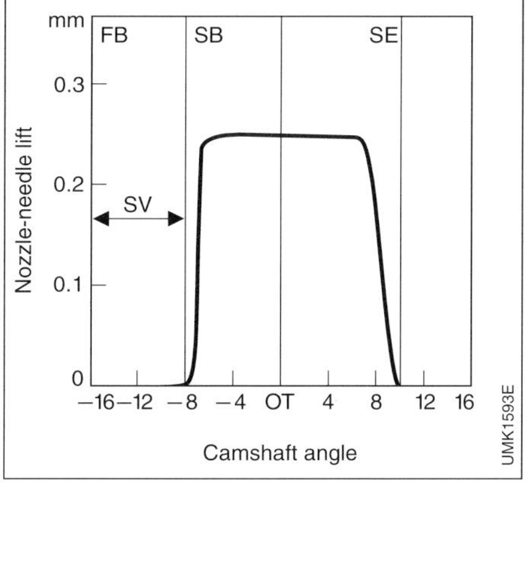

Design and construction

The hydraulically controlled timing device is installed in the bottom of the radial-piston distributor pump's housing, at right angles to the pump's longitudinal axis (Fig. 5). Through a ball pivot (2), the cam ring (1) engages with a transverse passage in the timing-device piston (3). This means that axial movement of the timing-device piston is converted to rotational movement of the cam ring. Located in the center of the timing-device piston, is a control collar (5) which opens and closes the spill ports in the piston. A spring-loaded hydraulic control plunger (12) is located along the same axis, and serves to define the control collar's set position.

The timing-device solenoid valve is arranged at right angles to the timing-device plunger axis. It is shown as Item 15 in the schematic diagram Fig. 5, in which it has been pivoted into the timing-device plane. When triggered by the pump ECU, it influences the pressure at the control plunger.

Fig. 5

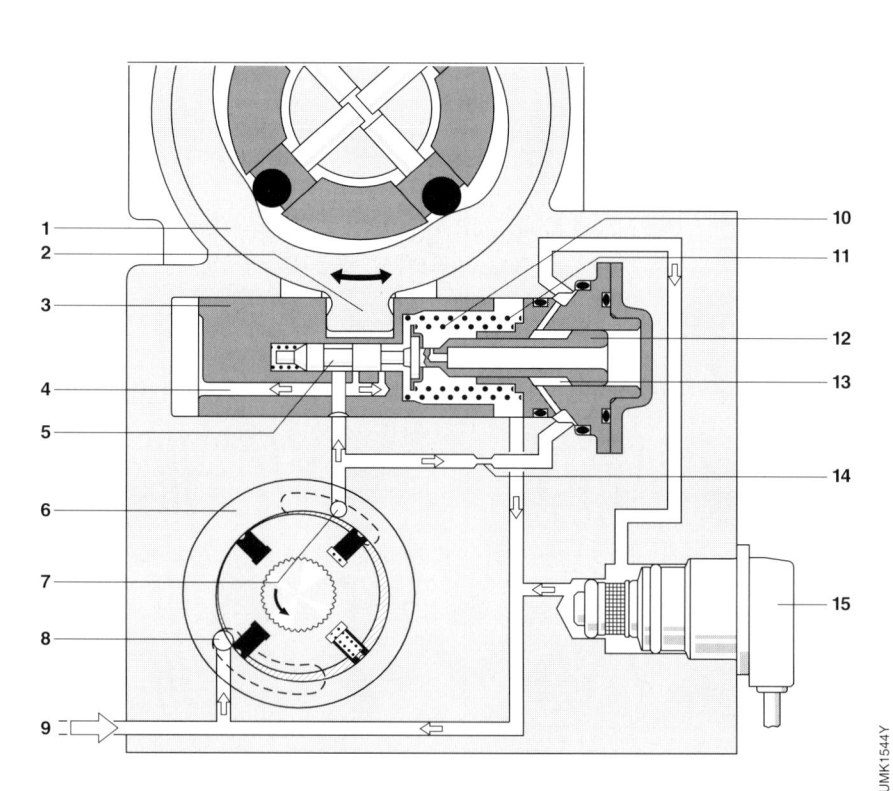

Fig. 5: Timing device with solenoid valve
(shown schematically in one plane).

1 Cam ring,
2 Ball pivot,
3 Timing-device piston,
4 Inlet passage/outlet passage,
5 Control collar,
6 Vane-type supply pump,
7 Pump outlet (pressure side),
8 Pump inlet (suction side),
9 Input from fuel tank,
10 Control plunger spring,
11 Return spring,
12 Control plunger,
13 Ring chamber of the hydraulic stop,
14 Throttle,
15 Timing-device solenoid valve.

UMK1544Y

Method of operation

Start-of-injection control

Depending upon the engine's operating status (load, engine speed, engine temperature), the engine ECU accesses the start-of-injection map and selects a start of injection which it stipulates as the setpoint value. The start-of-injection controller in the pump ECU continually compares the actual start of injection with the setpoint figure and, in case of deviation, triggers the solenoid with an appropriate on-off ratio. The signal from an angle-of-rotation sensor, or from a needle-motion sensor in the nozzle holder, provides the information on the actual start-of-injection value.

Injection timing in the "advance" direction

The timing-device piston (3) is held in its neutral position ("retard setting") by a return spring (11). During operation, the fuel pressure inside the pump is regulated by the pressure-control valve as a function of pump speed. This fuel pressure is applied as control pressure (through a throttle) to the ring chamber of the hydraulic stop (13), and with the timing-device solenoid valve (15) closed, shifts the control plunger (12) in the "advance" direction against the force of the spring (to the right in Fig. 5). This causes the control collar to shift in the "advance" direction as well so that the inlet passage in the timing-device piston is opened and fuel can flow into the space behind the timing-device piston and shift the piston to the right in the "advance" direction. The ball pivot (2) converts the piston's axial shift to a rotational movement at the high-pressure pump's cam ring (1). In the case of injection-timing shift in the "advance" direction, rotating the cam ring relative to the direction of rotation of the pump driveshaft leads to the rollers reaching the cam lobes sooner so that start of injection takes place earlier. Maximum injection advance is 20° camshaft (corresponding to 40° crankshaft).

Injection timing in the "retard" direction

The timing-device solenoid valve (15) opens upon receiving time-pulsed signals from the pump ECU, and the pressure in the ring chamber of the hydraulic stop (13) sinks as a result. The spring force shifts the control plunger (12) in the "retard" direction (to the left in Fig. 5). Initially, the timing-device piston (3) remains where it is. The control collar (5) must open the spill port for the control passage before the fuel can leave the space behind the timing-device piston. The spring (11) can now shift the piston in the "retard" direction and into its neutral position.

Control of the control pressure

The timing-device solenoid valve acts as a variable throttle when its needle is opened and closed rapidly. It can continuously vary the control pressure so that the control plunger can be shifted to any position required between advance and retard. The pump ECU is responsible for defining the valve needle's on/off ratio, that is, the ratio of the open time to the total duration of a working cycle.

If, for instance, the control plunger is to be adjusted further in the "advance" direction, the on/off ratio in changed by the pump ECU so that the proportion of time that the valve remains open is reduced. Less fuel flows via the timing-device solenoid valve, and the piston is free to move towards "advance".

System control using EDC

System blocks

The electronic diesel control (EDC) for the radial-piston distributor pump (Fig. 3) comprises three major system blocks:

1. Sensors and setpoint generators for registration of the operating conditions and the desired values. These convert a variety of physical parameters into electrical signals.
2. The engine ECU and a pump ECU for processing the information using specified arithmetic operations (control algorithms) in order to generate electrical output signals.
3. Actuators to convert the ECU's electrical output signals into mechanical parameters.

The ECU's apply their signals directly to the actuators through driver stages, or send them to other systems in the vehicle.

Sensors

Temperature sensors
Temperature sensors are installed at a number of different points:

– In the coolant circuit, in order to establish engine temperature by way of the coolant temperature (Fig. 1),
– In the intake manifold to measure the temperature of the intake air,
– In the engine lube oil to measure the oil temperature, and
– In the fuel-return line to measure the fuel temperature.

The sensors are all equipped with a temperature-dependent resistor having a Negative Temperature Coefficient (NTC). This is part of a voltage-divider circuit across which 5 V are applied. The voltage drop across the resistor is inputted into the ECU through an analog-to-digital converter (ADC) and is a measure for the temperature. A characteristic curve is stored in the engine-ECU microcomputer which defines the temperature as a function of the given voltage value (Fig. 2).

Crankshaft-speed sensor
The piston position in the combustion chamber is decisive in defining the instant of injection. The rotational speed defines the number of crankshaft rotations per minute. This important input variable is calculated in the engine ECU using the signal from the inductive crankshaft-speed sensor.

Fig. 1

Coolant temperature sensor
1 Electrical connection, **2** Housing,
3 NTC resistor, **4** Coolant.

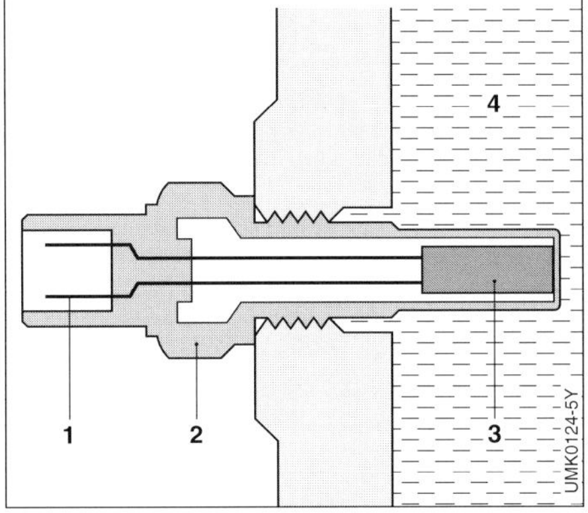

Fig. 2

Temperature-sensor characteristic (NTC)

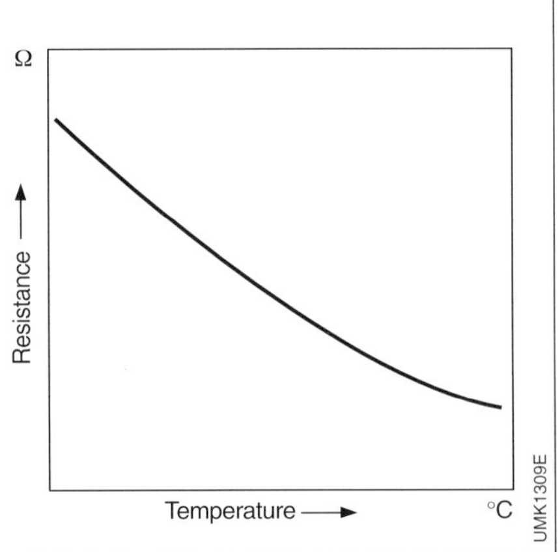

Fig. 3

System overview of a diesel injection system with VR radial-piston distributor pump and a variety of different system components

1 Fuel tank,
2 Fuel filter,
3 Injection pump,
4 Pump ECU,
5 High-pressure solenoid valve,
6 Timing-device solenoid valve,
7 Timing device,
8 Engine ECU,
9 Nozzle-and-holder assembly, 3 with needle-motion sensor,
10 Sheathed-element glow plug,
11 Glow control unit,
12 Coolant temperature sensor,
13 Crankshaft-speed sensor,
14 Intake-air temperature,
15 Air-mass meter,
16 Boost-pressure sensor,
17 Turbocharger,
18 EGR positioner,
19 Charge-pressure actuator,
20 Vacuum pump,
21 Battery,
22 Instrument panel with display of fuel consumption, engine speed etc.,
23 Pedal-travel sensor,
24 Clutch switch,
25 Brake contacts,
26 Vehicle-speed sensor,
27 Operator unit for the Cruise Control,
28 Air-conditioner compressor with switch,
29 Diagnosis display with connections for diagnostic tester.

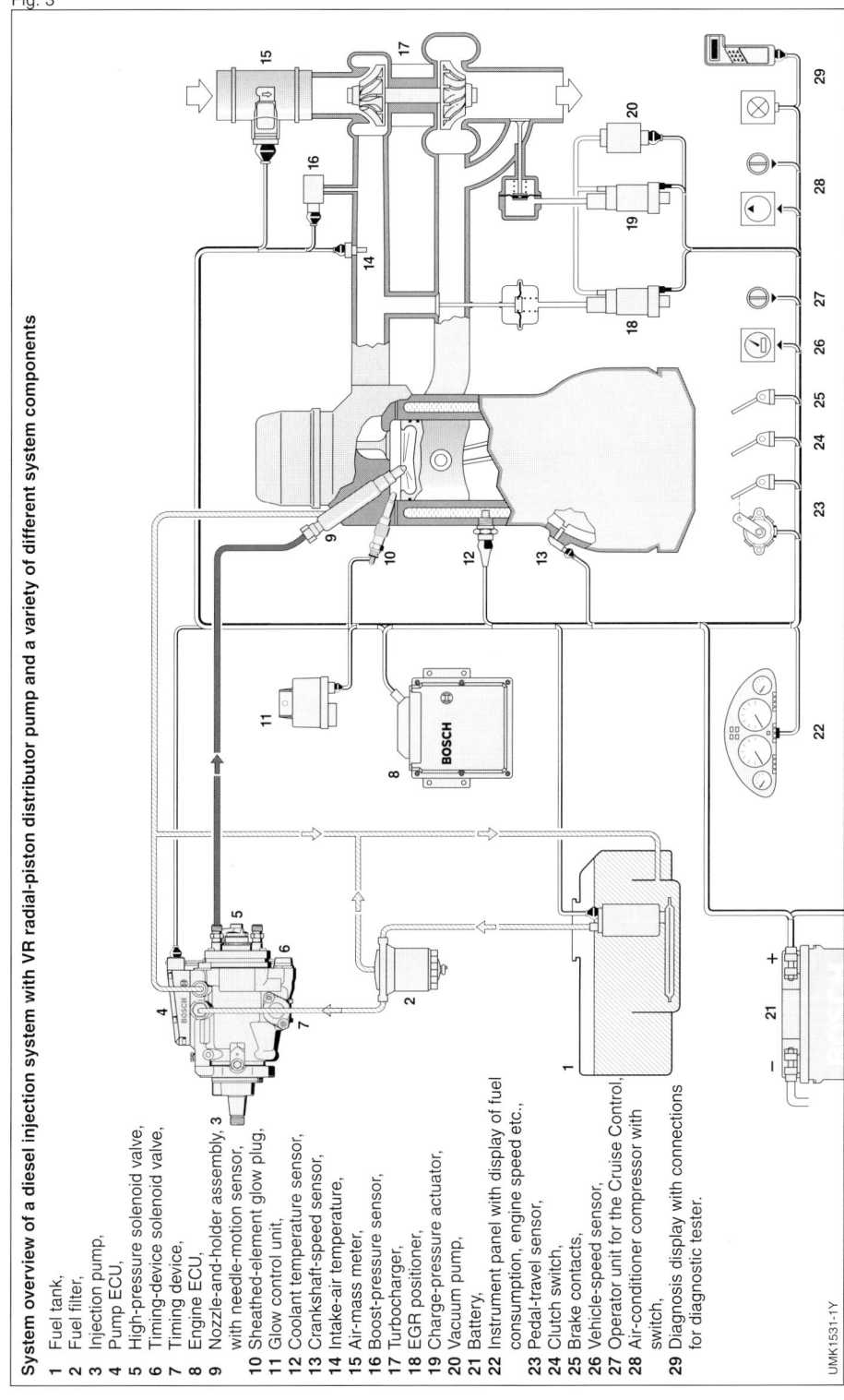

UMK1531-1Y

237

Signal generation

A ferromagnetic trigger wheel is attached to the crankshaft. Around its circumference, it has 1 tooth (segment) for each engine cylinder. The crankshaft-speed sensor (Fig. 4) registers the trigger wheel's tooth sequence. It comprises a permanent magnet and a soft-iron core with a copper winding. The magnetic flux in the sensor changes as the teeth and gaps pass by, and a sinusoidal AC voltage is generated, the amplitude of which increases sharply in response to higher engine (crankshaft) speeds. Adequate amplitude is already available as from a minimum speed of 50 min^{-1}.

Fig. 4

Crankshaft-speed sensor

1 Permanent magnet, 2 Housing, 3 Engine block,
4 Soft-iron core, 5 Winding,
6 Trigger wheel with 1 tooth per cylinder.

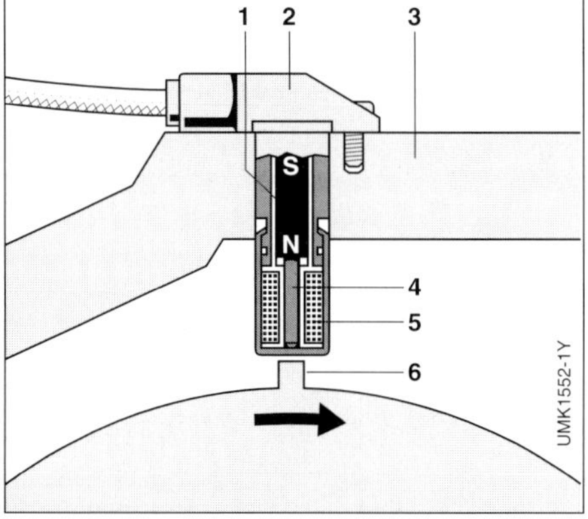

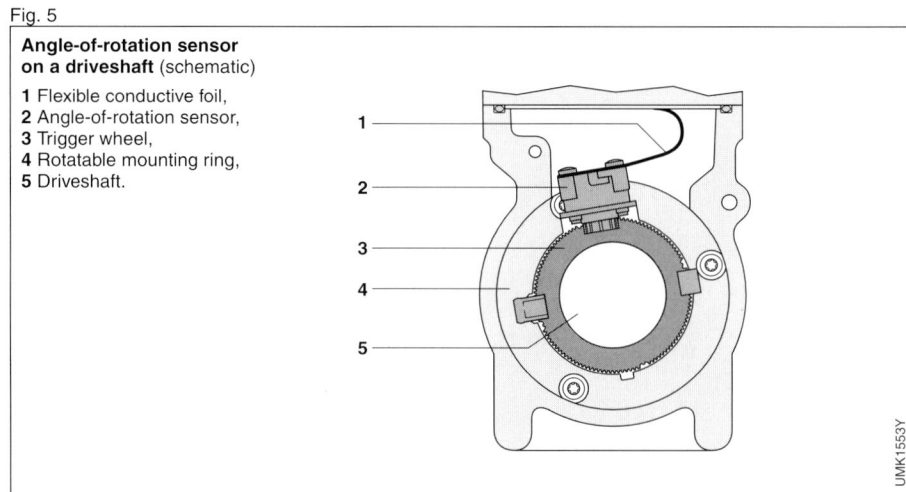

Calculation of engine speed

The angular relationship (offset) between the cylinder pistons is such that two full crankshaft rotations (720°) elapse before the start of each new working cycle at a given cylinder. If the pistons are offset uniformly with respect to each other, this means that

$$\text{angular ignition spacing [°]} = \frac{720°}{\text{No. of cylinders}}$$

On a 4-cylinder engine, the trigger wheel has 4 teeth (segments). This means that the crankshaft-speed sensor receives eight pulses for two revolutions of the crankshaft. The period of time between 2 pulses is termed the segment time, and the corresponding angle corresponds to half the angular ignition spacing.

Angle-of-rotation sensor (DWS)

A finely-toothed trigger wheel is firmly attached to the injection pump's driveshaft. It features a number of particularly large tooth gaps, spaced evenly around its circumference. The number of gaps corresponds to the number of engine cylinders. An angle-of-rotation sensor scans the succession of teeth and gaps (Fig. 5).

It must generate a signal which is a function of the cam ring's angular setting. Therefore, instead of being rigidly at-

Fig. 5

Angle-of-rotation sensor on a driveshaft (schematic)

1 Flexible conductive foil,
2 Angle-of-rotation sensor,
3 Trigger wheel,
4 Rotatable mounting ring,
5 Driveshaft.

tached to the driveshaft like the trigger wheel, it is mounted on the pump drive-shaft in such a manner that it can be turned relative to the shaft. When the timing device comes into action, the angle of rotation sensor is turned with the cam ring (the complete configuration is referred to as an incremental angle/time measuring system or IWZ).

Inside the injection pump, the signal from the angle-of-rotation sensor is transferred to the pump ECU via a flexible conductive foil.

The DWS signal is used for the following assignments:

- Defining the actual angular position,
- Measuring the actual pump speed, and
- Defining the timing-device setting (actual position).

The cam ring's current angular position is used to generate the triggering signal for the high-pressure solenoid valve. For the high-pressure solenoid valve to open and close at the corresponding cam lift, it is imperative that triggering takes place at the correct angle (Fig. 6).

The injection pump's current speed is the input variable for the pump ECU. It also serves as a substitute speed input for the engine ECU should the crankshaft-speed sensor fail.

The timing device's actual position is defined by comparing the signal from the crankshaft speed sensor with the angular setting of the angle-of-rotation sensor. It is needed for control of the timing device.

Needle-motion sensor

The needle-motion sensor is required on systems with start-of-injection control (for details, see the "Nozzles and nozzle holders" section). It determines the instant in time at which the needle in the injection nozzle opens, in other words the start of injection. The engine ECU is responsible for processing the signal from the needle-motion sensor.

Hot-film air-mass meter

In order to comply with the emission limits imposed by legislation, it is imperative that the desired A/F ratio is complied with exactly. This applies particularly when the IC engine is operating in the dynamic mode. To this end, sensors are required which precisely measure the air mass flow actually drawn in by the engine. The sensor's measuring accuracy must remain unaffected by pulsation, return flows, EGR, and variable camshaft control, as well as by changes in the intake-air temperature.

In the hot-film air-mass meter, heat is taken from a heated sensor element by heat transfer to the incoming air flow (Figs. 7 and 8). A micromechanical measuring system is used and, in combination with a hybrid circuit, registers the air mass flow and its direction. Reverse flows which take place during pronounced pulsation of the air flow are detected.

The micromechanical sensor element is located in the flow passage of the plug-in

Fig. 6

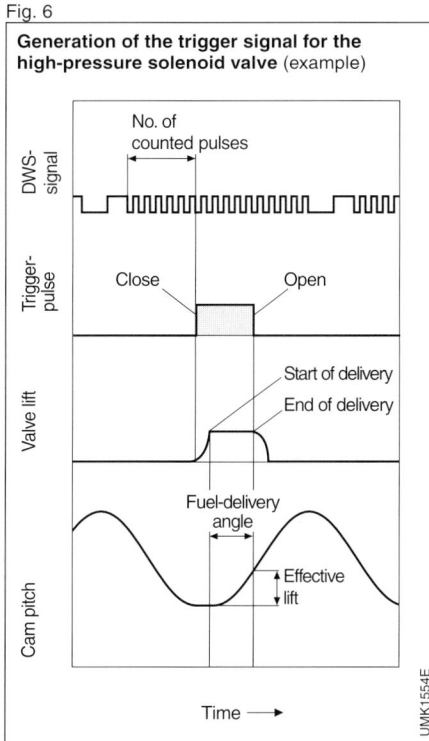

Generation of the trigger signal for the high-pressure solenoid valve (example)

UMK1554E

Sensor in the hot-film air-mass meter
1 Electrical terminations,
2 Electrical connections,
3 Evaluation electronics (hybrid circuit),
4 Air inlet,
5 Sensor element,
6 Air outlet,
7 Housing.

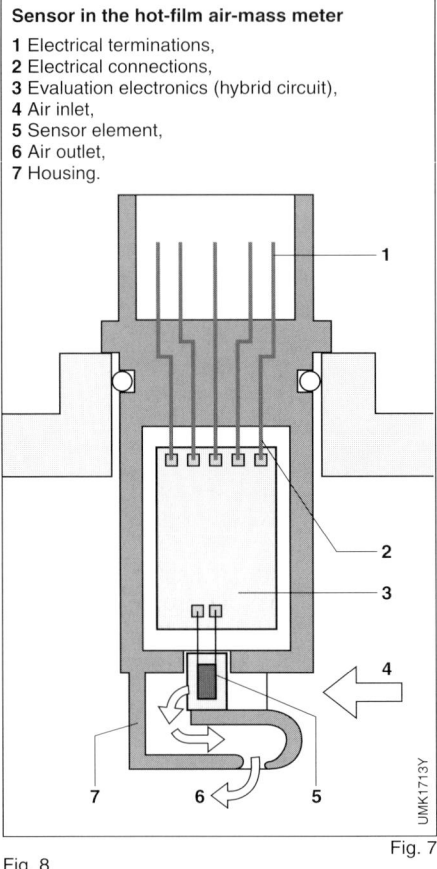

Fig. 7

Fig. 8

Hot-film air-mass meter
(measuring principle)
1 Temperature profile (no contact with air flow),
2 Temperature profile (contact with air flow),
3 M_1, M_2 measuring points, T_1, T_2 temperature signals, ΔT temperature difference results in measurement signal.

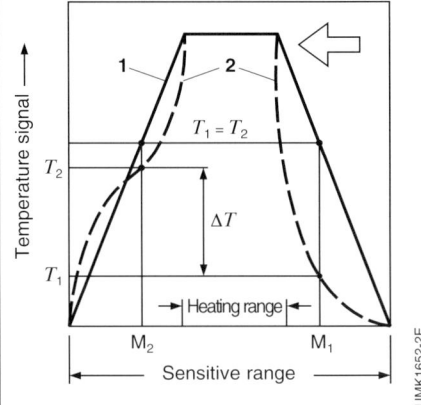

sensor (Fig. 7, Pos. 5). The plug-in sensor itself can be located in the air filter or in a measuring tube in the air inlet.

There are various sizes of measuring tube available depending upon the maximum air flow to be measured. The curve of the signal voltage as a function of incoming air flow is divided into signal sectors for forward and reverse flow. In order to increase the measuring accuracy, the measuring signal is referred to a reference voltage generated by the engine management. The characteristic curve is shaped so that in the workshop the engine management can be used for instance to diagnose an open circuit. A temperature sensor can be included to measure the temperature of the incoming air.

Accelerator-pedal sensor
In contrast to conventional distributor and in-line injection pumps, with EDC the driver's acceleration input is no longer transmitted to the injection pump by Bowden cable or mechanical linkage, but instead is registered by an accelerator-pedal sensor and transmitted to the ECU (this is also known as drive-by-wire). A voltage is generated across the potentiometer in the accelerator-pedal sensor and is a function of the accelerator-pedal setting. The pedal's position is then calculated using a programmed characteristic curve.

Boost-pressure sensor
The boost-pressure sensor (BPS) is pneumatically connected to the intake manifold and measures the intake manifold's absolute pressure between 0.5 and 3 bar.

The sensor is sub-divided into a pressure cell with two sensor elements, and a chamber for the evaluation circuit. The sensor elements and the evaluation circuit are mounted on a common ceramic substrate.

Each sensor element comprises a bell-shaped thick-film diaphragm, containing a reference volume with defined internal pressure. The diaphragm is displaced to a greater or lesser degree as a function of charge pressure.

"Piezoresistive" resistors are located on the diaphragm's surface, whose resistance changes when mechanical stress is applied. These resistors are connected as a bridge so that when the diaphragm moves this causes a change in the bridge balance. This means that the bridge voltage is a measure for the boost pressure.

The evaluation circuit is responsible for amplifying the bridge voltage, compensating for temperature influences, and linearization of the pressure characteristic. The evaluation circuit's output signal is inputted to the ECU where, with the help of a programmed characteristic curve, it is used for calculating the boost pressure.

Electronic control units (ECU's)

Operating conditions

A diesel fuel-injection installation with VR radial-piston distributor pump has two ECU's for electronic diesel control: A pump ECU and an engine ECU. This configuration is necessary so that on the one hand overheating of certain electronic components is avoided, and on the other to prevent the interference signals resulting from the very high currents (up to 20 A) which can flow in the injection pump.

The pump ECU registers the pump-internal sensor signals for the angle of rotation and fuel temperature and, together with the input from the engine ECU, processes these in order to adapt the instant of injection and the injected fuel quantity (Fig. 9). The engine ECU on the other

Fig. 9

Fig. 9: Control of the injection pump

To facilitate ease of understanding, a number of components have been turned with respect to their real position. **1** Engine ECU, **2** Pump ECU, **3** Angle-of-rotation sensor, **4** High-pressure solenoid valve, **5** Timing-device solenoid valve, **6** Needle-motion sensor in the injection nozzle.

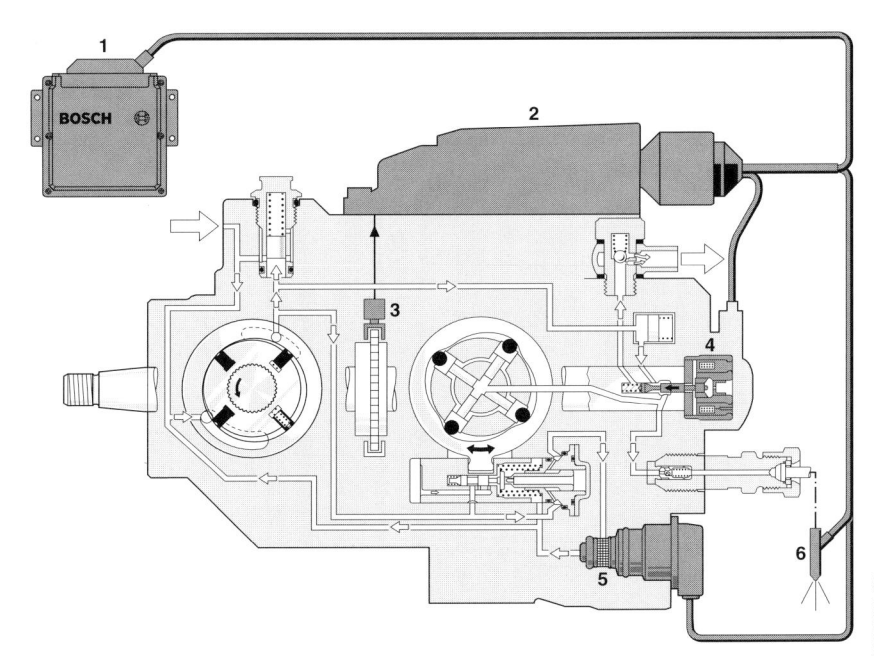

UMK1534-4Y

hand processes all the engine and environmental data registered by the external sensors, and uses this data to calculate the adjustments to be made at the engine. Each ECU is provided with specific maps for the respective operations.

The ECU input circuits condition the sensor signals, and while taking the operating conditions into account, the microprocessors use these to calculate the command signals for optimum vehicle operation.

Data exchange between engine and pump ECU uses the CAN bus system.

High demands are made upon the ECU regarding

– The surrounding (ambient) temperatures (in normal cases from – 40...+85 °C),
– Resistance to fuels and lubricants etc.,
– Resistance to humidity, and
– Mechanical loading.

Very high demands are also made upon electromagnetic compatibility (EMC) and upon the radiation of HF interference signals.

Pump ECU

Assignment

The pump ECU's prime assignment is that of an "intelligent" fuel-quantity adjuster. It triggers the timing device so that this adjusts to the correct start of delivery (closed-loop start-of-delivery control). In order to control the timing-device setting, the pump ECU uses the pulses from the engine-speed sensors or angle-of-rotation sensors as reference marks. The injected fuel quantity specified by the engine ECU is converted to a triggering period for the high-pressure solenoid valve. The rate of discharge (injected fuel quantity per degree camshaft) is also taken into account when the high-pressure solenoid is triggered. Due to the communication with the engine ECU, the pump ECU is included in the EDC safety and reliability concept.

Design and construction

The pump ECU is directly mounted on the pump and uses microhybrid techniques. It is provided with a 9-pole plug which connects it to the engine ECU, and through which all the communication between the two devices takes place. The pump ECU is cooled by fuel which runs through a passage underneath the ECU housing.

The only direct sensor inputs from the injection pump are the measuring signals from the angle-of-rotation sensor (DWS signal) and the input from the fuel-temperature sensor. In addition, the signal from the crankshaft-speed sensor, which has already been evaluated by the engine ECU, is also available for processing.

The housing of the pump ECU is sealed due to the ECU's exposed position on the injection pump.

Engine ECU

Assignment

The engine ECU evaluates the signals from the external sensors, and uses them to calculate the triggering signals for the actuators. It transmits the following parameters (which it has either received or calculated) to the pump ECU:
– The instantaneous crankshaft speed,
– the injected fuel quantity,
– the start of delivery, and
– the correct camshaft setting for the required rate of injection.

The engine ECU monitors the complete fuel-injection system within the framework of a safety and reliability concept.

Design and construction

The ECU is in a metal housing, and is connected to the sensors, power supply, and actuators through a multi-pole plug-in connection. Depending upon the type of ECU and its functional scope, the plug has between 105 and 134 pins.

The power components which directly trigger the actuators are integrated in the engine ECU so that they can efficiently dissipate their heat to the ECU housing. Both sealed and non-sealed versions of the ECU are available.

Operating-state control

In order that the engine operates with optimum combustion in every operating state, the engine ECU in each case calculates the appropriate injected fuel quantity. In the process, a number of parameters must be taken into account (Fig. 10).

Start quantity

For starting, the injected fuel quantity is calculated as a function of temperature and cranking speed. At low temperatures, in order to start reliably the engine needs far more fuel than is the case at operating temperature. At low temperatures, part of the fuel is deposited on the cold manifold and cylinder walls and is therefore not available for combustion. The start quantity is injected from the moment the starting switch is turned to "Start" (Fig. 10, Pos. A) until the engine has reached a given minimum speed. The driver has no influence upon the start quantity.

Drive mode

When the vehicle is being driven normally (Fig. 10, starting switch in Pos. B), the injected fuel quantity is calculated from the accelerator-pedal setting (accelerator-pedal sensor) and the engine speed. Calculation utilises the driving map so that the driver input and the engine O/P power are optimally matched to each other.

Idle-speed control

At idle, fuel consumption depends for the most part on engine efficiency and idle speed. Since a considerable portion of a vehicle's fuel consumption in dense traffic conditions is attributable to this operating state, it is obvious that idle speed must be kept to a minimum. It must be taken into account though that the idle speed must be set such that no matter what the operating conditions, it does not drop so far under load that the engine runs roughly or even stops. This applies for instance when the vehicle electrical system is loaded, when the air-conditioner is switched on, when a gear is engaged on

Fig. 10

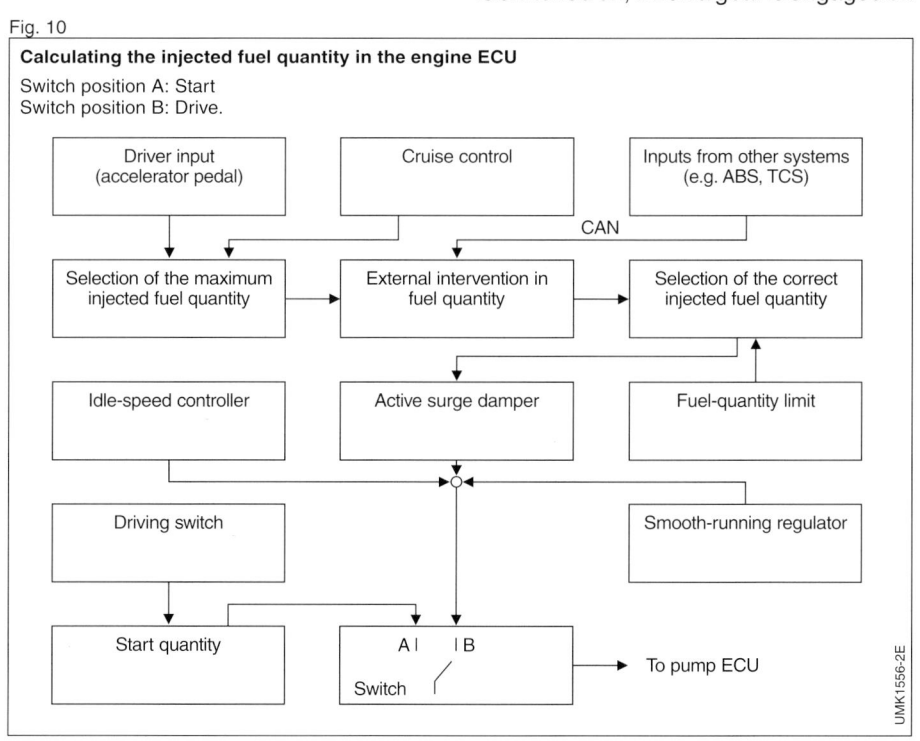

Calculating the injected fuel quantity in the engine ECU

Switch position A: Start
Switch position B: Drive.

an automatic transmission, or when the power steering is in operation.

In order to regulate to the desired idle speed, the idle controller varies the injected fuel quantity until the actual engine speed equals the desired idle speed. Here, the desired idle speed and the control characteristic are influenced by the selected gear and by the engine temperature (coolant-temperature sensor).

In addition to the external load moments, the internal friction moments must also be taken into account and compensated for by the idle-speed control. These change minimally but steadily throughout the vehicle's service life, as well as being highly dependent upon temperature.

Smooth-running control

Due to mechanical tolerances and ageing, there are differences in the torque generated by the engine's individual cylinders. Particularly at idle, this leads to rough or irregular running. The smooth-running (cylinder-balancing) control measures the engine-speed changes every time a cylinder has "fired" and compares them with each other. The injected fuel quantity for each cylinder is then adjusted in accordance with the measured differences in engine speed between the individual cylinders, so that each cylinder makes the same contribution to the torque generated by the engine. The smooth-running control is only operative in the lower engine-speed range.

Vehicle-speed controller

The vehicle-speed controller (Cruise Control) comes into operation when the vehicle is to be driven at a constant speed. It controls the vehicle speed to that selected by the driver at the operator unit in the instrument panel.

The injected fuel quantity is increased or reduced until the actual speed equals the set speed. The control process is interrupted if the driver depresses the clutch, or applies the brakes, while the Cruise Control is in operation. If the accelerator pedal is pressed, the vehicle can be accelerated beyond the speed which has been set with the Cruise Control. As soon as the accelerator pedal is released, the Cruise Control regulates the speed back down again to the previous set speed. Similarly, if the Cruise Control has been switched off, the driver only needs to press the reactivate key in order to again select the last speed which had been set.

Controlling the injected fuel quantity limit

For a number of reasons, it is undesirable that the injected fuel quantity as selected by the driver (or the maximum physically possible quantity) is always injected. The reasons include:

– Excessive pollutant emissions,
– Excessive soot emissions due to the injection of too much fuel,
– Mechanical overloading due to excessive torque or excessive engine speed, or
– Thermal overload as a result of excessive coolant, lube-oil, or turbocharger temperature.

The limit for the injected fuel quantity is formed from a number of input variables, for instance intake air mass, engine speed, and coolant temperature.

Fig. 11

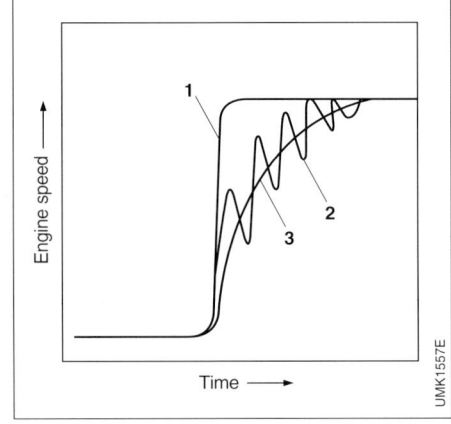

Active surge damper

1 Sudden operation of accelerator pedal (driver input), **2** Engine-speed curve without active surge control, **3** Speed curve with active surge control.

Engine speed ⟶

Time ⟶

UMK1557E

Surge-damping control

When the accelerator pedal is abruptly depresssed or released, this causes the injected fuel quantity to change rapidly with the result that there is also a rapid change in the torque developed by the engine. These abrupt load changes lead to the resilient engine mountings and the drivetrain generating bucking oscillations which result in fluctuations of engine speed (Fig. 11).

The surge-damping control reduces these periodic speed fluctuations by varying the injected fuel quantity at the same frequency as the periodic speed fluctuations: Less fuel is injected when the speed increases, and more when it decreases. This effectively damps the surge movements.

Start-of-injection control

The start of injection has considerable influence upon power output, fuel consumption, noise, and exhaust emissions. Its desired (or setpoint) value is stored in the engine ECU as a function of engine speed and injected fuel quantity. Correction can also be made as a function of coolant temperature. For optimum start of injection, the scatter bands of the nitrogen-oxide emissions (NO_x) and hydrocarbon emissions (HC) must also be taken into account (Fig. 12). In order to ascertain the actual start-of-injection point, the needle-motion sensor's signal is evaluated. If the actual value (of start of injection) deviates from its desired value, the engine ECU changes the desired value of the start of delivery accordingly. The start of delivery is defined as that instant at which the high-pressure solenoid valve in the distributor head closes. Its desired value is calculated from the desired value for the start of injection, whereby the time needed for the pressure wave to pass through the line and reach the injectors (the delay) is taken into account.

If the needle-motion sensor is defective, or not installed in the system in question, it is still possible to operate the radial-piston distributor injection pump, although start-of-injection tolerances are not so tight.

Engine shutoff

The diesel engine operates according to the "auto-ignition" principle. This means that it can only be switched off by interrupting its supply of fuel.

In the case of the electronic diesel control (EDC), the engine is switched off by the ECU stipulating "injected fuel quantity zero".

Actuators

High-pressure solenoid valve

For fuel-quantity metering, a high-pressure solenoid valve is integrated in the injection pump's high-pressure stage. At the beginning of each injection process, the coil is energized and the solenoid, together with the valve needle, is forced in the direction of the valve seat. As soon as the valve seat has been closed off completely by the valve needle, no more fuel can flow. The fuel pressure in the high-pressure stage increases sharply as a result and opens the injection nozzle in question. As soon as the required amount of fuel has been injected, the solenoid is de-energized, the high-pressure solenoid valve opens again, and the pressure in the high-pressure stage collapses.

Fig. 12

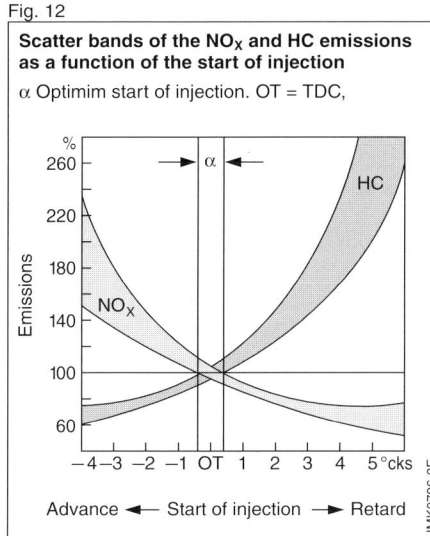

Scatter bands of the NO_X and HC emissions as a function of the start of injection

α Optimim start of injection. OT = TDC,

The reduction in injection pressure causes the injection nozzle to close again, and fuel injection is terminated.

In order to control this process even more precisely, the pump ECU can use the current curve to define the actual instant of closing of the high-pressure solenoid valve (Fig. 13).

Timing-device solenoid valve

The pump ECU controls the timing-device piston through the timing-device solenoid valve (Fig. 14) which is permanently pulsed by a constant-frequency control current.

Here, the flow quantity is determined by the ratio of pulsed time to non-pulsed time (on/off ratio). It is possible to vary the flow quantity so that the timing device moves to its desired position.

Glow control unit

The glow control unit is responsible for ensuring efficient cold starting. It also shortens the warm-up period, a fact which is highly relevant for exhaust emissions.

The preheating time is a function of the coolant temperature. The further glow phases during engine start or when the engine is actually running are determined by a number of parameters which include engine speed and injected fuel quantity. Glow control utilises a power relay.

Electropneumatic transducer

The valves or flaps of the swirl controller, EGR positioner, and boost-pressure actuator are actuated mechanically using overpressure or negative pressure. Here, the engine ECU generates an electrical signal which is converted to overpressure or negative pressure by an electropneumatic transducer.

Boost-pressure actuator
Passenger-car engines with exhaust-gas turbocharging must develop high torques even at low engine speeds. The turbocharger housing is therefore designed for a low exhaust-gas mass flow. But in order that excessive charge-air pressure is not developed when larger exhaust-gas masses flow, part of this flow must be

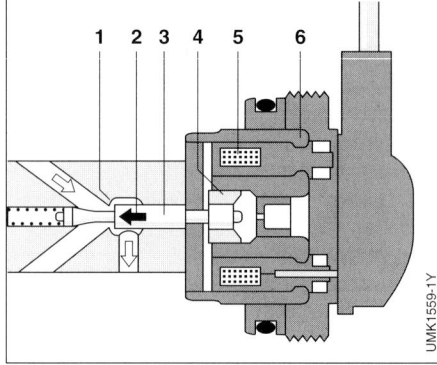

High-pressure solenoid valve

1 Valve seat, 2 Direction of closing,
3 Valve needle, 4 Solenoid armature,
5 Coil, 6 Solenoid.

Fig. 13

Fig. 14

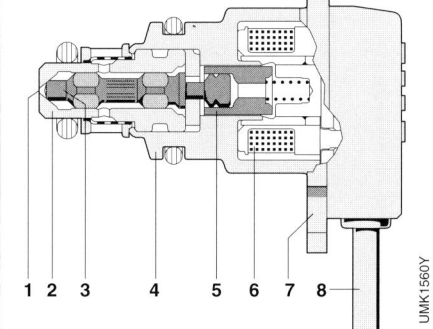

Timing-device solenoid valve

1 Throttle bore, 2 Valve body, 3 Valve needle,
4 Valve housing, 5 Solenoid armature,
6 Solenoid coil, 7 Mounting flange,
8 Electrical connection.

Fig. 15

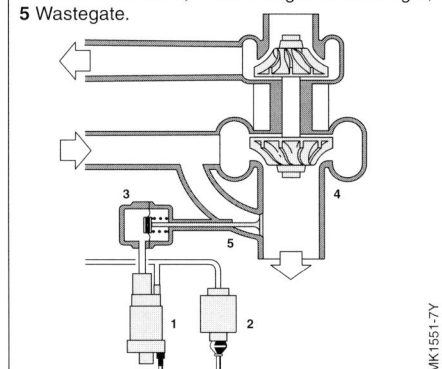

Boost-pressure actuator

1 Boost-pressure actuator, 2 Vacuum pump,
3 Pressure actuator, 4 Exhaust-gas turbocharger,
5 Wastegate.

diverted past the turbine by means of a bypass valve ("wastegate") and into the exhaust system. To do so, the boost-pressure actuator (Fig. 15) changes the cross section at the wastegate as a function of engine speed and injected fuel quantity etc. Variable turbine geometry (VTG) can be applied instead of the wastegate. This varies the approach angle of the turbine wheel and as a result influences the charge-air pressure.

Swirl controller

Swirl control serves to influence the swirl movement of the intake air. The swirl itself is usually generated by spiral-shaped inlet passages and determines the mixing of the fuel and the air in the combustion chamber. It therefore has considerable influence upon the combustion quality. As a rule, a pronounced swirl is generated at low engine speeds, and a weak swirl at high speeds. The swirl can be modified by means of the swirl controller (flap or slide valve) in the vicinity of the intake valve.

EGR positioner

With exhaust-gas recirculation (EGR) a portion of the exhaust gas is led into the intake tract. Up to a certain degree, an increasing portion of the residual exhaust-gas content has a positive effect upon energy conversion and therefore upon the exhaust-gas emissions. Depending upon the engine's operating point, the air/gas mass drawn into the cylinders can be composed of up to 40% exhaust gas (Figs. 16 and 17). For ECU control, the actual drawn-in fresh-air mass is measured and compared at each operating point with the air-mass setpoint value. Using the signal generated by the control circuit, the EGR positioner (a valve) opens so that exhaust gas can flow into the intake tract.

Throttle-valve control

The throttle valve in the diesel engine fulfils a completely different function to that in the spark-ignition (SI) engine. It serves to increase the exhaust-gas recirculation rate by reducing the overpressure in the intake manifold. Throttle-valve control is only operative in the lower speed range.

Fig. 16

Influence of exhaust-gas recirculation rate (EGR rate) upon pollutant emissions

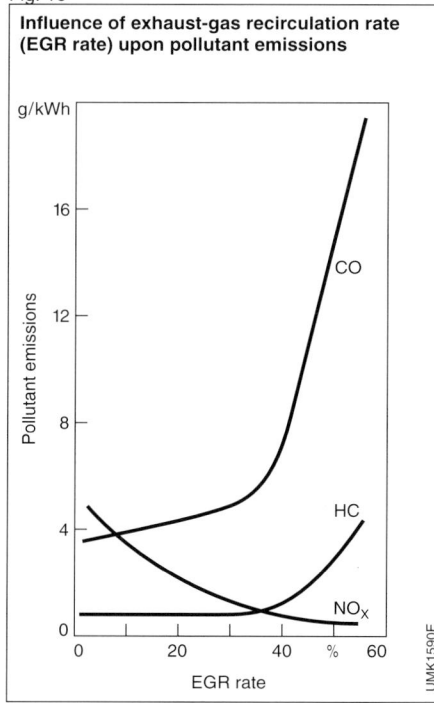

Fig. 17

Influence of exhaust-gas recirculation rate (EGR rate) upon excess-air factor λ, soot emissions, and fuel consumption

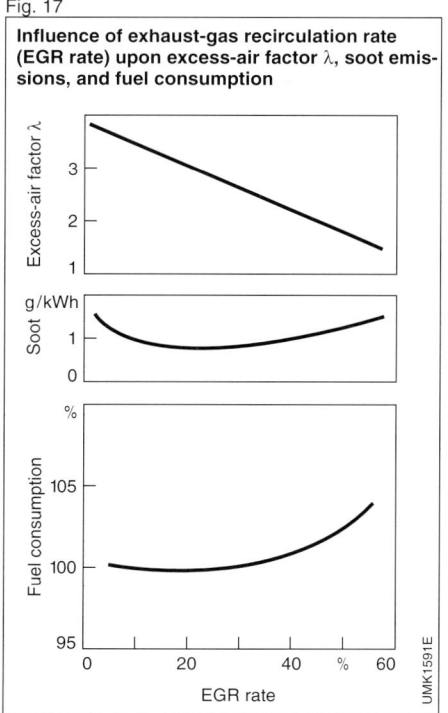

Information exchange

Communication between the ECU's

Communication between the engine ECU and the pump ECU takes place using CAN-Bus (Controller Area Network). This is used to transmit the desired values, operating data, and status information needed for operation and for fault monitoring.

There are three messages defined from the engine ECU to the pump ECU (MSG1 to MSG3), and three from the pump ECU to the engine ECU (PSG1 to PSG3).

Engine ECU messages

The MSG1 message from engine ECU to pump ECU contains

– The injected fuel quantity
  (desired value)
– The start of injection
  (referred to camshaft and crankshaft)
– The engine speed.

Using the desired value for injected fuel quantity, the pump ECU defines the triggering time for the high-pressure solenoid valve. The start of delivery referred to the crankshaft is needed for the calculation of the start of injection. On the other hand, the start of delivery referred to the camshaft is used to vary the fuel's rate of discharge.

The engine speed is used for monitoring purposes and to this end is compared with the fuel-injection pump's rotational speed.

The engine ECU is provided with a diagnosis plug in order that the data from both ECU's can be called up (e.g. for access to stored fault messages), for instance during vehicle inspection or customer service. The engine ECU passes on the data enquiries to the pump ECU using the MSG2 message.

The MSG3 message provides the pump ECU with the information on the position of the crankshaft-speed sensor and on the engine ECU configuration.

Messages from the pump ECU

The PSG1 message from the pump ECU to the engine ECU contains data on
– The duration of triggering of the high-pressure solenoid valve,
– The injection-pump speed,
– The injection-pump temperature, and
– Fault messages.

The information on the duration of triggering of the high-pressure solenoid valve and on the injection-pump speed are applied in the engine ECU for monitoring purposes. The injection-pump temperature is an additional actuating variable applied in the calculation of the start of delivery and of the duration of triggering. Using the message PSG 2, the pump ECU transmits back to the engine ECU the information it demanded with the message MSG2. At the engine ECU the pump ECU's data can be read out through the diagnosis plug.

A self-test is performed every time the pump ECU is "reset". The results are sent with the message PSG3 to the engine ECU which answers with the message MSG3.

Exchange of information with other systems

External intervention in injected fuel quantity

In the case of external intervention, the injected fuel quantity is influenced by another ECU (e.g. ABS, TCS). This informs the engine ECU whether the engine torque (and therefore the injected fuel quantity) is to be changed, and if so by how much.

Electronic immobilizer

For the purposes of theft deterrence, the vehicle can be protected by means of an extra immobilizer ECU which prevents the engine being started. Using remote control for instance, the driver can signal to the vehicle that he is authorized to use it. The immobilizer ECU then informs the engine ECU that fuel is required so that the vehicle can be started and driven.

Air-conditioner

When outside temperatures are high, the air-conditioner uses a refrigeration compressor to cool the air inside the vehicle to a pleasant temperature.

Depending upon the engine and the particular driving situations, the air-conditioners power consumption can be between 1 and 30% of the engine's output power. The target is to make maximum use of the engine's torque and not of the temperature control's power.

As soon as the driver accelerates heavily (in other words he wants maximum engine torque), the EDC switches of the refrigeration compressor for a brief period.

Integrated diagnosis

Sensor monitoring

In sensor monitoring the sensors are checked by means of the integrated diagnosis facility to ensure that they are adequately supplied and that their output signals are within the permissible range (for instance, temperature between $-40°C$ and $+150°C$). As far as possible, important signals are available twice (redundant principle), which means that there is always the possibility of switching to another similar signal in case of fault.

Monitoring module

In addition to the microcontroller, the engine ECU also has a monitoring module. ECU and monitoring module monitor each other. If an error is detected in the process, each of them can switch off the vehicle independent of the other.

Fault monitoring

Fault monitoring is only possible within the monitoring range of the particular sensor. A signal path is identified as faulty as soon as a fault is present for longer than a predefined period. The fault is then stored in the engine ECU's fault memory, together with the surrounding conditions which prevailed when it occurred (e.g. coolant temperature, speed etc.). For a number of faults, it is possible for the signal path to be assessed as serviceable again in such cases in which it remains without fault for a defined period of time.

Fault procedure

One differentiates between a variety of different system reactions depending upon the severity of the fault:
− Switch-over to a substitute value,
− Reversible shutoff,
− Irreversible shutoff.

Switch-over to a substitute value

As soon as a given sensor's permissible signal range is exceeded, switch-over takes place to a substitute value.

This procedure is applied for the following input signals:
− Battery voltage,
− Coolant, air, lube-oil temperature,
− Boost pressure,
− Atmospheric pressure, and
− Intake-air quantity.

In addition, an accelerator-pedal-sensor substitute signal is applied in case of plausibility violation of the signals from accelerator-pedal sensor and brake.

Reversible shutoff

The MAB line (fuel shutoff) enables the engine ECU to intervene directly in the output stage of the high-pressure solenoid valve and suppress its triggering so that no more fuel can be injected. This measure is reversible.

This means that fuel can be made available again for injection as soon as the condition which led to its shutoff is no longer present (for instance, in case of permanent trigger monitoring during overrun). When double the injection-pump speed is compared with the speed of the engine, and the deviation exceeds a given threshold, the vehicle is also reversibly shutoff.

Irreversible shutoff

Main-relay shutoff is irreversible. Shutoff takes place only when the fault "High-pressure solenoid valve permanently triggered" is reported. This method must be used since vehicle shutoff using "Zero fuel quantity" or via the MAB line is in this case impossible.

Nozzles and nozzle holders

Hole-type nozzles

Assignments and designs

The injection nozzles and their respective nozzle holders are vitally important components situated between the radial-piston pump and the diesel engine. Their assignments are as follows:
– Metering the injection of fuel,
– Management of the fuel mixture,
– Defining the rate-of-discharge curve,
– Sealing-off against the combustion chamber.

Considering the wide variety of combustion processes and the different forms of combustion chamber, it is necessary that the shape, "penetration force", and atomization of the fuel spray injected by the nozzle are adapted to the prevailing conditions. This also applies to the injection time, and the injected fuel quantity per degree camshaft.

Since the design of the nozzle holder makes maximum use of standardized components and assemblies, this means that the required flexibility can be achieved with a minimum of components.

Fig. 1

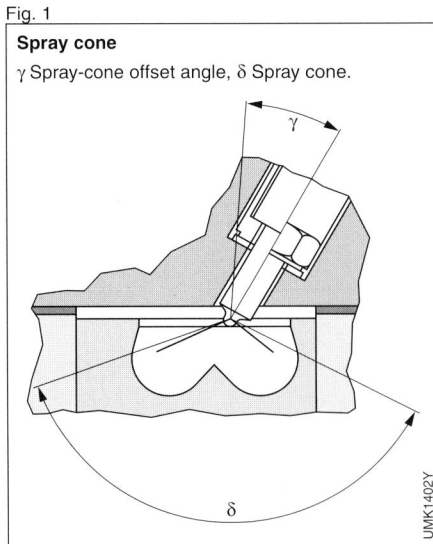

Spray cone

γ Spray-cone offset angle, δ Spray cone.

Application

Hole-type nozzles are used for direct-injection (DI) engines. One differentiates between sac-hole and seat-hole nozzles. Hole-type nozzles also vary according to their size:
– <u>Type P</u> with 4 mm needle diameter, and
– <u>Type S</u> with 5 and 6 mm needle diameters.

Design and construction

The spray holes are located on the envelope of a spray cone (Fig. 1). The number of the spray holes and their diameter depend upon:
– The injected fuel quantity,
– The combustion-chamber shape, and
– The air swirl in the combustion chamber.

The input edges of the spray holes can be rounded by hydro-erosive (HE) machining.
At those points where high flow rates occur (spray-hole entrance), the abrasive particles in a hydro-erosive (HE) medium cause material loss.

The so-called HE-rounding process can be applied to both sac-hole and seat-hole nozzles, whereby the target is:
– Prevent in advance the edge wear caused by the abrasive particles in the fuel and/or
– Tighten the flow tolerance.

To achieve low hydrocarbon emissions, it is highly important that the volume filled with fuel (residual volume) below the edge of the nozzle-needle seat is kept to a minimum. This is achieved by using seat-hole nozzles.

Designs

Sac-hole nozzles

The spray holes of the sac-hole nozzle (Fig. 2) are arranged in the sac hole.
In the case of a round nozzle tip (Fig. 3), depending upon the design the spray holes are drilled mechanically or by means of electrical-discharge machining (EDM electrical particle removal).

Sac-hole nozzles with conical tip are always with spray-hole lengths of 0.6 mm.

Sac-hole nozzles with conical tip are always drilled using EDM.

The sac-hole nozzles are available with cylindrical and conical sac holes in a variety of different dimensions.

The sac-hole nozzle with cylindrical sac hole and round tip (Fig. 3a), comprising a cylindrical and hemispherical portion, permits a high level of design freedom with respect to number of spray holes, spray-hole length, and injection angle. The nozzle tip is hemispherical, and together with the shape of the sac hole, ensures that the spray holes are of identical length.

Sac-hole nozzle with cylindrical sac hole and conical tip (Fig. 3b) is used exclu-

sively with spray-hole lengths of 0.6 mm. The tip's conical shape enables the wall thickness to be increased between the throat radius and the nozzle-body seat. This results in an improvement of nozzle-tip strength.

In the sac-hole nozzle with conical sac hole and conical tip (Fig. 3c), the sac-hole volume is less than that for a nozzle with cylindrical sac hole. The volume is between that for a seat-hole nozzle and a sac-hole nozzle with cylindrical sac hole. In order to achieve uniform tip-wall thickness, the tip's conical design corresponds to that of the sac hole.

Fig. 2

Sac-hole nozzle

1 Pressure pin, 2 Needle-lift stop face,
3 Inlet passage, 4 Exposed annular area,
5 Needle shaft, 6 Nozzle tip,
7 Nozzle-body shaft, 8 Nozzle-body shoulder,
9 Pressure chamber, 10 Needle guide,
11 Nozzle-body collar, 12 Locating hole,
13 Sealing surface,
14 Pressure-pin contact surface.

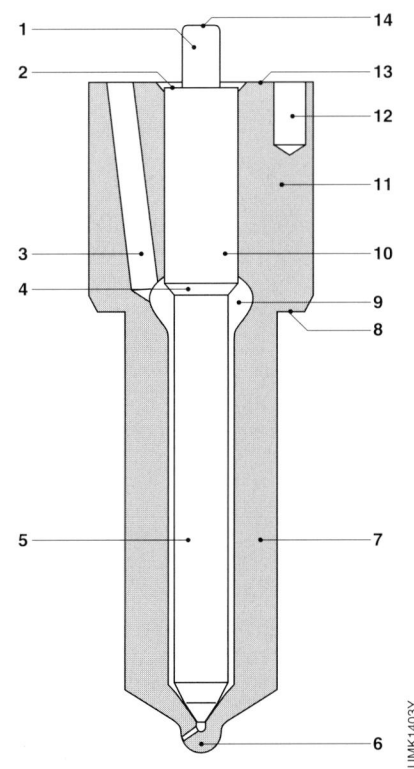

UMK1403Y

Fig. 3

Sac-hole shapes

a Cylindrical sac hole with round tip,
b Cylindrical sac hole with conical tip,
c Conical sac hole with conical tip.
1 Shoulder, 2 Seat entrance, 3 Needle seat,
4 Needle tip, 5 Injection orifice,
6 Injection-orifice entrance, 7 Sac hole,
8 Throat radius, 9 Nozzle-tip cone,
10 Nozzle-body seat, 11 Damping cone.

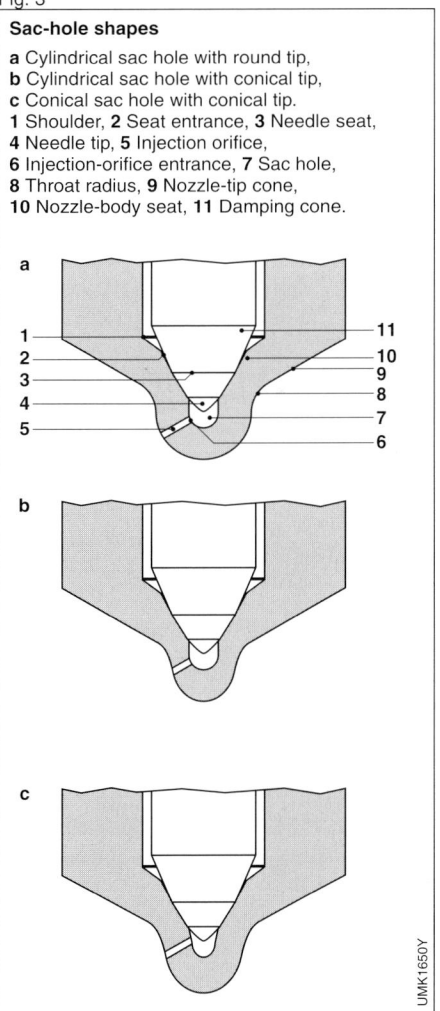

UMK1650Y

Seat-hole nozzle

In order to minimise the residual volume – and therefore the HC emissions – the start of the spray hole is located in the seat taper, and with the nozzle closed it is covered by the nozzle needle. This means that there is no direct connection between the sac hole and the combustion chamber (Figs. 4 and 5). The sac-hole volume here is much lower than that in the sac-hole nozzle. Compared to sac-hole nozzles, seat-hole nozzles have a much lower loading limit and are therefore only manufactured as Size P with a spray-hole length of 1 mm.

For reasons of strength, the nozzle tip is conically shaped. The spray holes are always formed by electrical discharge machining (EDM) methods.

Fig. 4

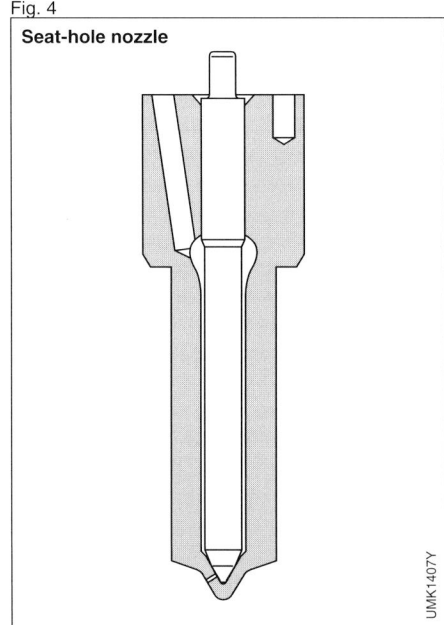

Seat-hole nozzle

UMK1407Y

Fig. 5

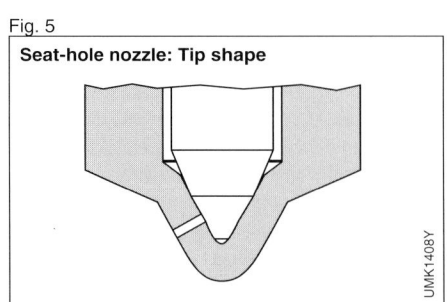

Seat-hole nozzle: Tip shape

UMK1408Y

Standard nozzle holders

Designs

Nozzle holders with hole-type nozzles in combination with a radial-piston pump are used for the injection on direct-injection engines.

With regard to the nozzle holders, one differentiates between
– Standard nozzle holders
  (single-spring nozzle holders) with and without needle-motion sensor, and
– Two-spring nozzle holders, with and without needle-motion sensor.

Application

The nozzle holders described here have the following characteristics:

– Cylindrical external shape with diameters between 17 and 21 mm,
– Bottom-mounted springs
  (leads to low moving masses),
– Fixed nozzles for direct-injection engines, and
– Standardised components (springs, pressure pin, nozzle-retaining nut) make combinations a practical matter.

Design

The nozzle-and-holder assembly comprises the injection nozzle and the nozzle holder.

The nozzle holder comprises the following components (Fig. 6):

– Nozzle-holder body,
– Intermediate element,
– Nozzle-retaining nut,
– Pressure pin,
– Spring,
– Shim, and
– Locating pins.

The nozzle is centered in the nozzle body and fastened using the nozzle-retaining nut. When nozzle body and retaining nut are screwed together the intermediate element is forced up against the sealing surfaces of nozzle body and retaining nut. The intermediate element serves as

the needle-lift stop, and centers the nozzle in the nozzle-holder body with its locating pins.
The nozzle-holder body contains the
– Pressure pin,
– Spring, and
– Shim.

The spring is centered by the pressure pin, whereby pressure-pin guidance is taken over by the nozzle-needle's pressure pintle.
The nozzle is connected to the injection pump's high-pressure line via the nozzle-holder feed passage, the intermediate element, and the nozzle-body feed passage. If required, an edge-type filter can be installed in the nozzle holder.

Method of operation
The spring applies pressure to the nozzle needle through the pressure pin. The spring's initial tension defines the nozzle's opening pressure which can be adjusted using a shim.
On its way to the nozzle seat, the fuel passes through the nozzle-holder inlet passage, the intermediate element, and the nozzle body. When injection takes place, the nozzle needle is lifted by the injection pressure and fuel is injected through the injection orifices into the combustion chamber. Injection terminates as soon as the injection pressure drops far enough for the nozzle spring to force the nozzle needle back onto its seat.

Two-spring nozzle holders

Application
The two-spring nozzle holder is a further development of the standard nozzle holder, and serves to reduce combustion noise particularly in the idle and part-load ranges.

Design
The two-spring nozzle holder features two springs located one behind the other. At first, only one of these springs has an influence on the nozzle needle and as

Fig. 6

Standard nozzle holder
1 Edge-type filter, **2** Inlet passage,
3 Pressure pin, **4** Intermediate element,
5 Nozzle-retaining nut, **6** Wall thickness,
7 Nozzle, **8** Locating pins, **9** Spring,
10 Shim, **11** Leak-fuel passage,
12 Leak-fuel connection thread,
13 Nozzle-holder body, **14** Connection thread,
15 Sealing cone.

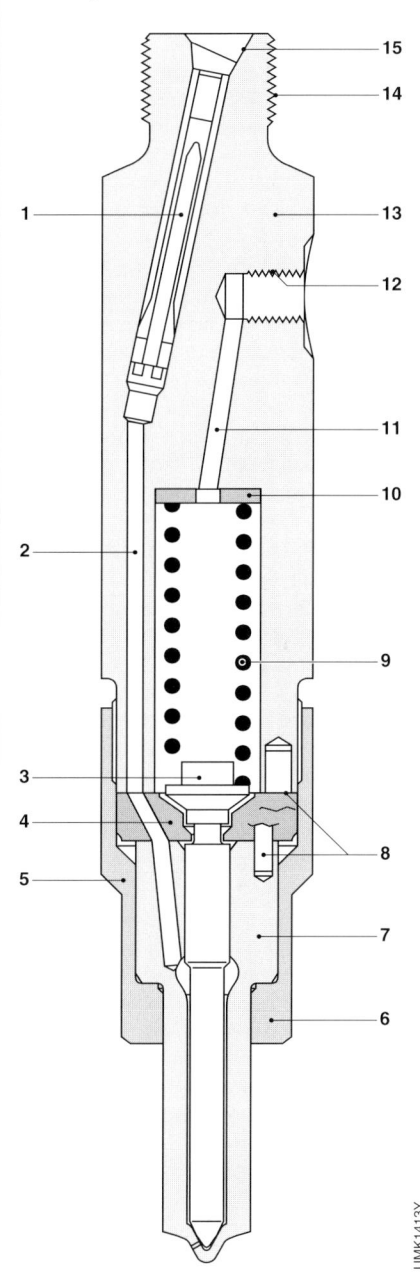

UMK1413Y

Fig. 7

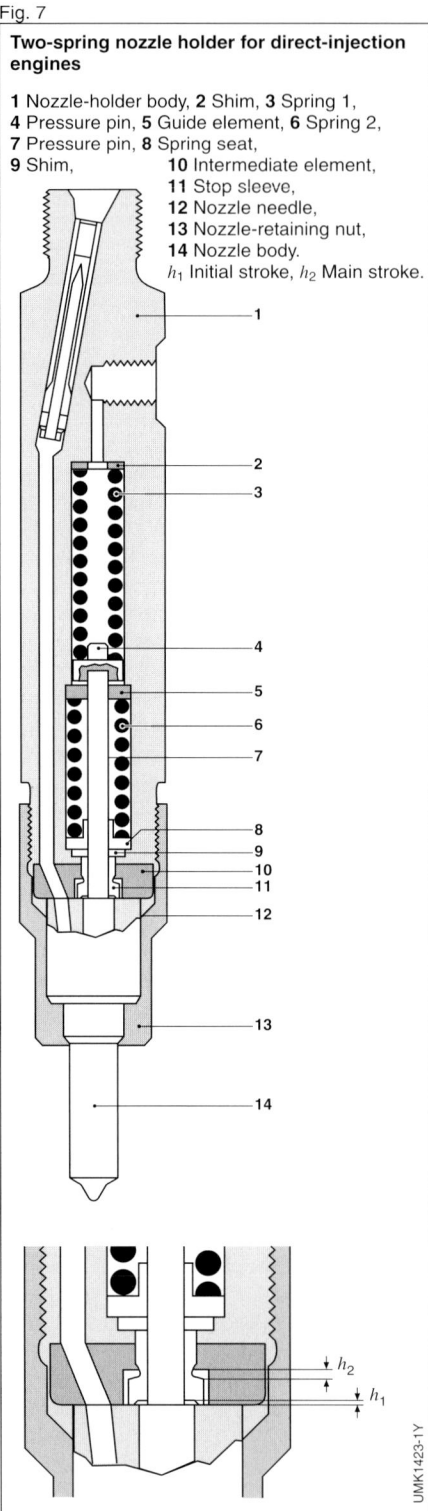

Two-spring nozzle holder for direct-injection engines

1 Nozzle-holder body, **2** Shim, **3** Spring 1,
4 Pressure pin, **5** Guide element, **6** Spring 2,
7 Pressure pin, **8** Spring seat,
9 Shim,
10 Intermediate element,
11 Stop sleeve,
12 Nozzle needle,
13 Nozzle-retaining nut,
14 Nozzle body.
h_1 Initial stroke, h_2 Main stroke.

UMK1423-1Y

such defines the initial opening pressure. The second spring is in contact with a stop sleeve which limits the needle's initial stroke.

When strokes take place in excess of the initial stroke, the stop sleeve lifts and both springs have an effect upon the nozzle needle (Fig. 7).

Method of operation

During the actual injection process, the nozzle needle opens only the initial amount. This permits only a small amount of fuel to enter the combustion chamber.

Along with increasing injection pressure in the nozzle holder though, the nozzle needle opens completely and the principle quantity is injected (Fig. 8). This 2-stage rate-of-discharge curve leads to "softer" combustion and to a reduction in noise.

Nozzle holder with needle-motion sensor

Application

The start-of-injection point is an important parameter for the optimum operation of the diesel engine. For instance, its

Fig. 8

Comparison of needle-stroke curves

a Standard nozzle holder (single-spring nozzle holder), **b** Two-spring nozzle holder.
h_1 Initial stroke, h_2 Main stroke.

Nozzle-needle lift

a
mm

0.4
0.2
0

b
mm

0.4
0.2
0

0   1   2 ms
h_1 Time

h_2

UMK1422E

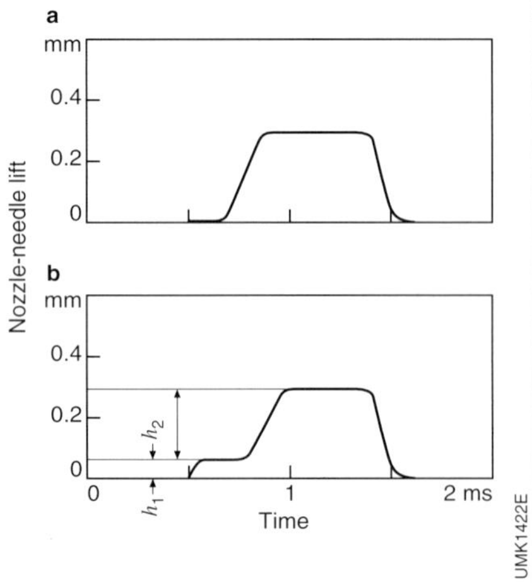

evaluation permits load and speed-dependent injection timing, and/or control of the exhaust-gas recirculation (EGR) rate.

This necessitates a nozzle holder with needle-motion sensor (Fig. 10) which outputs a signal as soon as the nozzle needle opens.

Design

The extended pressure pin enters the current coil.

The degree to which it enters the coil (overlap length "X" in Fig. 11) determines the strength of the magnetic flux.

Method of operation

Due to nozzle-needle movement the magnetic flux in the coil changes and induces a signal voltage which is proportional to the speed of movement but not to the distance travelled. This signal is processed directly in an evaluation circuit (Fig. 9).

When a given threshold voltage is exceeded, this serves as the signal for the start of injection.

Two-spring nozzle holder with needle-motion sensor for direct-injection engines

1 Nozzle-holder body, 2 Needle-motion sensor, 3 Spring 1, 4 Guide element, 5 Spring 2, 6 Pressure pin, 7 Nozzle-retaining nut.

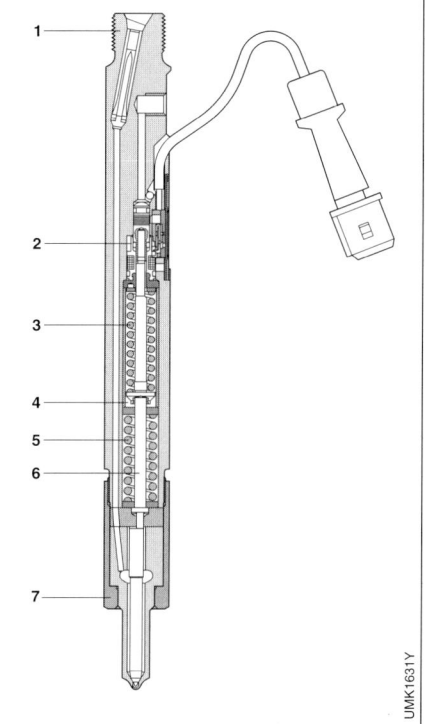

Fig. 10

Fig. 9

Comparison between a needle-stroke curve and the corresponding signal-voltage curve of the needle-motion sensor

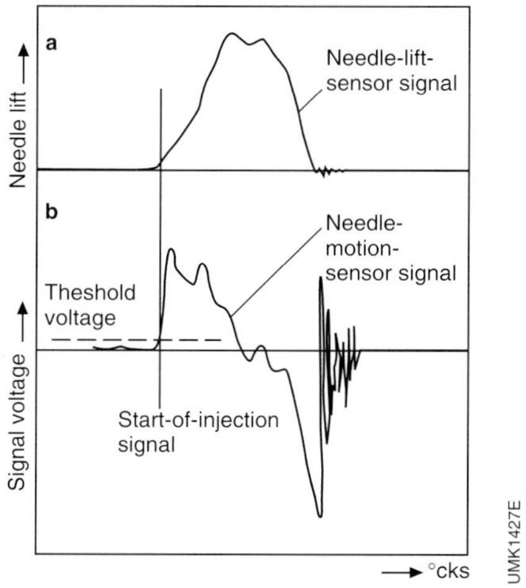

a Needle-lift-sensor signal

b Needle-motion-sensor signal

Threshold voltage

Start-of-injection signal

Needle lift

Signal voltage

→ °cks

UMK1427E

Fig. 11

Needle-motion sensor in a two-spring nozzle holder for direct-injection engines

1 Adjusting pin, 2 Terminal, 3 Sensor coil, 4 Pressure pin, 5 Spring seat. X Overlap length.

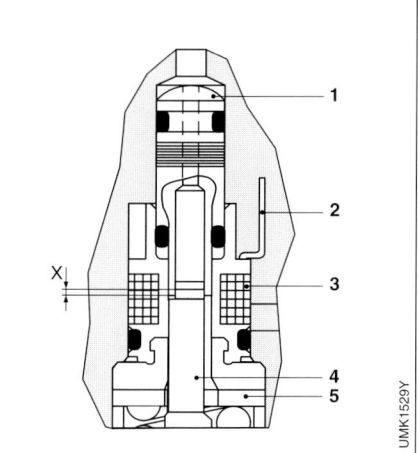

UMK1529Y

Common Rail accumulator fuel-injection system

System overview

Field of application

The introduction of the first series-production in-line fuel-injection pump in 1927 marked the beginning of diesel fuel-injection system manufacture at Bosch. The in-line fuel-injection pump's main area of application is still in all sizes of commercial-vehicle diesel engines, stationary diesel engines, locomotives and ships. Injection pressures of up to approx. 1350 bar are used to generate output powers of up to about 160 kW per cylinder.

Over the years, a wide variety of different requirements, such as the installation of direct-injection (DI) engines in small delivery vans and passenger cars, have led to the development of various diesel fuel-injection systems which are aligned to the requirements of a particular application. Of major importance in these developments are not only the increase in specific power, but also the demand for reduced fuel consumption, and the call for lower noise and exhaust-gas emissions. Compared to conventional cam-driven systems, the Bosch "Common Rail" fuel-injection system for direct-injection (DI) diesel engines provides for considerably higher flexibility in the adaptation of the injection system to the engine, for instance:

– Extensive area of application (for passenger cars and light commercial vehicles with output powers of up to 30 kW/cylinder, as well as for heavy-duty vehicles, locomotives, and ships with outputs of up to approx. 200 kW/cylinder,
– High injection pressures of up to approx. 1400 bar,
– Variable start of injection,
– Possibility of pilot injection, main injection, and post injection,
– Matching of injection pressure to the operating mode.

Functions

Pressure generation and fuel injection are completely decoupled from each other in the "Common Rail" accumulator injection system. The injection pressure is generated independent of engine speed and injected fuel quantity. The fuel is stored under pressure in the high-pressure accumulator (the "Rail") ready for injection. The injected fuel quantity is defined by the driver, and the start of injection and injection pressure are calculated by the ECU on the basis of the stored maps. The ECU then triggers the solenoid valves so that the injector (injection unit) at each engine cylinder injects accordingly. The ECU and sensor stages of such a CR fuel-injection system comprise:
– ECU,
– Crankshaft-speed sensor,
– Camshaft-speed sensor,
– Accelerator-pedal sensor,
– Boost-pressure sensor,
– Rail-pressure sensor,
– Coolant sensor and
– Air-mass meter.
Using the input signals from the above sensors, the ECU registers the driver's requirements (accelerator-pedal setting) and defines the instantaneous operating performance of the engine and the vehicle as a whole. It processes the signals which have been generated by the sensors and which it receives via data lines. On the basis of this information, it can then intervene with open and closed-loop controlling action at the vehicle and particularly at the engine. The engine speed is measured by the crankshaft-speed sensor, and the camshaft-speed sensor determines the firing sequence (phase length). The electrical signal generated across a potentiometer in the accelerator-pedal module informs the ECU about how far the driver has depressed the

pedal, in other words about his (her) torque requirement.

The air-mass meter provides the ECU with data on the instantaneous air flow in order that combustion can be adapted so as to comply with the emissions regulations. Insofar as the engine is equipped with an exhaust-gas turbocharger and boost-pressure control, the boost-pressure sensor also measures boost-pressure. At low outside temperatures and with the engine cold, the ECU applies the data from the coolant-temperature and air-temperature sensors to adapt the setpoint values for start of injection, post injection, and further parameters to the particular operating conditions. Depending upon the vehicle in question, in order to comply with the increasing demands for safety and comfort, further sensors and data lines provide inputs to the ECU.

Fig. 1 shows an example of a 4-cylinder diesel engine fitted with a fuel-injection installation using the "Common Rail" accumulator injection system. Various components are shown

Basic functions

The basic functions control the injection of the diesel fuel at the right moment, in the right quantities, and with the correct injection pressure. They ensure that the diesel engine not only runs smoothly, but also economically.

Auxiliary functions

Auxiliary closed and open-loop control functions serve to improve both the exhaust-gas emission and fuel-consumption figures, or are used for increasing safety, comfort, and convenience. Examples here are Exhaust-Gas Recirculation (EGR), boost-pressure control, vehicle-speed control (cruise control), and electronic immobilizer etc.

The CAN bus system permits the exchange of data with other electronic systems in the vehicle (e.g. ABS and electronic transmission-shift control). During vehicle inspection in the workshop, a diagnosis interface permits evaluation of the stored system data.

Bild 1

Common Rail accumulator injection system on a 4-cylinder diesel engine

1 Air-mass meter, **2** ECU, **3** High-pressure pump, **4** High-pressure accumulator (rail), **5** Injectors, **6** Crankshaft-speed sensor, **7** Coolant-temperature sensor, **8** Fuel filter, **9** Accelerator-pedal sensor.

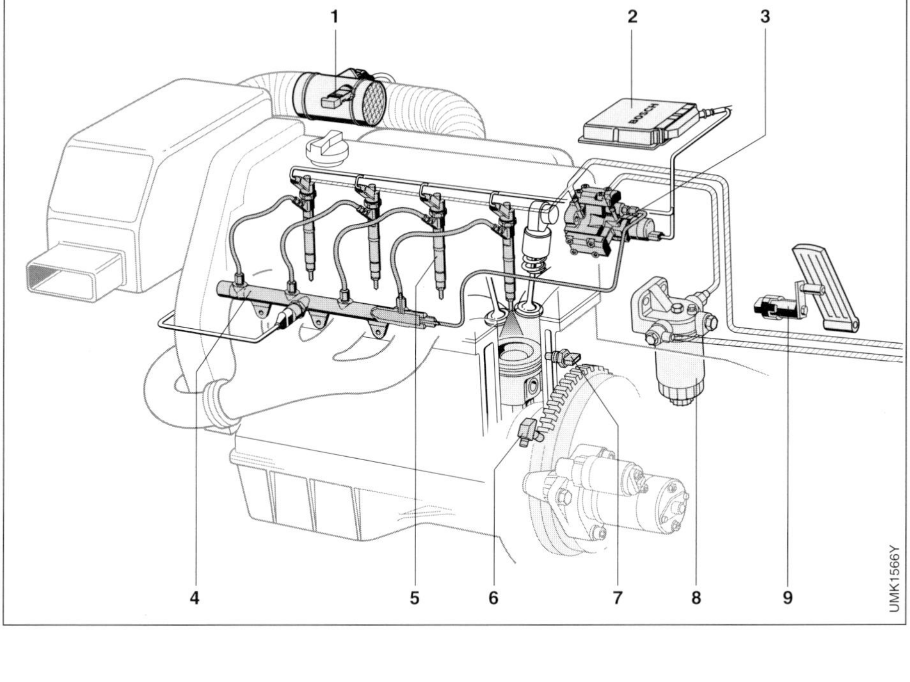

UMK1556Y

Injection characteristics

Conventional injection characteristics

With conventional injection systems, using distributor and in-line injection pumps, fuel injection today comprises only the main injection phase – without pilot and post-injection phases (Fig. 1). On the solenoid-valve-controlled distributor pump though, developments are progressing towards the introduction of a pilot-injection phase. In conventional systems, pressure generation and the provision of the injected fuel quantity are coupled to each other by a cam and a pump plunger. This has the following effects upon the injection characteristics:
– The injection pressure increases together with increasing speed and injected fuel quantity,
– During the actual injection process, the injection pressure increases and then drops again to the nozzle closing pressure at the end of injection,
The consequences are as follows:
– Smaller injected fuel quantities are injected with lower pressures than larger injected fuel quantities (refer to Fig. 1),
– The peak pressure is more than double that of the mean injection pressure, and
– In line with the requirements for efficient combustion, the rate-of-discharge curve is practically triangular.

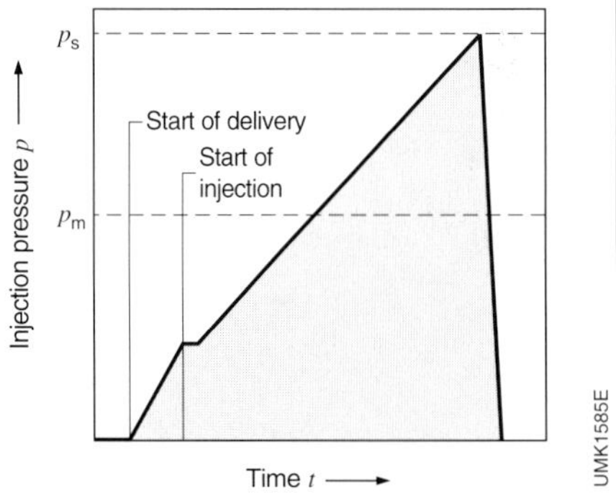

The peak pressure is decisive for the mechanical loading of a fuel-injection pump's components and drive. On conventional fuel-injection systems it is decisive for the quality of the A/F mixture formation in the combustion chamber.

Injection characteristics with Common Rail

Compared to conventional injection characteristics, the following demands are made upon an ideal injection characteristic:
– Independently of each other, injected fuel quantity and injection pressure should be definable for each and every engine operating condition (provides more freedom for achieving ideal A/F mixture formation)
– At the beginning of the injection process, the injected fuel quantity should be as low as possible (that is, during the ignition lag between the start of injection and the start of combustion).
These requirements are complied with in the Common Rail accumulator injection system with its pilot and main-injection features (Figs. 2 and 4).
The Common Rail system is a modular system, and essentially the following components are responsible for the injection characteristic:
– Solenoid-valve-controlled injectors which are screwed into the cylinder head,
– Pressure accumulator (rail), and
– High-pressure pump.

Fig. 1

Rate-of-discharge curve for conventional fuel injection

p_m Mean injection pressure, p_s Peak pressure.

Injection pressure p

p_s

Start of delivery
Start of injection

p_m

Time t

UMK1585E

Fig. 2

Rate-of-discharge curve for Common Rail fuel injection

p_m Mean injection pressure, p_R Rail pressure.

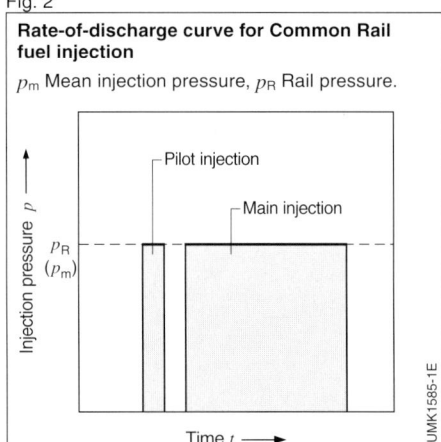

Injection pressure p

Pilot injection
Main injection

p_R
(p_m)

Time t

UMK1585-1E

The following components also also required in order to operate the system:
– Electronic control unit (ECU),
– Crankshaft-speed sensor, and
– Camshaft-speed sensor (phase sensor).
For passenger-car systems, a radial-piston pump is used as the high-pressure pump for pressure generation. Pressure is generated independently of the injection process. The speed of the high-pressure pump is coupled directly to the engine speed with a non-variable transmission ratio. In comparison with conventional injection systems, the fact that delivery is practically uniform, means that not only is the Common Rail high-pressure pump much smaller, but also that its drive is not subject to such high pressure-loading peaks.

The injectors are connected to the rail by short lines and, essentially, comprise a nozzle, and a solenoid valve which is energized by the ECU to switch it on (start of injection). When the solenoid valve is switched off (de-energized) injection ceases. Presuming constant pressure, the injected fuel quantity is directly proportional to the length of time the solenoid valve is energized. It is completely independent of the engine or pump speed (time-controlled fuel injection).

The required high-speed solenoid switching is achieved by using high voltages and currents. This means that the solenoid-valve triggering stage in the ECU must be designed accordingly.

The start of injection is controlled by the angle-time control system of the EDC (Electronic Diesel Control). This uses a sensor on the crankshaft to register engine speed, and a sensor on the camshaft for phase detection (working cycle).

Pilot injection

Pilot injection can be advanced by up to 90° crankshaft (90°cks) referred to TDC. If the start of injection occurs less than 40°cks BTDC, fuel can be deposited on the surface of the piston and the cylinder walls, and can lead to unwanted dilution of the lube-oil.

With pilot injection, a small amount of diesel fuel (1...4 mm^3) is injected into the cylinder to "precondition" the combustion chamber. Combustion efficiency can be improved as a result, and the following effects are achieved:
– The compression pressure is increased slightly due to pilot reaction and partial combustion, this in turn leads to
– The main-injection ignition delay being reduced, and
– A reduction of combustion-pressure rise and of the combustion-pressure peaks (softer combustion).

These effects reduce the combustion noise, the fuel consumption, and in many cases the exhaust-gas emissions as well. In the case of a rate-of-discharge curve without pilot injection (Fig. 3), in line with the compression only a slight, flat pressure rise is evident just before TDC, after which it peaks relatively sharply at the point of maximum pressure. The steep pressure increase together with the sharp peak contribute considerably to the diesel engine's combustion noise. As shown by the rate-of-discharge curve with pilot injection, (Fig. 4), pressure in the vicinity of TDC reaches a somewhat higher value, and the combustion-pressure increase is less rapid.

Since it reduces the ignition delay, pilot injection makes an indirect contribution to the generation of engine torque.

The specific fuel consumption can increase or decrease as a function of the start

Fig. 3

Needle lift in the injector nozzle, and rate-of-discharge curve without pilot injection

h_{HE} Needle lift for main injection.

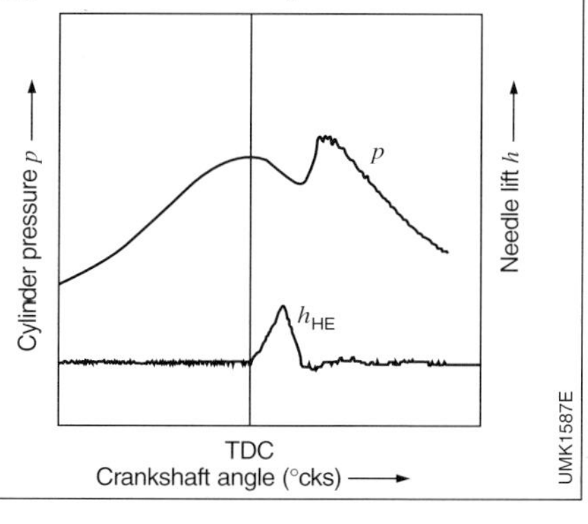

Cylinder pressure p → | Needle lift h →

p

h_{HE}

TDC
Crankshaft angle (°cks) →

UMK1587E

of main injection and the time between the pilot and main injection sequences.

Main injection

The energy for the engine's output work comes from the main injection sequence. This means that essentially the main injection is responsible for the development of the engine's torque. With the Common Rail accumulator fuel-injection system, the injection pressure remains practically constant throughout the whole of the injection process.

Secondary injection

With certain versions of NO_X catalytic converter, secondary injection can be applied for NO_X combustion (reduction). It follows the main injection process and is timed to take place during the expansion or exhaust cycle up to 200°cks after TDC. Secondary injection introduces a precisely metered quantity of fuel to the exhaust gas.

In contrast to the pilot and main injection processes, the injected fuel does not combust but instead vaporises due to the residual heat in the exhaust gas. During the exhaust cycle, the resulting mixture of exhaust gas and fuel is forced out through the exhaust valves and into the exhaust-gas system. Part of the fuel though is returned for combustion via the EGR system and has the same effects as very advanced pilot injection. Provi-

ded suitable NO_X catalytic converters are fitted, these utilise the fuel in the exhaust gas as a reduction agent to lower the NO_X content in the exhaust gas.

Since very late secondary injection leads to dilution of the engine lube oil, it must be approved by the engine manufacture.

Exhaust-gas reduction

Mixture formation and combustion behavior

Compared to SI engines, diesel engines burn low-volatility (high-boiling) fuel, and not only prepare the air/fuel mixture in the period between injection and start of combustion, but also during the actual combustion process. The result is a less homogenous mixture. The diesel engine always operates with excess air ($\lambda > 1$). Fuel consumption, and the emissions of soot, CO, and HC increase if there is insufficient excess air.

The A/F mixture formation is defined by the following parameters:
- Injection pressure,
- Rate of discharge (injection time),
- Spray distribution (number of spray jets, spray cross-section, spray direction),
- Start of injection,
- Air movement, and
- Air mass.

These quantities all have an effect upon the engine's emissions and fuel consumption. High combustion temperatures and high levels of oxygen concentration lead to increased NO_X generation. Soot emissions rise due to lack of air and poor A/F mixture formation.

Measures at the engine

The configuration of the combustion chamber and air-intake tract can have a positive effect upon the exhaust-gas emissions. If the air movement in the combustion chamber is carefully matched to the fuel jets leaving the nozzle, this promotes efficient mixing of air and fuel and

Fig. 4

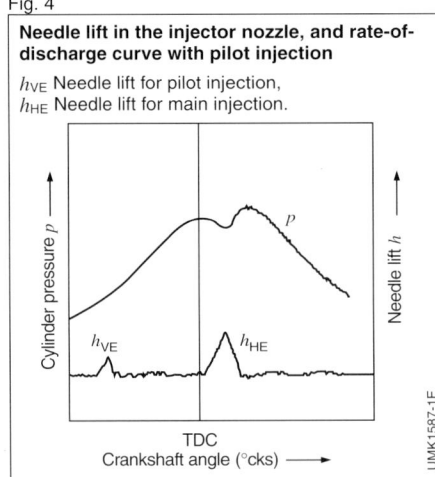

Needle lift in the injector nozzle, and rate-of-discharge curve with pilot injection

h_{VE} Needle lift for pilot injection,
h_{HE} Needle lift for main injection.

Cylinder pressure p

Needle lift h

p

h_{VE}

h_{HE}

TDC
Crankshaft angle (°cks) ⟶

UMK1587-1E

thus complete combustion of the injected fuel. In addition, positive effects are achieved with a homogenous mixture of air and exhaust gas and a cooled EGR tract. Four-valve techniques and turbochargers with variable-turbine geometry (VTG) also contribute to lower emissions and higher power density.

Exhaust-gas recirculation (EGR)

Without EGR, NO_X emissions are excessive from the emission-control legislation standpoint, whereas soot emissions are within limits. Exhaust-gas recirculation (EGR) is a method for reducing the emissions of NO_X without drastically increasing the engine's soot output. This can be implemented very efficiently with the CR system thanks to the excellent A/F mixture formation resulting from the high injection pressures. With EGR, a portion of the exhaust gases are diverted into the intake tract during part-load operation. This not only reduces the oxygen content, but also the rate of combustion and the peak temperature at the flame front, with the result that NO_X emissions drop. If too much exhaust gas is recirculated though (exceeding 40% of the intake air volume), the soot, CO, and HC emissions, as well as the fuel consumption rise due to the lack of oxygen.

Influence of fuel injection

Start of injection, rate-of-discharge curve, and atomization of the fuel also have an influence upon fuel consumption and upon exhaust-gas emissions.

Start of injection
Due to lower process temperatures, retarded fuel-injection reduces the NO_X emissions. But if it is too far retarded, HC emissions and fuel consumption increase, as do soot emissions under high loading conditions. If the start of injection deviates by only 1°cks (crankshaft) from the desired value, NO_X emissions can increase by as much as 5%. Whereas a deviation of 2°cks in the advance (early)

direction can lead to a 10 bar increase in the cylinder peak pressure, a deviation of 2°cks in the retarded (late) direction can increase the exhaust-gas temperature by 20°C. Such high sensitivity demands utmost accuracy when adjusting the start of injection.

Rate-of-discharge curve
The rate-of-discharge curve defines the variations in fuel mass flow during a single injection cycle (from start of injection till end of injection). The rate-of-discharge curve determines the mass of fuel delivered during the combustion lag (between start of injection and start of combustion). Furthermore, since it also influences the distribution of the fuel in the combustion chamber it also has an effect upon the efficiency of the air utilization. The rate-of-discharge curve must climb slowly in order that fuel injection during the combustion lag is kept to a minimum. This fuel, namely, combusts suddenly as soon as combustion is initiated with the attendant negative effects upon engine noise and NO_X emissions. The rate-of-discharge curve must drop-off sharply in order to prevent poorly atomized fuel leading to high HC and soot emissions, and increased fuel consumption during the final phase of combustion.

Fuel atomization
Finely atomized fuel promotes the efficient mixing of air and fuel. It contributes to a reduction in HC and soot emissions. High injection pressures and optimal geometrical configuration of the nozzle injection orifices lead to good atomization. To prevent visible soot emission, the injected fuel quantity must be limited in accordance with the intake air quantity. This necessitates excess air in the order of at least 10...40% ($\lambda = 1.1...1.4$).

Once the nozzle needle has closed, the fuel in the injection orifices can vaporize (in the case of sac-hole (blind-hole) nozzles the fuel vaporizes in the sac-hole volume) and in the process increase the HC emissions. This means that such (harmful) volumes must be kept to a minimum.

Fuel system

The fuel system in a "Common Rail" fuel-injection system (Fig. 1) comprises a low-pressure stage for the low-pressure delivery of fuel, a high-pressure stage for the high-pressure delivery, and the ECU (11).

Low-pressure delivery

The low-pressure stage of the Common Rail fuel system incorporates:
– Fuel tank (1) with pre-filter (2),
– Presupply pump (3),
– Fuel filter (4), and
– Low-pressure fuel lines (5).

Fuel tank
(Excerpt from § 45 StVZO (FMVSS/CUR))
The fuel tank must be of non-corroding material, and must remain free from leaks at double the operating pressure, and in any case at 0.3 bar. Suitable openings or safety valves must be provided, or appropriate measures taken, to permit excess pressure to escape of its own accord. Fuel must not leak past the filler cap or through pressure-compensation devices when the vehicle is subjected to minor shocks, as well as when cornering and when standing or driving on an incline.

Fuel tank and engine must be so far apart from each other that in case of accident there is no danger of fire. This does not apply to tractor vehicles (with open cab), nor to motorcycles and mopeds etc.

In open-cab vehicles, tractors, and buses, special regulations apply regarding the fuel tank's protective shielding and its height.

Fuel lines for the low-pressure stage
(§ 46 StVZO (FMVSS/CUR))
As an alternative to steel pipes, flame-inhibiting steel-braid-armoured flexible fuel lines can be used for the low-pressure stage. They must be routed so that they cannot be damaged mechanically, and fuel which has dripped or evaporated must not be able to accumulate, nor must it be able to ignite.

When the vehicle twists, or the engine

Fig. 1

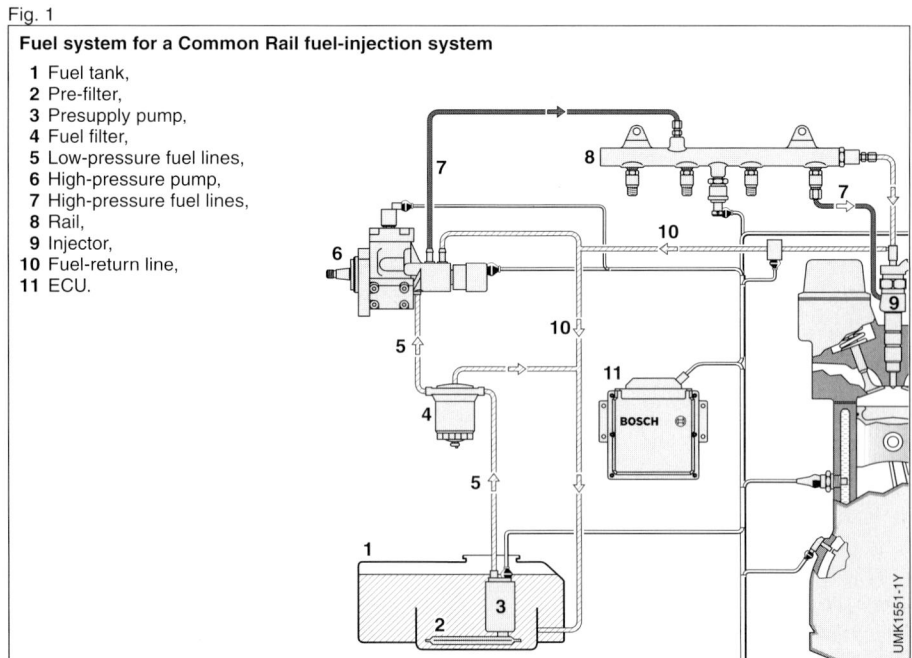

Fuel system for a Common Rail fuel-injection system
1 Fuel tank,
2 Pre-filter,
3 Presupply pump,
4 Fuel filter,
5 Low-pressure fuel lines,
6 High-pressure pump,
7 High-pressure fuel lines,
8 Rail,
9 Injector,
10 Fuel-return line,
11 ECU.

moves etc., this must have no derogatory effects upon fuel-line function. All parts which carry fuel must be protected against the effects of heat. In the case of buses, fuel lines must not be located in the passenger compartment or in the driver's cab, nor may fuel be delivered by force of gravity.

Low-pressure system components

Presupply pump
The presupply pump is either an electric fuel pump with pre-filter, or a gear-type fuel pump. The pump draws the fuel from the fuel tank and continually delivers the required quantity of fuel in the direction of the high-pressure pump.

Fuel filter
Inadequate filtering can lead to damage at the pump components, delivery valves, and injector nozzles. The fuel filter cleans the fuel before it reaches the high-pressure pump, and thereby prevents premature wear at the pump's sensitive components.

High-pressure delivery

The high-pressure stage of the fuel system in a Common Rail installation comprises:
– High-pressure pump (6) with pressure-control valve,
– High-pressure fuel lines (7),
– The rail as the high-pressure accumulator (8) with rail-pressure sensor, pressure-limiting valve, and flow limiter,
– Injectors (9), and
– Fuel-return lines (10).

High-pressure-system components

High-pressure pump
The high-pressure pump pressurises the fuel to a system pressure of up to 1,350 bar. This pressurized fuel then passes through a high-pressure line and into the tubular high-pressure fuel accumulator (rail).

High-pressure accumulator (rail)
Even after an injector has taken fuel from the rail in order to inject it, the fuel pressure inside the rail remains practically constant. This is due to the accumulator effect arising from the fuel's inherent elasticity. Fuel pressure is measured by the rail-pressure sensor and maintained at the desired level by the pressure-control valve. It is the job of the pressure-limiter valve to limit the fuel pressure in the rail to maximum 1,500 bar. The highly pressurized fuel is directed from the rail to the injectors by a flow limiter, which prevents excess fuel reaching the combustion chamber.

Injectors
The nozzles of these injectors open when the solenoid valve is triggered and permit the flow of fuel. They inject the fuel directly into the engine's combustion chamber.

The excess fuel which was needed for opening the injector nozzles flows back to the tank through a collector line. The return fuel from the pressure-control valve and from the low-pressure stage is also led into this collector line together with the fuel used to lubricate the high-pressure pump.

Fuel lines in the high-pressure section
These fuel lines carry the high-pressure fuel. They must therefore be able to permanently withstand the maximum system pressure and, during the pauses in injection, the sometimes high-frequency pressure fluctuations which occur. They are therefore manufactured from steel tubing. Normally, they have an outside diameter of 6 mm and an internal diameter of 2.4 mm.
The injection lines between the rail and the injectors must all be of the same length. The differences in length between the rail and the individual injectors are compensated for by using slight or pronounced bends in the individual lengths of tubing. Nevertheless, the injection lines should be kept as short as possible.

Design and function of the components

Low-pressure stage

The low-pressure stage (Fig. 1) provides enough fuel for the high-pressure section. The most important components are:

- Fuel tank (1),
- Pre-supply pump (3) with prefilter (2),
- Low-pressure fuel lines for supply and return (5, 7),
- Fuel filter (4) and
- Low-pressure area of the high-pressure pump (6).

Pre-supply pump

It is the pre-supply pump's job to maintain an adequate supply of fuel to the high-pressure pump. This applies

- In every operating state,
- At the necessary pressure, and
- Throughout the complete service life.

At present, there are two possible versions. An electric roller-cell fuel pump is the standard solution. An alternative is the mechanically driven gear-type fuel pump.

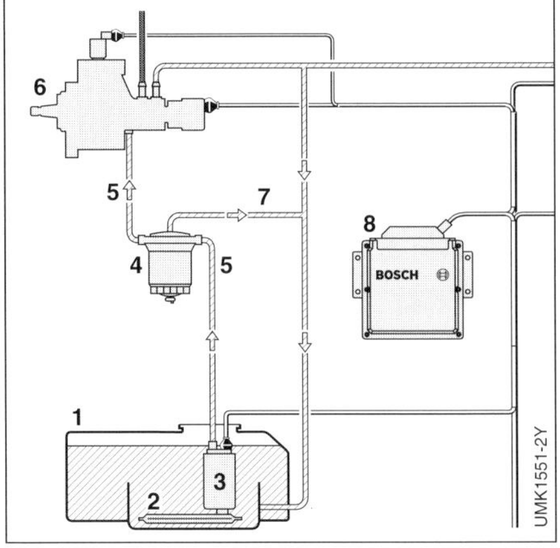

Electric fuel pump

The electric fuel pump (Figs. 2 and 3) is only used in passenger cars and light commercial vehicles. It is not only responsible for delivering the fuel to the high-pressure pump, but within the framework of system monitoring it must also interrupt the flow of fuel in case of an emergency.

Beginning with the engine cranking process, the electric fuel pump runs continuously independent of engine speed. This means that the pump permanently delivers fuel from the fuel tank, and through the filter to the high-pressure pump. Excess fuel can flow back to the tank through an overflow valve.

A safety circuit is provided to prevent the delivery of fuel should the ignition be on with the engine stopped.

Electric fuel pumps are available as in-line or in-tank versions. In-line fuel pumps are installed outside the tank in the fuel line between the tank and the fuel filter. They are attached to the vehicle's floor assembly. In-tank fuel-pump versions on the other hand are installed in the fuel tank itself using a special mounting. Apart from the electrical and hydraulic connections to the outside, this mounting usually incorporates a fuel

Fig. 1

Low-pressure stage

1 Fuel tank, **2** Prefilter, **3** Pre-supply pump,
4 Fuel filter, **5** Low-pressure fuel lines,
6 Low-pressure stage of high-pressure pump,
7 Fuel return line, **8** ECU.

Fig. 2

Electric fuel pump (schematic)

A Pumping element, **B** Electric motor,
C End cover.
1 Pressure end, **2** Motor armature,
3 Roller-cell pump, **4** Pressure limiter,
5 Suction end.

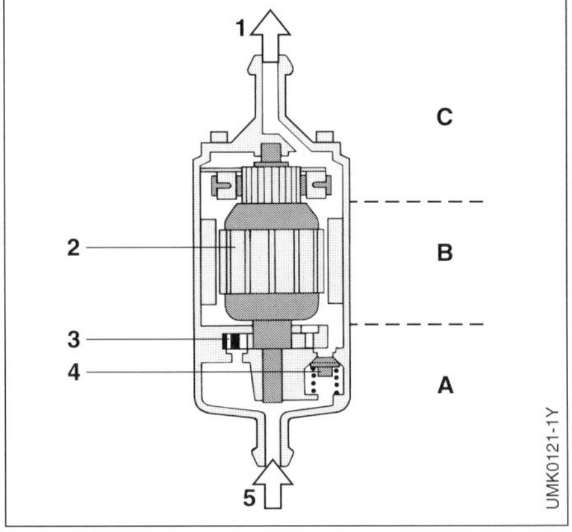

strainer, a fuel-level indicator, and a swirl pot which acts as a fuel reservoir. An electric fuel pump comprises the three function elements:
– Pumping element (Fig. 2, A),
– Electric motor (Fig. 2, B), and
– End cover (Fig. 2, C).

There are a variety of different pumping-element versions available depending upon the pump's particular field of application. The fuel pump for Common Rail is of the roller-cell type (positive-displacement pump). This form of pump comprises an eccentrically located chamber in which a slotted rotor is fitted. A movable roller is located in each slot. Rotor rotation, together with the pressure of the fuel, causes the rollers to move outward against the outside roller path and against the driving flanks of the slots. The result is that the rollers act as rotating seals, whereby a chamber is formed between the rollers of adjacent slots and the roller path. Pumping itself depends upon the fact that once the kidney-shaped inlet opening has closed, the chamber volume reduces continuously. When the outlet opening opens, fuel flows through the electric motor and leaves the pump through the pressure-side cover.

The electric motor comprises a permanent-magnet system, and an armature whose design is determined by the required delivery quantity at the given system pressure. The electric motor and pumping element are located in a common housing. With the pump operating, they are permanently flushed by gasoline so that they are always kept cool. This design permits high motor performance without the necessity for complicated sealing elements between the pumping element and the electric motor.

The end cover contains the electrical connections as well as the pressure-side hydraulic connection. Interference-suppression units can also be integrated in the end cover.

Gear-type fuel pump
On passenger cars, commercial vehicles, and off-road vehicles, a gear-type fuel pump is used to supply the Common Rail's high-pressure pump with fuel. It is either integrated in the high-pressure pump with which it shares a common drive, or it is directly attached to the engine and has its own drive.
Common forms of drive are coupling, gearwheel, or toothed belt.

Fig. 3

Roller-cell pump of the electric fuel-pump (schematic)

1 Suction end, **2** Rotor, **3** Roller, **4** Base plate, **5** Pressure end.

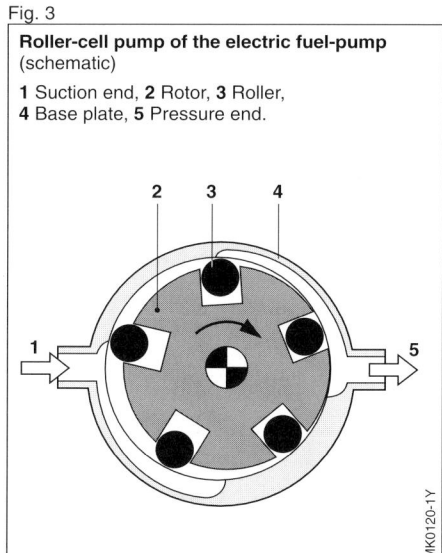

UMK0120-1Y

Fig. 4

Gear-type fuel pump (schematic)

1 Suction end,
2 Drive gear,
3 Pressure end.

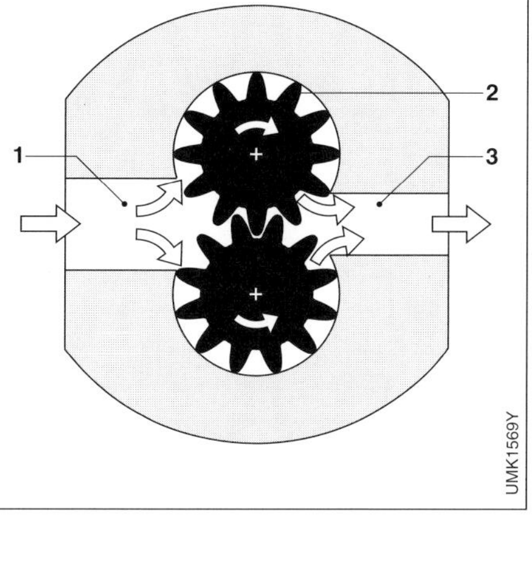

UMK1569Y

The main components are two counter-rotating gear wheels (Fig. 4) which mesh with each other when rotating, whereby fuel is trapped in the chambers formed between the gearwheels and the pump wall and transported to the outlet (pressure side). The line of contact between the rotating gearwheels provides the seal between suction and pressure ends of the pump, and prevents fuel flowing back again.

The gear-type fuel pump's delivery quantity is practically proportional to the engine speed. This is why the gear pump's delivery quantity is reduced by a suction throttle at the inlet (suction) end, or limited by an overflow valve at the outlet (pressure) end.

The gear-type pump is maintenance-free. To bleed the fuel system before the first start, or when the tank has been driven "dry", a hand pump can be installed directly on the gear-type pump or in the low-pressure lines.

Fig. 5

Fuel filter (schematic)

1 Filter cover, **2** Fuel inlet, **3** Paper filter element, **4** Case, **5** Water reservoir, **6** Water drain screw, **7** Fuel outlet.

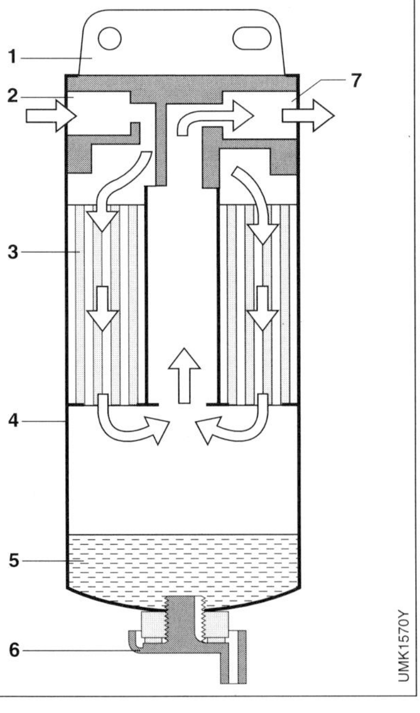

Fuel filter

Contaminants in the fuel can lead to damage at the pump components, delivery valves, and injection nozzles. This, therefore, necessitates the use of a fuel filter which is specifically aligned to the requirements of the particular injection system, otherwise faultless operation and a long service life cannot be guaranteed. Diesel fuel can contain water either in bound form (emulsion) or in free form (e.g. condensation of water due to temperature change). If this water enters the injection system, it can lead to damage as a result of corrosion.

Similar to other injection systems, the Common Rail also needs a fuel filter with water reservoir (Fig. 5), from which the water must be drained at regular intervals. The increasing number of diesel engines used in passenger cars, has led to the demand for an automatic water warning device which indicates by means of a warning lamp when water must be drained (this is binding in those countries in which there is a high level of water in the fuel).

High-pressure stage

In addition to high-pressure generation, fuel distribution and fuel-metering also take place in the high-pressure stage (Fig. 6). The most important components are:
– High-pressure pump (1) with element shutoff valve (2) and pressure-control valve (3),
– High-pressure accumulator (5, rail),
– Rail-pressure sensor (6),
– Pressure-limiter valve (7),
– Flow limiter (8) and
– Injectors (9).

High-pressure pump

Assignments
The high-pressure pump (Figs. 7 and 8) is the interface between the low-pressure and the high-pressure stages. Under all operating conditions, it is responsible for providing adequate high-

UMK1570Y

High-pressure stage of the Common Rail accumulator injection system

1 High-pressure pump,
2 Element shutoff valve,
3 Pressure-control valve,
4 High-pressure fuel lines,
5 High-pressure accumulator (rail),
6 Rail-pressure sensor,
7 Pressure-limiter valve,
8 Flow limiter,
9 Injector,
10 ECU.

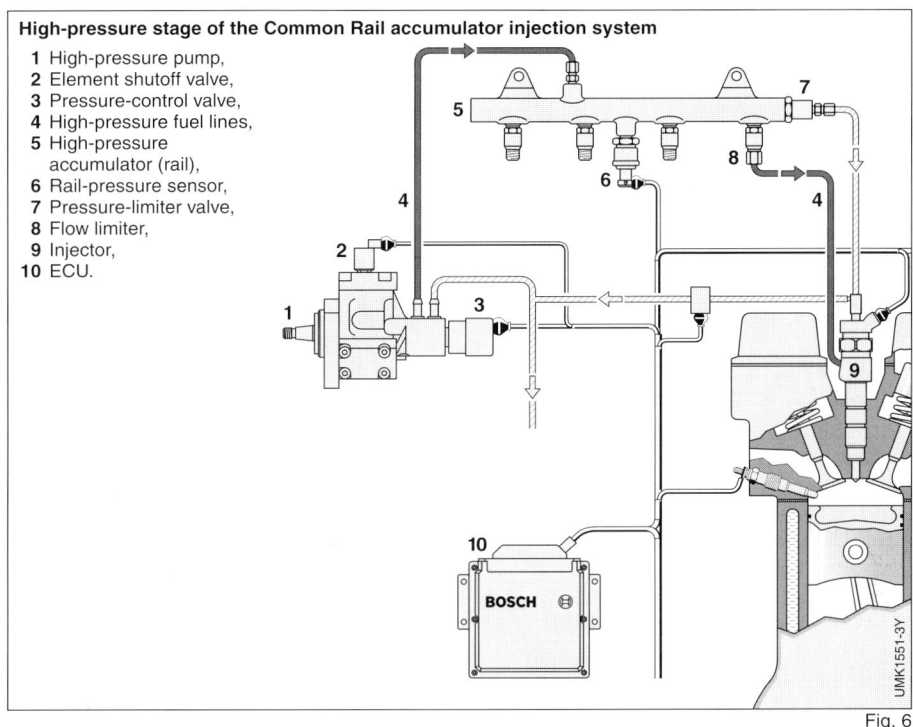

Fig. 6

Fig. 7

High-pressure pump (schematic, longitudinal section)

1 Driveshaft, 2 Eccentric cam, 3 Pumping element with pump plunger, 4 Pumping-element chamber, 5 Suction valve, 6 Element shutoff valve, 7 Outlet valve, 8 Seal, 9 High-pressure connection to the rail, 10 Pressure-control valve, 11 Ball valve, 12 Fuel return, 13 Fuel inlet from the presupply pump, 14 Safety valve with throttle bore, 15 Low-pressure passage to the pumping element.

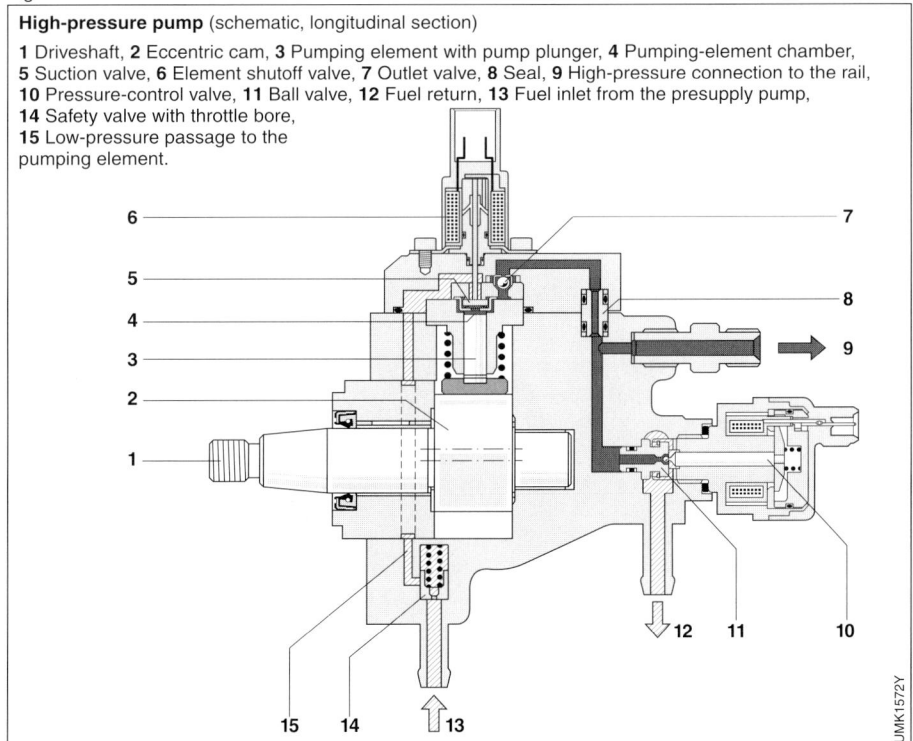

pressure fuel throughout the vehicle's complete service life. This also includes the provision of extra fuel as needed for rapid starting and for rapid build-up of pressure in the rail.

The high-pressure pump continually generates the system pressure as needed in the high-pressure accumulator (rail). This means therefore, that in contrast to conventional systems, the fuel does not have to be specially compressed for each individual injection process.

Design and construction

The high-pressure pump is installed preferably at the same point on the diesel engine as a conventional distributor pump. It is driven by the engine (at half engine speed, but max. 3000 min^{-1}) through a coupling, gearwheel, chain, or toothed belt, and lubricated by the diesel fuel which it pumps.

Depending on available space, a pressure-control valve is installed directly on the high-pressure pump or remote from it.

Inside the high-pressure pump, the fuel is compressed with three radially arranged pump pistons which are at an angle of 120° to each other. Since three delivery strokes take place for every revolution, only low peak drive torques are generated so that the stress on the pump drive remains uniform. With 16 Nm, the torque is only about 1/9 of that required to drive a comparable distributor pump. This means that Common Rail places less loading on the pump drive than is the case with conventional injection systems. The power required to drive the pump climbs in proportion to the pressure set in the rail and to the pump's speed (delivery quantity). For a 2-liter engine turning at rated speed, and with a set pressure of 1,350 bar in the rail, the high-pressure pump requires 3.8 kW presuming a mechanical efficiency of approx. 90%. The higher power demand (higher than theoretically necessary) results from the leak-fuel and control quantities at the injector, and

Fig. 8

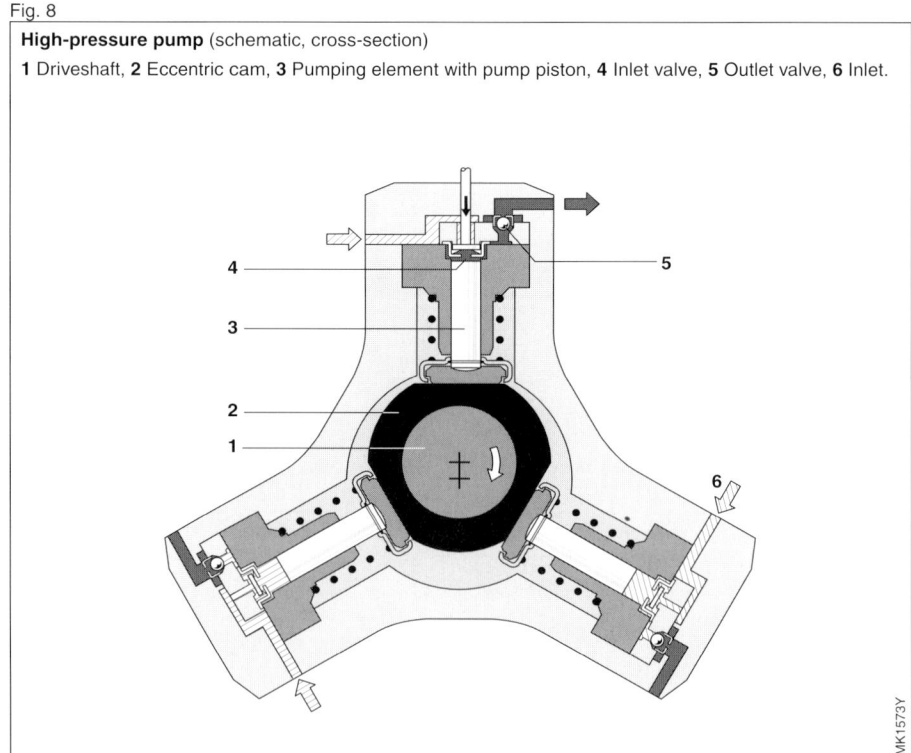

High-pressure pump (schematic, cross-section)

1 Driveshaft, **2** Eccentric cam, **3** Pumping element with pump piston, **4** Inlet valve, **5** Outlet valve, **6** Inlet.

UMK1573Y

from the fuel return through the pressure-control valve.

Method of operation
Via a filter with water separator, the pre-supply pump pumps fuel from the tank to the high-pressure pump through the fuel inlet (Fig. 7, Item 13) and the safety valve. It forces the fuel through the safety valve's throttle bore (14) and into the high-pressure pump's lubrication and cooling circuit. The driveshaft (1) with its eccentric cams (2) moves the three pump plungers (3) up and down in accordance with the shape of the cam.

As soon as the delivery pressure exceeds the safety valve's opening pressure (0.5...1.5 bar), the pre-supply pump can force fuel through the high-pressure pump's inlet valve into the pumping-element chamber (4) whose pump piston is moving downwards (suction stroke). The inlet valve closes when the pump piston passes through BDC and, since it is impossible for the fuel in the pumping-element chamber to escape, it can now be compressed beyond the delivery pressure. The increasing pressure opens the outlet valve (7) as soon as the rail pressure is reached, and the compressed fuel enters the high-pressure circuit.

The pump piston continues to deliver fuel until it reaches TDC (delivery stroke), after which the pressure collapses so that the outlet valve closes. The fuel remaining in the pumping-element chamber relaxes and the pump piston moves downwards again.

As soon as the pressure in the pumping-element chamber drops below the presupply-pump pressure, the inlet valve opens and the pumping process starts again.

Fuel-delivery rate
Since the high-pressure pump is designed for large delivery quantities, excess high-pressure fuel is delivered during idle and part-load operation. This excess fuel is returned to the tank via the pressure-control valve. The compressed fuel relaxes in the tank, and the energy is lost which was used for compressing the fuel in the first place. In addition to the unnecessary heating up of the fuel, overall efficiency is also reduced.

To a certain extent, this loss of efficiency can be compensated for by switching off one of the pumping elements.

Element switch-off:
When one of the pumping elements (3 in Fig. 7) is switched off, this leads to a reduction of the amount of fuel which is pumped into the rail. Switch-off involves the suction valve (5 in Fig. 7) remaining open permanently. When the solenoid valve of the pumping-element switchoff is triggered, a pin attached to its armature continually holds the inlet valve open. The result is that the fuel drawn into this pumping element cannot be compressed during the delivery stroke. No pressure is generated in the element chamber since the fuel flows back into the low-pressure passage again. With one of its pumping elements switched off when less power is needed, the high-pressure pump no longer delivers the fuel continuously but rather with brief interruptions in delivery.

Transmission ratio:
The high-pressure pump's delivery rate is proportional to its rotational speed. And this, in turn, is a function of the engine speed. During the injection-system application-engineering work on the engine, the transmission ratio is defined so that on the one hand the amount of excess fuel is not too high, and on the other, the fuel requirements can still be satisfied during WOT operation. Referred to the crankshaft, transmission ratios of 1:2 and 2:3 are possible.

Pressure-control valve

Assignment

The pressure-control valve sets the correct pressure in the rail as a function of engine loading, and maintains it at this level.

– If the rail pressure is excessive, the pressure-control valve opens and a portion of the fuel returns from the rail to the fuel tank via a collector line.
– If the rail pressure is too low, the pressure-control valve closes and seals off the high-pressure stage from the low-pressure stage.

Design and construction

The pressure-control valve (Fig. 9) is provided with a mounting flange for attachment to the high-pressure pump or to the high-pressure accumulator (rail).

In order to seal off the high-pressure and low-pressure stages from each other, the armature forces a ball against the seal seat. There are two forces acting upon the armature. Firstly it is pushed down by a spring, and secondly a force is exerted on it by an electromagnet. For lubrication and heat-dissipation, the complete armature assembly is permanently surrounded by fuel.

Method of operation

The pressure-control valve incorporates two control loops:

– An slow-response electrical control loop for setting a variable mean pressure in the rail, and
– A fast-response mechanical control loop to compensate for the high-frequency pressure fluctuations.

Pressure-control valve non-energized:
The high pressure at the rail or at the high-pressure pump's outlet is applied to the pressure-control valve via the high-pressure input. Since the non-energized electromagnet exerts no force, the force of the high-pressure fuel exceeds the spring force so that the control valve opens and remains open to a degree depending upon the delivery quantity. The spring is designed so that a maximum pressure of approx. 100 bar is achieved.

Pressure-control valve energized:
If the pressure in the high-pressure circuit is to be increased, the force of the electromagnet must be generated in addition to the spring force. The pressure-control valve is energized so that is closes and remains closed until equilibrium is reached between the high-pressure forces on the one side and the combined forces of the

Fig. 9

Pressure-control valve

1 Valve ball,
2 Armature,
3 Electromagnet,
4 Spring,
5 Electrical connection.

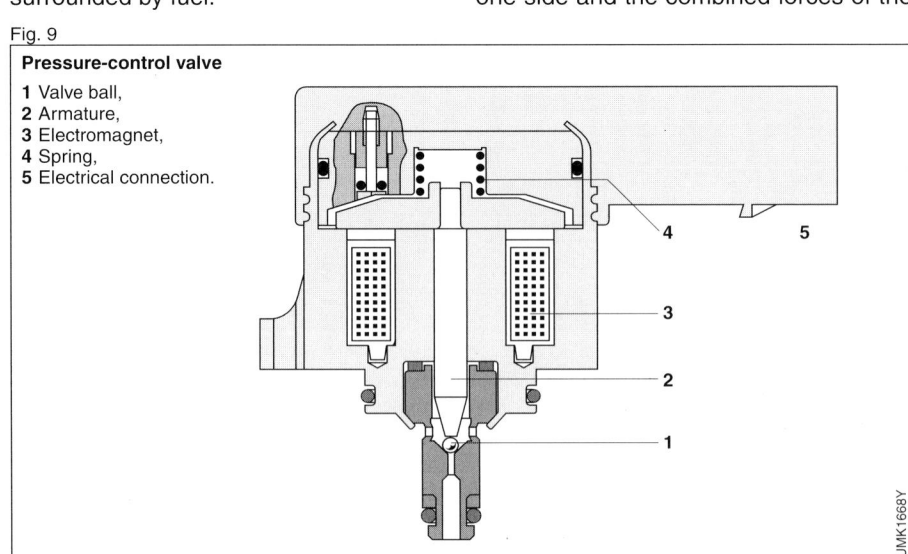

UMK1668Y

spring and the electromagnet on the other. The valve then remains open and maintains the fuel pressure constant. A change in the pump's delivery quantity, or the removal of fuel from the high-pressure stage, is compensated for by the valve assuming a different setting. The electromagnet's forces are proportional to its energizing current which is varied by pwm pulsing (pwm = pulse-width modulation). The 1 kHz pulsing frequency is high enough to prevent unwanted electromagnet-armature motion and/or pressure fluctuations in the rail.

High-pressure accumulator (rail)

Assignments
The high-pressure accumulator (the Rail in Fig. 10) stores the fuel at high pressure. At the same time, the pressure oscillations which are generated due to the high-pressure pump delivery and the injection of fuel are damped by the rail volume.

This high-pressure accumulator is common to all cylinders, hence its name "common rail". Even when large quantities of fuel are extracted, the common rail maintains its inner pressure practically constant. This ensures that the injection pressure remains constant from the moment the injector opens.

Design and construction
In order to comply with the wide variety of engine installation conditions, the rail with its flow limiters and the provisions for attaching rail-pressure sensor, pressure-control valve, and pressure-limiter valve is available in a number of different designs.

Function
The available rail volume is permanently filled with pressurized fuel. The compressibility of the fuel resulting from the high pressure is utilised to achieve the accumulator effect. When fuel leaves the rail for injection, the pressure in the high-pressure accumulator remains practically constant. Similarly, the pressure variations resulting from the pulsating fuel supply from the high-pressure pump are compensated for.

Fig. 10

High-pressure accumulator (rail)

1 Rail, **2** Inlet from the high-pressure pump, **3** Rail-pressure sensor, **4** Pressure-limiter valve, **5** Return from the rail to the fuel tank, **6** Flow limiter, **7** Line to the injector.

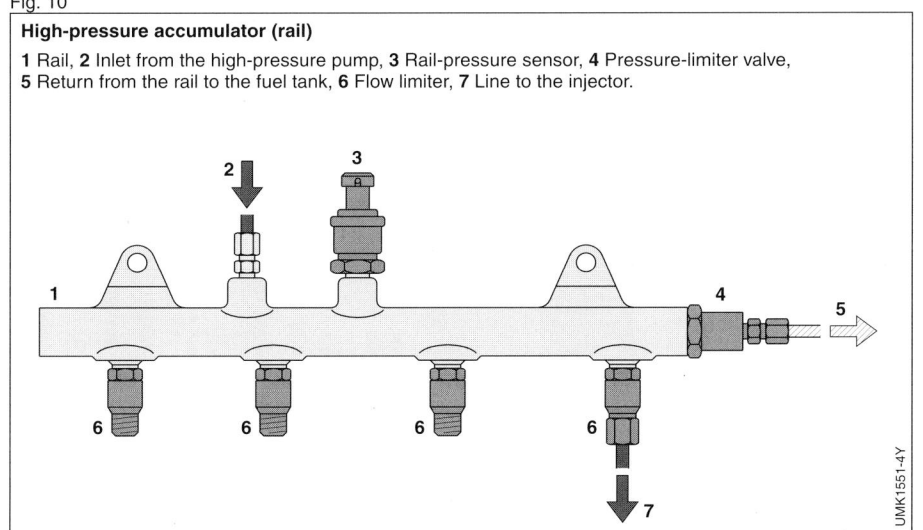

UMK1551-4Y

Rail-pressure sensor

Assignment

In order to output a voltage signal to the
ECU which corresponds to the applied
pressure, the rail-pressure sensor must
measure the instantaneous pressure in
the rail
– With adequate accuracy, and
– As quickly as possible

Design and construction

The rail-pressure sensor (Fig. 12) comprises the following components:
– An integrated sensor element welded
  to the pressure fitting,
– A printed-circuit board (pcb) with electrical evaluation circuit, and
– A sensor housing with electrical plug-in
  connection.

The fuel flows to the rail-pressure sensor
through an opening in the rail, the end
of which is sealed off by the sensor
diaphragm. Pressurized fuel reaches the
sensor's diaphragm through a blind hole.
The sensor element (semiconductor
device) for converting the pressure to
an electric signal is mounted on this
diaphragm. The signal generated by the

sensor is inputted to an evaluation circuit
which amplifies the measuring signal and
sends it to the ECU.

Function

The rail-pressure sensor (Fig. 12) operates as follows:
When the diaphragm's shape changes,
the electrical resistance of the layers
attached to the diaphragm also change.
The change in shape (approx. 1 mm at
1500 bar) which results from the build-
up of system pressure, changes the
electrical resistance and causes a voltage
change across the 5 V resistance bridge.

This voltage change is in the range 0...70
mV (depending upon the applied pressure) and is amplified by the evaluation
circuit to 0.5...4.5 V.
The precise measurement of rail
pressure is imperative for correct system
functioning. This is one of the reasons for
the very tight tolerances which apply to
the rail-pressure sensor during pressure
measurement. In the main operating
range, the measuring accuracy is approx.
± 2% of full-scale reading. If the rail-
pressure sensor should fail, the pressure-control valve is triggered "blind"
using an emergency (limp-home)
function and fixed values.

Fig. 11
**Common Rail injection system on the
engine test bench**

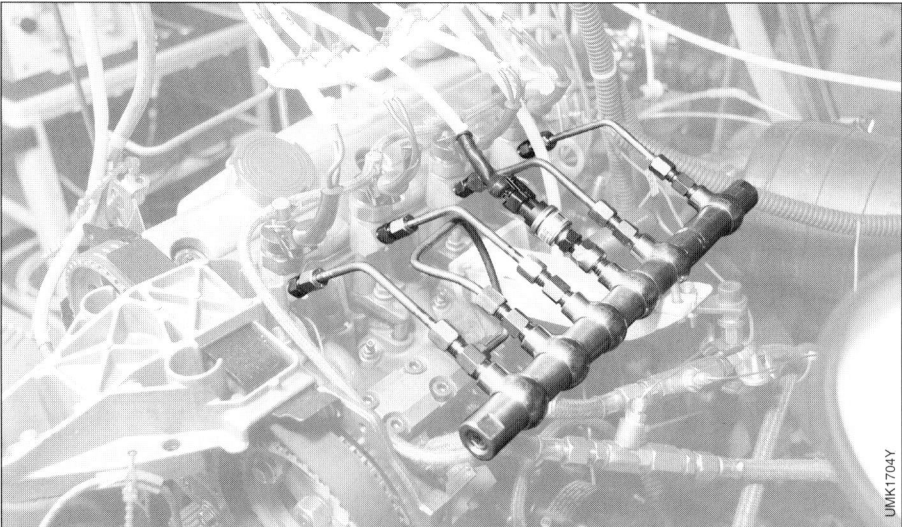

Pressure limiter valve

Assignment

The pressure limiter valve has the same job as an overpressure valve. In case of excessive pressure, the pressure limiter valve limits the rail pressure by opening an escape passage. The pressure limiter permits a short-time maximum rail pressure of 1500 bar.

Design and construction

The pressure-limiter valve (Fig. 13) is a mechanical device comprising the following components:

– Housing with external thread for screwing to the rail,
– A connection to the fuel-tank return line,
– A movable plunger, and
– A spring.

At the connection end to the rail, the housing is provided with a passage which is closed by the cone-shaped end of the plunger coming up against the sealing seat inside the housing. At normal operating pressures (up to 1350 bar), a spring forces the plunger against the seat and the rail remains closed. As soon as the maximum system pressure is exceeded, the plunger is forced up by the rail pressure against the force of the spring. The fuel under high pressure can now escape, whereby it flows through passages into the plunger's interior from where it is led through a collector line back to the fuel tank. When the valve opens, fuel leaves the rail so that the rail pressure drops.

Flow limiter

Assignment

It is the job of the flow limiter to prevent continuous injection in the very unlikely case that one of the injectors remains open permanently. To comply with this task, as soon as the amount of fuel leaving the rail exceeds a defined level, the flow limiter closes the line to the injector in question.

Design and construction

The flow limiter (Fig. 14) comprises a metal housing with external thread for screwing onto the rail (high pressure) and an external thread for screwing into the injector lines. The housing has a passage at each end which provides the hydraulic connection to the rail and to the injector lines.

There is a plunger inside the flow limiter which is forced in the direction of the fuel accumulator by a spring. This plunger

Fig. 12

Rail-pressure sensor (schematic)

1 Electric connections, **2** Evaluation circuit,
3 Metal diaphragm with sensor element,
4 High-pressure connection, **5** Mounting thread.

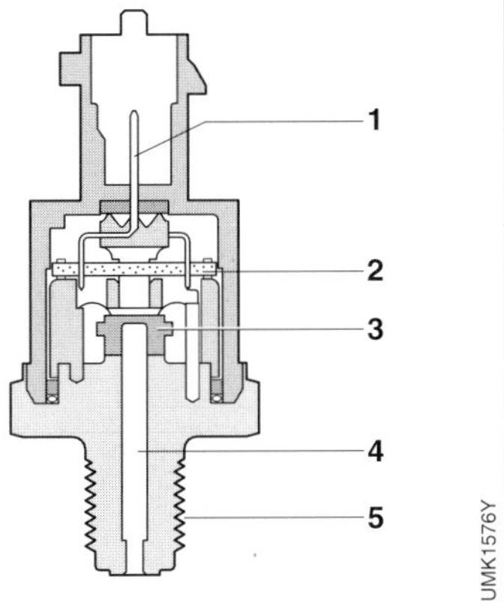

Fig. 13

Pressure limiter valve (schematic)

1 High-pressure connection, **2** Valve,
3 Flow passages, **4** Plunger, **5** Spring, **6** Stop,
7 Valve body, **8** Fuel return.

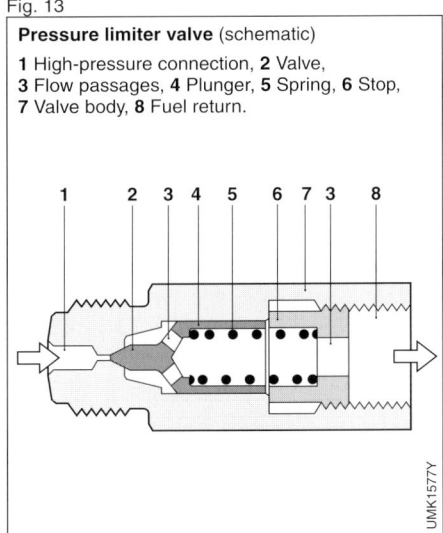

seals off to the housing walls, and the longitudinal passage through its center is the hydraulic connection between inlet and outlet.

The diameter reduces at the end of this longitudinal passage, and acts as a throttle with precisely defined flow rate.

Function
Normal operation (Fig. 15):
The plunger is in its at-rest position, in other words up against the stop at the flow limiter's rail end. When fuel is injected, the injection pressure drops at the injector end and causes the plunger to shift in the direction of the injector. The flow limiter compensates for the fuel volume taken from the rail by the injector by means of the fuel volume displaced by its plunger, and not by the throttle since this is too small. At the end of the injection process, the plunger takes up a mid-position away from its seat without closing off the outlet completely. The spring forces it back into its at-rest position and the fuel can now flow through the throttle.

Spring and throttle bore are dimensioned so that even at maximum injected fuel quantity (plus safety reserve), it is possi-

Fig. 14

Flow limiter (schematic)

1 Connection to the rail, **2** Sealing washer, **3** Plunger, **4** Spring, **5** Housing, **6** Connection to the injector, **7** Seat, **8** Throttle.

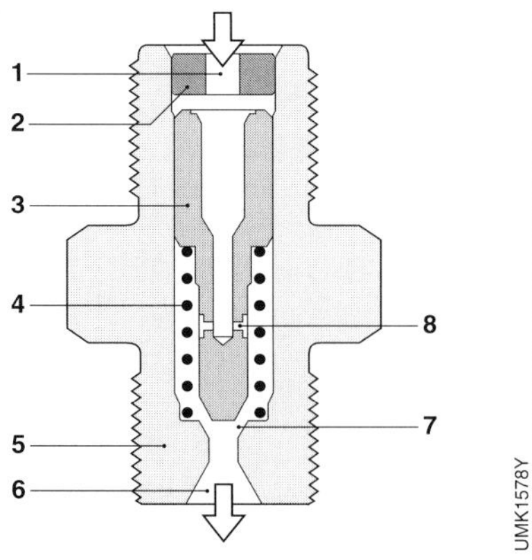

UMK1578Y

ble for the plunger to move back again to the stop at the rail end of the flow limiter, where it remains until the next injection.

Malfunction operation with large amount of leakage:
Due to the large quantity of fuel leaving the rail, the flow-limiter plunger is forced away from its at-rest position and up against the seal seat at the outlet. It remains in this position up against the stop at the injector side of the flow limiter and prevents fuel reaching the injector.

Malfunction operation with only slight leakage (Fig. 15):
Due to the leakage quantity, the flow-limiter plunger is unable to reach its at-rest position. After a number of in-jections have taken place, the plunger moves into the seal seat at the outlet bore.
It remains in this position up against the stop at the injector side of the flow limiter until the engine is switched off, and closes the input to the injector.

Injectors

Assignment
The start of injection and the injected fuel quantity are adjusted by electrically triggered injectors. These injectors super-sede the nozzle-and-holder assembly (nozzle and nozzle-holder).

Similar to the already existing nozzle-hol-der assemblies in direct-injection (DI) diesel engines, clamps are preferably used for installing the injectors in the cylinder head. This means that the Common Rail injectors can be installed in already existing DI diesel engines with-out major modifications to the cylinder head.

Design and construction
The injector (Fig. 16) can be sub-divided into a number of function blocks:
– The hole-type nozzle ,
– The hydraulic servo-system, and
– The solenoid valve.

Referring to Fig. 16, fuel is fed from the high-pressure connection (4), to the nozzle through the passage (10), and to the control chamber (8) through the feed orifice (7). The control chamber is connected to the fuel return (1) via a bleed orifice (6) which is opened by the solenoid valve.

With the bleed orifice closed, the hydraulic force applied to the valve control plunger (9) exceeds that at the nozzle-needle pressure shoulder (11). As a result, the needle is forced into its seat and seals off the high-pressure passage from the combustion chamber.

When the injector's solenoid valve is triggered, the bleed orifice is opened. This leads to a drop in control-chamber pressure and, as a result, the hydraulic pressure on the plunger also drops. As soon as the hydraulic force drops below the force on the nozzle-needle pressure shoulder, the nozzle needle opens and fuel is injected through the spray holes into the combustion chamber. This indirect control of the nozzle needle using a hydraulic force-amplification system is applied because the forces which are necessary for opening the needle very quickly cannot be directly generated by the solenoid valve. The so-called control quantity needed for opening the nozzle needle is in addition to the fuel quantity which is actually injected, and it is led back to the fuel-return line via the control chamber's orifices.

In addition to the control quantity, fuel is also lost at the nozzle-needle and valve-plunger guides. These control and leak-off fuel quantities are returned to the fuel tank via the fuel return and the collector line to which overflow valve, high-pressure pump, and pressure-control valve are also connected.

Method of operation

The injector's operation can be subdivided into four operating states with the engine running and the high-pressure pump generating pressure:
− Injector closed (with high pressure applied),
− Injector opens (start of injection),
− Injector opened fully, and
− Injector closes (end of injection).

These operating states result from the distribution of the forces applied to the injector's components. With the engine at standstill and no pressure in the rail, the nozzle spring closes the injector.

Fig. 15

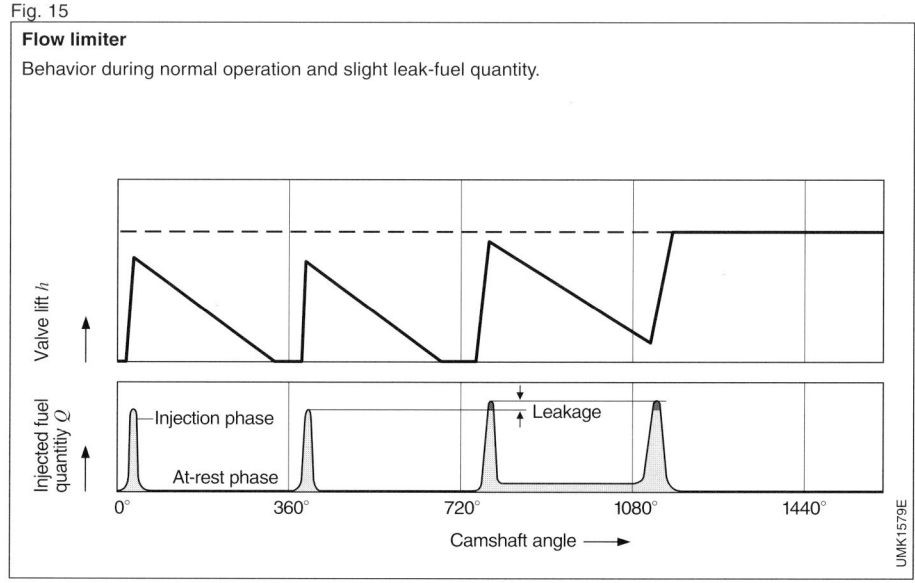

Flow limiter
Behavior during normal operation and slight leak-fuel quantity.

Valve lift *h*

Injected fuel quantity *Q*

Injection phase

At-rest phase

Leakage

0° 360° 720° 1080° 1440°

Camshaft angle ⟶

UMK1579E

Fig. 16

Injector (schematic)

a Injector closed
(at-rest status),
b Injector opened
(injection).
1 Fuel return,
2 Electrical connection,

3 Triggering element
(solenoid valve),
4 Fuel inlet (high pressure)
from the rail,
5 Valve ball,
6 Bleed orifice,

7 Feed orifice,
8 Valve control chamber,
9 Valve control plunger,
10 Feed passage
to the nozzle,
11 Nozzle needle.

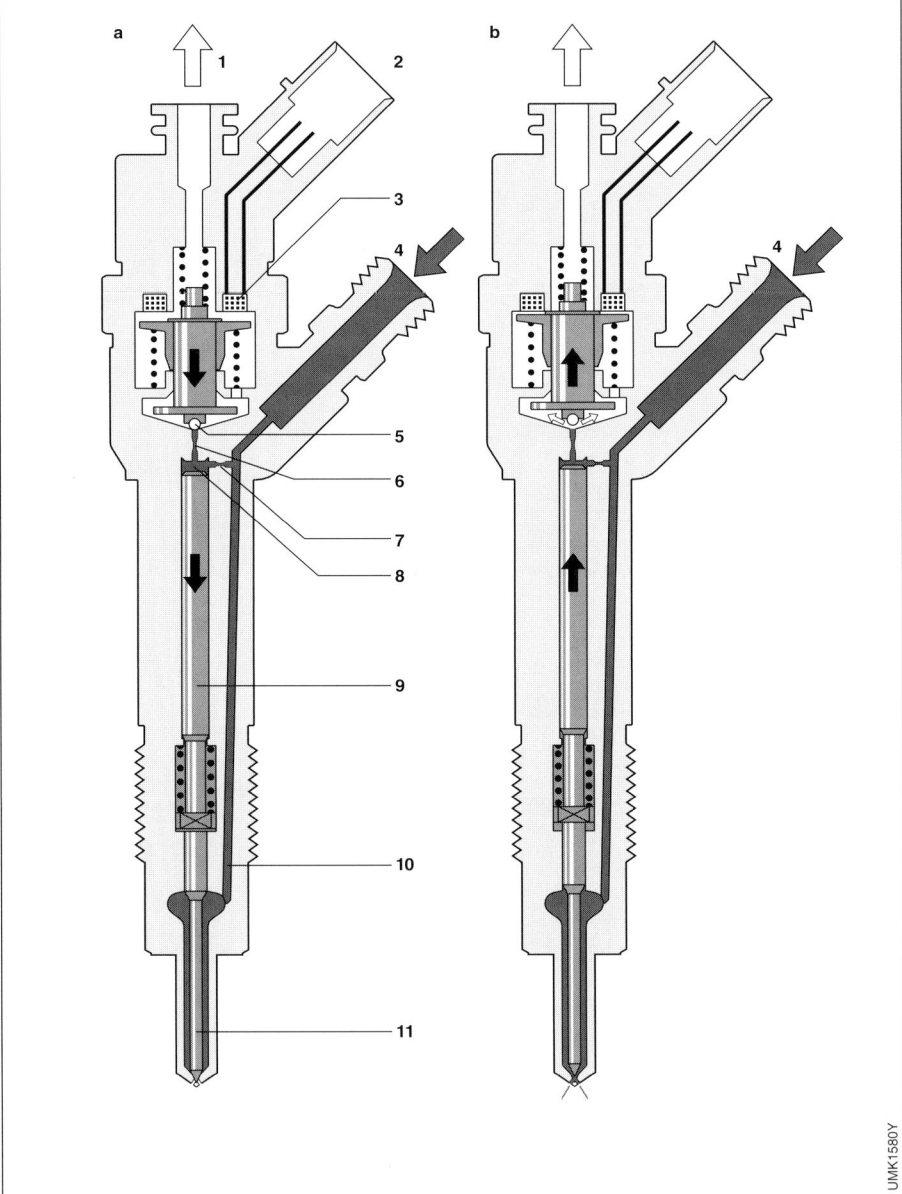

UMK1580Y

Injector closed (at-rest status):

In the at-rest state, the solenoid valve is not energized and is therefore closed (Fig. 16a).

With the bleed orifice closed, the valve spring forces the armature's ball onto the bleed-orifice seat. The rail's high pressure builds up in the valve control chamber, and the same pressure is also present in the nozzle's chamber volume. The rail pressure applied at the control plunger's end face, together with the force of the nozzle spring, maintain the nozzle in the closed position against the opening forces applied to its pressure stage.

Injector opens (start of injection):

The injector is in its at-rest position. The solenoid valve is energized with the pick-up current which serves to ensure that it opens quickly (Fig.16b). The force exerted by the triggered solenoid now exceeds that of the valve spring and the armature opens the bleed orifice. Almost immediately, the high-level pick-up current is reduced to the lower holding current required for the electromagnet. This is possible due to the magnetic circuit's air gap now being smaller. When the bleed orifice opens, fuel can flow from the valve-control chamber into the cavity situated above it, and from there via the fuel return to the fuel tank. The bleed orifice prevents complete pressure balance, and the pressure in the valve control chamber sinks as a result. This leads to the pressure in the valve-control chamber being lower than that in the nozzle's chamber volume which is still at the same pressure level as the rail. The reduced pressure in the valve-control chamber causes a reduction in the force exerted on the control plunger, the nozzle needle opens as a result, and injection starts.

The nozzle needle's opening speed is determined by the difference in the flow rate through the bleed and feed orifices. The control plunger reaches its upper stop where it remains supported by a cushion of fuel which is generated by the flow of fuel between the bleed and feed orifices. The injector nozzle has now opened fully, and fuel is injected into the combustion chamber at a pressure almost equal to that in the fuel rail. Force distribution in the injector is similar to that during the opening phase.

Injector closes (end of injection):

As soon as the solenoid valve is no longer triggered, the valve spring forces the armature downwards and the ball closes the bleed orifice. The armature is a 2-piece design. Here, although the armature plate is guided by a driver shoulder in its downward movement, it can "overspring" with the return spring so that it exerts no downwards-acting forces on the armature and the ball.

The closing of the bleed orifice leads to pressure buildup in the control chamber via the input from the feed orifice. This pressure is the same as that in the rail and exerts an increased force on the control plunger through its end face. This force, together with that of the spring, now exceeds the force exerted by the chamber volume and the nozzle needle closes.

The nozzle needle's closing speed is determined by the flow through the feed orifice. Injection ceases as soon as the nozzle needle comes up against its bottom stop again.

Hole-type nozzles

Assignments

The injection nozzles are installed in the Common Rail injectors which thus assume the role of the nozzle-holder assemblies. The nozzles must be matched carefully to the specific engine conditions.

Nozzle design is also decisive for
– Metering of injected fuel (injection time and injected fuel quantity per degree crankshaft),
– Fuel management (number of injection jets, spray shape, and atomization of the injection spray), distribution of the fuel in the combustion chamber,
– Sealing-off from the combustion chamber.

Application

Type P hole-type nozzles with 4 mm needle diameter are used for Common Rail direct-injection (DI) engines.
These nozzles are available in 2 types:
– Sac-hole nozzle, and
– Seat-hole nozzle.

Fig. 17

Spray cone

γ Spray-cone offset angle,
δ Spray cone.

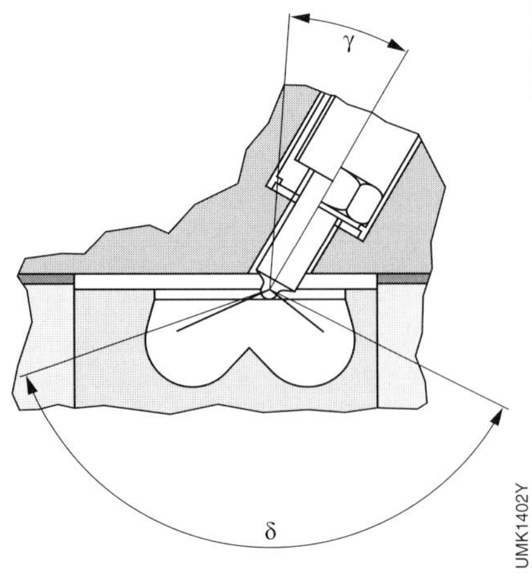

Design and construction

The spray holes are located on the envelope of a spray cone (Fig. 17). The number of spray holes and their diameter depend upon:
– The injected fuel quantity,
– The combustion-chamber shape, and
– The air swirl in the combustion chamber.

For both sac-hole and seat-hole nozzles, the input edges of the spray holes can be rounded by hydro-erosive (HE) machining. Such measures are aimed at:
– Preventing in advance the edge wear caused by the abrasive particles in the fuel and/or
– Reducing the flow tolerance.

To achieve low hydrocarbon emissions, it is important that the volume filled with fuel (residual volume) below the edge of the nozzle-needle seat is kept to a minimum. This is best done by using seat-hole nozzles.

Designs

Sac-hole nozzles

The spray holes of the sac-hole nozzle (Fig. 18) are arranged in the sac hole.
In the case of a round nozzle tip, depending upon the design, the spray holes are drilled mechanically or by means of electrical-discharge machining (EDM electrical particle removal).
Sac-hole nozzles with conical tip are always drilled using EDM.
The sac-hole nozzles are available with the following sac-hole shapes and in a variety of different dimensions:
– Cylindrical sac hole, and
– Conical sac hole.

1. Sac-hole nozzle with cylindrical sac hole and round tip:
This sac-hole shape, comprising a cylindrical and a semispherical portion, permits a high level of design freedom with respect to:
– Number of spray holes,
– Spray-hole length, and
– Injection angle.

The nozzle tip is semispherical, and together with the shape of the sac hole, ensures that the spray holes are of identical length.

2. Sac-hole nozzle with cylindrical sac hole and conical tip:
This form of nozzle is used exclusively with spray-hole lengths of 0.6 mm. The tip's conical shape enables the wall thickness to be increased between the throat radius and the nozzle-body seat. This results in an improvement of nozzle-tip strength.

3. Sac-hole nozzle with conical sac hole and conical tip:
With this version, as a result of its conical shape the sac-hole volume is less than that of the nozzle with cylindrical sac hole. It is between that for a seat-hole nozzle and a sac-hole nozzle with cylindrical sac hole. In order to achieve uniform tip-wall thickness, the tip's conical design corresponds to that of the sac hole

Seat-hole nozzle
In order to minimise the "harmful" volume – and therefore the HC emissions – the start of the spray hole is located in the seat taper, and with the nozzle closed it is covered by the nozzle needle. This means that there is no direct connection between the sac hole and the combustion chamber (Fig. 19). The "harmful" volume here is much lower than that in the sac-hole nozzle. Compared to sac-hole nozzles, seat-hole nozzles have a much lower loading limit and are therefore only manufactured as Size P with a spray-hole length of 1 mm.
For reasons of strength, the nozzle tip is conically shaped. The spray holes are always formed by electrical discharge machining (EDM) methods.

Sac-hole nozzle
1 Pressure pin, **2** Needle-lift stop face,
3 Inlet passage, **4** Exposed annular area,
5 Needle shaft, **6** Nozzle tip,
7 Nozzle-body shaft, **8** Nozzle-body shoulder,
9 Pressure chamber, **10** Needle guide,
11 Nozzle-body collar,
12 Locating hole, **13** Sealing surface,
14 Pressure-pin contact surface.

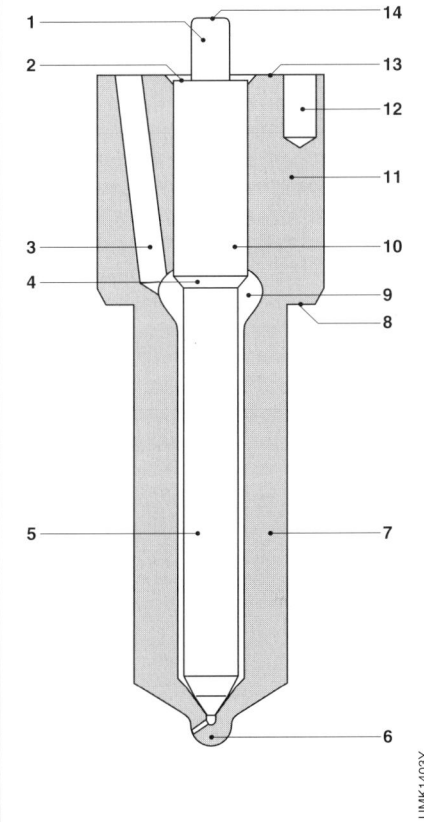

Fig. 18

Fig. 19

Seat-hole nozzle: Tip shape

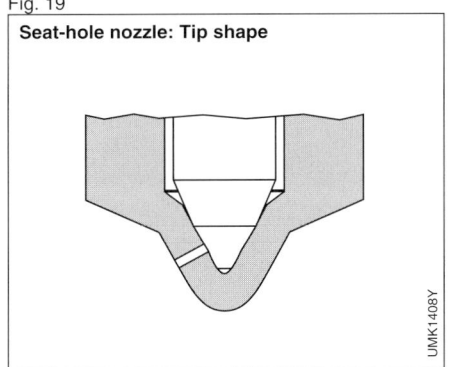

System control using EDC

System blocks

The Electronic Diesel Control (EDC) for Common Rail comprises three major system blocks:
1. Sensors and setpoint generators for registration of the operating conditions and the desired values. These convert a variety of physical parameters into electrical signals.
2. The ECU for generating the electrical output signals by processing the information using specified arithmetic operations (control algorithms)
3. Actuators to convert the ECU's electrical output signals into mechanical parameters.

Sensors (Fig. 2)

Crankshaft-speed sensor

The piston position in the combustion chamber is decisive in defining the start of injection. All the engine's pistons are connected to the crankshaft by connecting rods (conrods). A sensor on the crankshaft can therefore provide information on the position of all the pistons. The rotational speed defines the number of crankshaft rotations per minute.
This important input variable is calculated in the ECU using the signal from the inductive crankshaft-speed sensor.

Signal generation
A 60-tooth ferromagnetic trigger wheel is attached to the crankshaft. On the actually used trigger wheel 2 teeth are missing. This large gap is allocated to a defined crankshaft position for cylinder 1.
The crankshaft-speed sensor registers the trigger wheel's tooth sequence. It comprises a permanent magnet and a soft-iron core with a copper winding

(Fig. 1). The magnetic flux in the sensor changes as the teeth and gaps pass by, and a sinusoidal AC voltage is generated the amplitude of which increases sharply in response to higher engine (crankshaft) speeds. Adequate amplitude is already available from speeds as low as 50 min^{-1}.

Calculation of engine speed
The angular relationship (offset) between the cylinder pistons is such that two full crankshaft rotations (720°) elapse before the start of each new working cycle at cylinder 1. If the pistons are offset uniformly with respect to each other, this means that

$$\text{Angular ignition spacing [°]} = \frac{720°}{\text{No. of cylinders}}$$

On a 4-cylinder engine, the angular ignition spacing is 180°, in other words the crankshaft-speed sensor must scan 30 teeth between two ignitions. The period of time required is termed the segment time, and the mean crankshaft speed in the segment time is the engine speed.

Fig. 1

Crankshaft-speed sensor

1 Permanent magnet, **2** Housing,
3 Engine crankcase, **4** Soft-iron core, **5** Winding,
6 Trigger wheel.

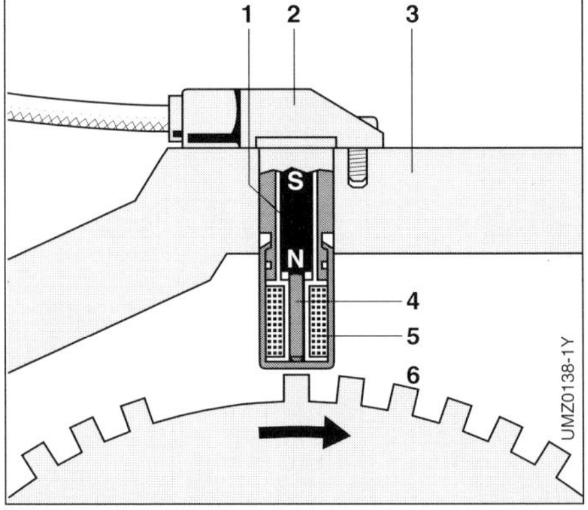

Fig.2

Sensors of a Common Rail injection system, together with various system components

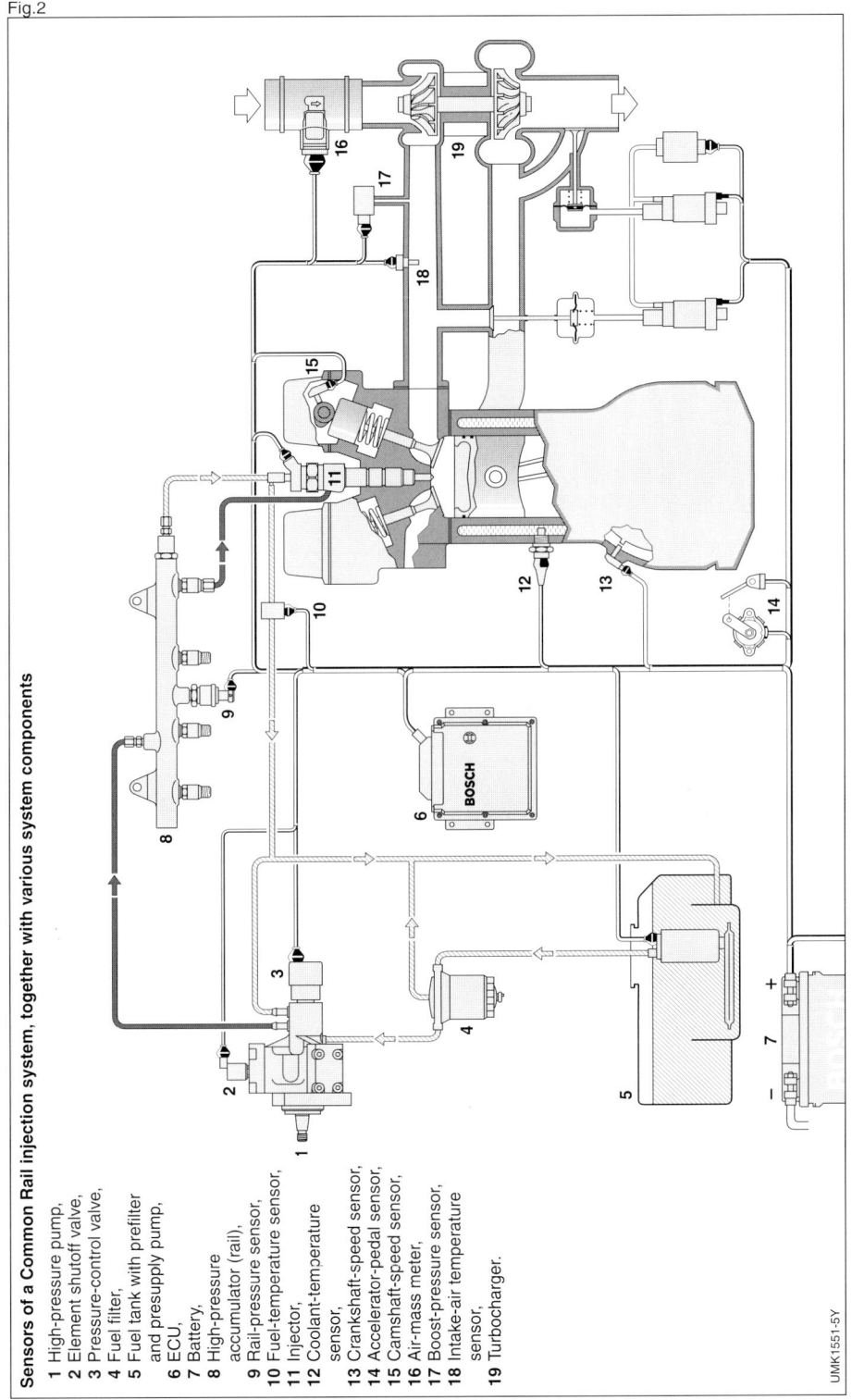

1 High-pressure pump,
2 Element shutoff valve,
3 Pressure-control valve,
4 Fuel filter,
5 Fuel tank with prefilter and presupply pump,
6 ECU,
7 Battery,
8 High-pressure accumulator (rail),
9 Rail-pressure sensor,
10 Fuel-temperature sensor,
11 Injector,
12 Coolant-temperature sensor,
13 Crankshaft-speed sensor,
14 Accelerator-pedal sensor,
15 Camshaft-speed sensor,
16 Air-mass meter,
17 Boost-pressure sensor,
18 Intake-air temperature sensor,
19 Turbocharger.

UMK1551-5Y

Camshaft-speed sensor

The camshaft controls the engine's intake and exhaust valves. It turns at half the speed of the crankshaft. When a piston travels in the direction of TDC, the camshaft position determines whether it is in the compression phase with subsequent ignition, or in the exhaust phase. This information cannot be generated from the crankshaft position during the starting phase. During normal engine operation on the other hand, the information generated by the crankshaft sensor suffices to define the engine status. In other words, this means that if the camshaft sensor should fail while the vehicle is being driven, the ECU still receives information on the engine status from the crankshaft sensor.

The camshaft sensor utilises the Hall effect when establishing the camshaft position. A tooth of ferromagnetic material is attached to the camshaft and rotates with it. When this tooth passes the semiconductor wafers of the camshaft sensor, its magnetic field diverts the electrons in the semiconductor wafers at right angles to the direction of the current flowing through the wafers. This results in a brief voltage signal (Hall voltage) which informs the ECU that cylinder 1 has just entered the compression phase.

Temperature sensors

Temperature sensors are installed at a number of different points:
– In the coolant circuit, to establish engine temperature by way of the coolant temperature (Fig. 3),
– In the intake manifold to measure the temperature of the intake air,
– In the engine lube oil to measure the oil temperature (optional), and
– In the fuel-return line to measure the fuel temperature (optional).

The sensors are equipped with a temperature-dependent resistor with a negative temperature coefficient (NTC) which is part of a voltage-divider circuit across which 5 V are applied.
The voltage drop across the resistor is inputted into the ECU through an analog-to-digital converter (ADC) and is a measure for the temperature. A characteristic curve is stored in the ECU microcomputer which defines the temperature as a function of the given voltage value (Fig. 4).

Hot-film air-mass meter

Particularly during dynamic operation, precise compliance with the correct A/F ratio is imperative in order to comply with the exhaust-gas limits as stipulated by law. This necessitates the use of sensors which precisely register the air-mass

Fig.3

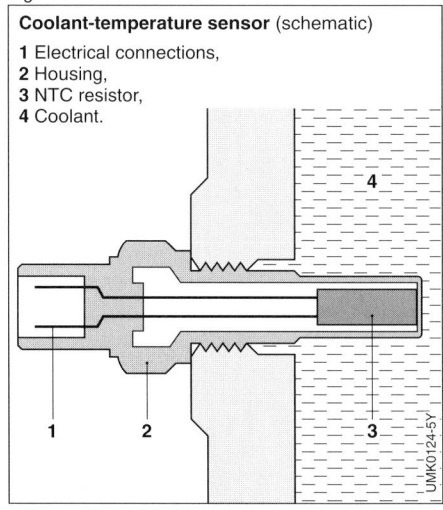

Coolant-temperature sensor (schematic)

1 Electrical connections,
2 Housing,
3 NTC resistor,
4 Coolant.

Fig.4

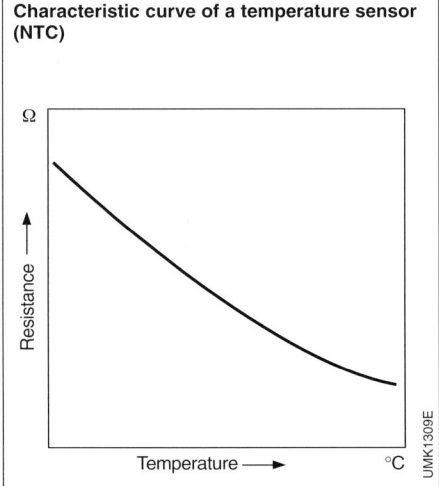

Characteristic curve of a temperature sensor (NTC)

flow actually being drawn in by the engine at a particular moment. This load sensor's measuring accuracy must be completely independent of pulsation, reverse flow, EGR, variable camshaft control, and changes in the intake-air temperature.

A hot-film air-mass meter was selected as being most suitable for complying with the above stipulations. The hot-film principle is based on the transfer of heat from a heated sensor element to the air-mass flow (Fig. 5). A micromechanical measuring system is utilised which permits registration of the air-mass flow and detection of flow direction. Reverse flows are also detected in case of strongly pulsating air flow.

The micromechanical sensor element is located in the plug-in sensor's flow passage (Fig. 5, Item 5). The plug-in sensor can be installed in the air filter or in a measuring tube in the engine's air-intake duct.

Fig.5

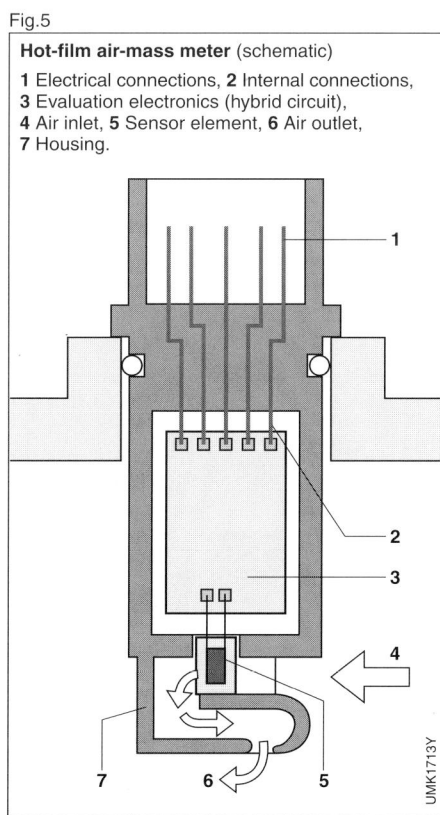

Hot-film air-mass meter (schematic)

1 Electrical connections, **2** Internal connections,
3 Evaluation electronics (hybrid circuit),
4 Air inlet, **5** Sensor element, **6** Air outlet,
7 Housing.

UMK1713Y

There are a variety of different-sized measuring tubes available, depending upon the maximum air throughput required by the engine. The signal-voltage curve, as a function of the air-mass flow, is divided into signal ranges for forward flow and reverse flow. In order to increase measuring accuracy, the measuring signal is referred to a reference voltage outputted by the engine management. The characteristic curve has been designed so that during diagnosis in the workshop an open-circuit conductor, for instance, can be detected with the help of the engine management.

A temperature sensor can be incorporated for measuring the intake-air temperature.

Accelerator-pedal sensor

In contrast to conventional distributor and in-line injection pumps, with EDC the driver's acceleration input is no longer transmitted to the injection pump by Bowden cable or mechanical linkage, but is registered by an accelerator-pedal sensor and transmitted to the ECU (this is also known as drive-by-wire).

A voltage is generated across the potentiometer in the accelerator-pedal sensor as a function of the accelerator-pedal setting. Using a programmed characteristic curve, the pedal's position is then calculated from this voltage.

Boost-pressure sensor

The boost-pressure sensor (BPS) is pneumatically connected to the intake manifold and measures the intake manifold's absolute pressure between 0.5 and 3 bar. The sensor is sub-divided into a pressure cell with two sensor elements, and a chamber for the evaluation circuit. The sensor elements and the evaluation circuit are mounted on a common ceramic substrate.

Each sensor element comprises a bell-shaped thick-film diaphragm, containing a reference volume with defined internal pressure. The diaphragm is displaced to a greater or lesser degree as a function of charge pressure.

Piezoresistive resistors are located on the diaphragm's surface whose resistance changes when mechanical stress is applied. These resistors are connected as a bridge so that when the diaphragm moves this causes a change in the bridge balance. This means that the bridge voltage is a measure for the boost presssure.

The evaluation circuit is responsible for amplifying the bridge voltage, compensating for temperature influences, and for linearization of the pressure characteristic. The evaluation circuit's output signal is inputted to the ECU where, with the help of a programmed characteristic curve, it is used for calculating the boost pressure.

ECU

Assignment and method of operation

The ECU evaluates the signals it receives from the external sensors and limits them to the permissible voltage level.

From this input data, and from stored characteristic maps, the ECU microprocessors calculate the injection times and the instants of injection, and convert these times to signal characteristics which are adapted to the movements of the engine pistons and crankshaft. The specified accuracy and the engine's high dynamic response demands high levels of computing power.

The output signals from the ECU microprocessors are used to trigger driver stages which provide adequate power for switching the actuators for rail-pressure control and element switch-off. In addition, actuators for engine function are triggered (e.g. EGR actuator, boost-pressure actuator, and the relay for the electric fuel pump), as well as those for further auxiliary functions such as blower relay, auxiliary-heater relay, glow relay, air-conditioner). The driver stages are proof against short-circuit and destruction due to brief electrical overloading. Errors of this type, and open-circuit or unplugged lines, are reported

to the microprocessor. Diagnosis functions in the injector driver stages detect faulty signal characteristics, and in addition a number of the output signals are transferred via interfaces for use in other systems in the vehicle. And within the framework of a special safety concept, the ECU monitors the complete fuel-injection system.

Injector triggering places particularly heavy demands on the driver stages. In the injector, the current from the driver stage generates a magnetic force in the triggering element which is applied to the injector's high-pressure system. In order to ensure very tight tolerances, and high reproducibility of the injected fuel quantity, this coil must be triggered with steep current flanks. This necessitates high voltages being made available in the ECU.

A current control circuit divides the energisation time (injection time) into a pickup-current phase and a hold phase. It must operate so accurately that the injector guarantees reproducible injection under all operating conditions. In addition, it must reduce the power loss in the ECU and the injectors.

Operating conditions

High demands are made upon the ECU regarding

– The surrounding (ambient) temperatures (in normal cases from $-40...+85°$ C),
– The resistance to fuels and lubricants etc.,
– The resistance to humidity, and
– Mechanical loading.

Very high demands are also made upon electromagnetic compatibility (EMC) and upon the radiation of HF interference signals. ·

Design and construction

The ECU has a metal housing. The sensors, the actuators, and power supply are connected to the ECU through a multi-pole plug-in connector. The power components which directly trigger the actuators are integrated in the ECU in such a manner that they can efficiently dissipate their heat to the

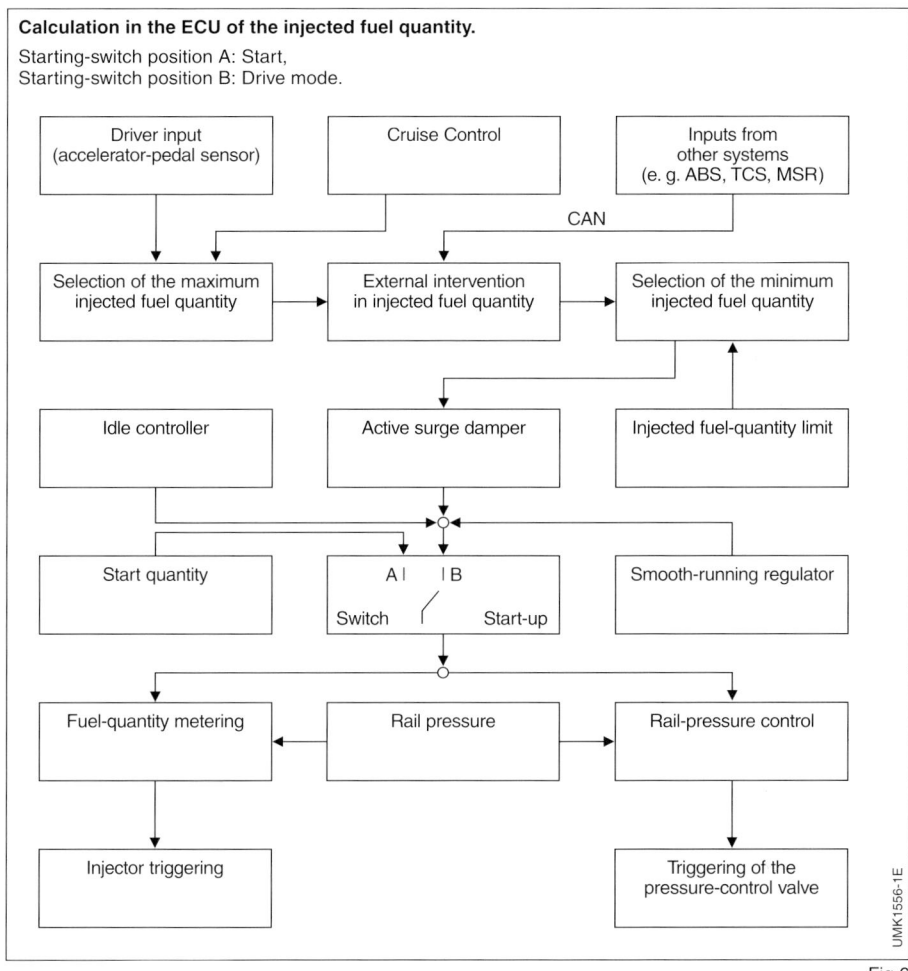

Calculation in the ECU of the injected fuel quantity.

Starting-switch position A: Start,
Starting-switch position B: Drive mode.

Fig.6

ECU housing. Both sealed and non-sealed versions of the ECU are available.

Operating-state control
In order that the engine operates with optimum combustion in every operating state, the ECU in each case calculates the appropriate injected fuel quantity. In the process, a number of parameters must be taken into account (Fig. 6).

Start quantity
For starting, the injected fuel quantity is calculated as a function of temperature and cranking speed. The start quantity is injected from the moment the starting switch is turned to "Start" (Fig. 6, Pos. A) until the engine has reached a given minimum speed. The driver has no influence upon the start quantity.

Drive mode
When the vehicle is being driven normally (Fig. 6, starting switch in Pos. B), the injected fuel quantity is calculated from the accelerator-pedal setting (accelerator-pedal sensor) and the engine speed. Calculation utilises the driving map so that the driver input and the engine O/P power are optimally matched to each other.

Idle-speed control
At idle, fuel consumption depends for the most part on engine efficiency and idle speed. Since a considerable portion of a

vehicle's fuel consumption in dense traffic conditions is attributable to this operating state, it is obvious that idle speed must be kept to a minimum. The idle speed, though, must be set so that no matter what the operating conditions, it does not drop so far under load that the engine runs roughly or even stops. This applies for instance when the vehicle electrical system is loaded, when the air-conditioner is switched on, when a gear is engaged on an automatic transmission, or when the power steering is in operation. In order to regulate to the desired idle speed, the idle controller varies the injected fuel quantity until the actual engine speed equals the desired idle speed. Here, the desired idle speed and the control characteristic are influenced by the selected gear and by the engine temperature (coolant-temperature sensor). In addition to the external load moments, the internal friction moments must also be taken into account and compensated for by the idle-speed control. These change minimally but steadily throughout the vehicle's service life, as well as being highly dependent upon temperature.

Smooth-running control
Due to mechanical tolerances and ageing, there are differences in the torques generated by the engine's individual cylinders. This leads to rough or irregular running, particularly at idle. The smooth-running (cylinder-balancing) control measures the engine-speed changes every time a cylinder has "fired" and compares them with each other. The injected fuel quantity for each cylinder is then adjusted in accordance with the measured differences in engine speed between the individual cylinders, so that each cylinder makes the same contribution to the torque generated by the engine. The smooth-running control is only operative in the lower engine-speed range.

Vehicle-speed controller
The vehicle-speed controller (Cruise Control) comes into operation when the vehicle is to be driven at a constant speed. It controls the vehicle speed to

that inputted by the driver at the operator unit in the instrument panel,
The injected fuel quantity is increased or reduced until the actual speed equals the set speed. While the Cruise Control is in operation, the control process is interrupted if the driver depresses the clutch, or applies the brakes. If the accelerator pedal is pressed, the vehicle can be accelerated beyond the speed which has been set with the Cruise Control. As soon as the accelerator pedal is released, the Cruise Control regulates the speed back down again to the previous set speed. Similarly, if the Cruise Control has been switched off, the driver only needs to press the re-activate key in order to again select the last speed which had been set.

Controlling the injected fuel quantity limit
There are a number of reasons why the fuel quantity desired by the driver (or the maximum physically possible quantity) must not be injected.
These include:
– Excessive pollutant emissions,
– Excessive soot emissions,
– Mechanical overloading due to excessive torque or engine speed, or
– Thermal overload as a result of excessive coolant, lube-oil, or turbocharger temperature.

Fig.7

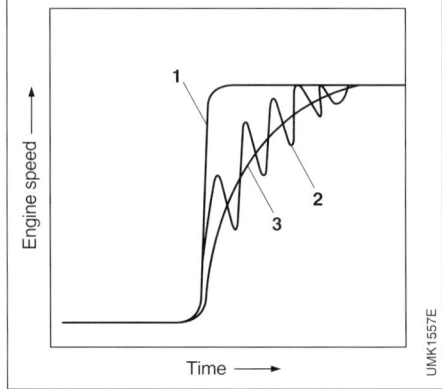

Active surge damper

1 Sudden accelerator-pedal movement (driver input), **2** Engine-speed curve without active surge-damping control, **3** With active surge-damping control.

Engine speed →

Time →

UMK1557E

The limit for the injected fuel quantity is formed from a number of input variables, for instance intake air mass, engine speed, and coolant temperature.

Active surge-damping control

When the accelerator pedal is abruptly depresssed or released, this causes the injected fuel quantity to change rapidly with the result that there is also a rapid change in the torque developed by the engine. These abrupt load changes lead to the resilient engine mountings and the drivetrain generating bucking oscillations which result in fluctuations of engine speed (Fig. 7).

The active surge-damper reduces these periodic speed fluctuations by varying the injected fuel quantity at the same frequency as the periodic speed fluctuations: Less fuel is injected when the speed increases, and more when it decreases. This effectively damps the surge movements.

Engine shutoff

The diesel engine operates according to the "auto-ignition" principle. This means that it can only be switched off by interrupting its supply of fuel.

In the case of the "Electronic Diesel Control (EDC)", the engine is switched off by the ECU stipulating "injected fuel quantity zero". The system also features a number of additional (redundant) switch-off paths.

Actuators (Fig. 8)

Injector

Special injectors with hydraulic servo-system and electrical triggering element

Fig.8

Actuators and other components of the Common Rail system

1 Glow control unit,
2 ECU,
3 Instrument panel with displays for fuel consumption, engine speed etc.,
4 Battery,
5 Glow plug,

6 Injector,
7 EGR positioner,
8 Charge-pressure actuator,
9 Vacuum pump,
10 Turbocharger.

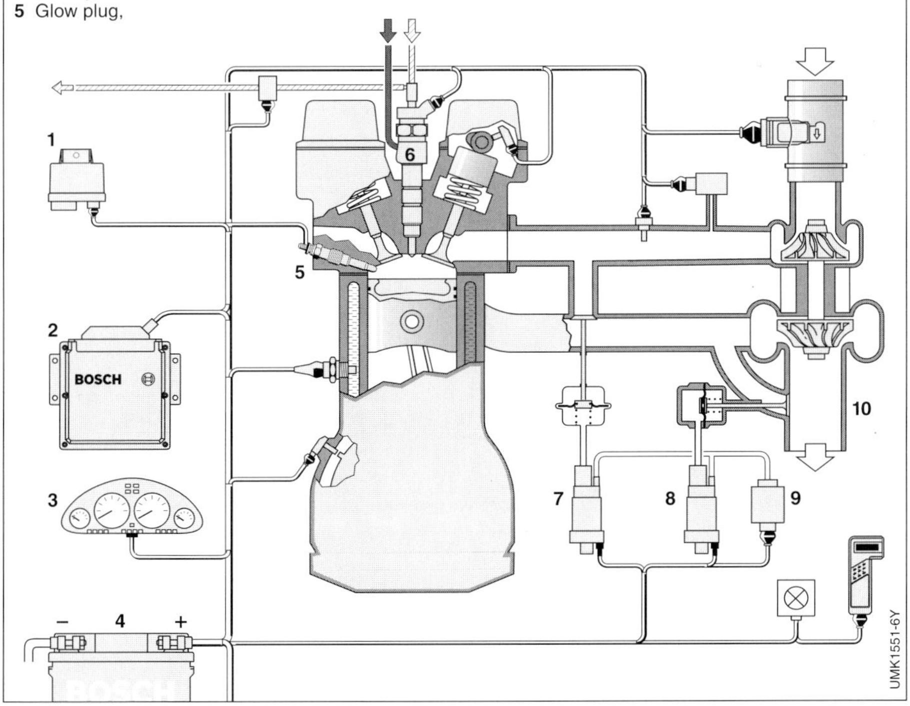

UMK1551-6Y

(solenoid valve) are used with the Common Rail system in order to achieve efficient start of injection and precise injected fuel quantity. At the start of injection, a high pickup current is applied to the injector so that the solenoid valve opens quickly. As soon as the nozzle needle has travelled its complete stroke, and the nozzle has opened completely, the energizing current is reduced to a lower holding value. The injected fuel quantity is now defined by the injector opening time and the rail pressure. Injection is terminated when the solenoid valve is no longer triggered and closes as a result.

Pressure-control valve

The ECU uses the pressure-control valve to control the rail pressure. When the pressure-control valve is triggered, the energized electromagnet forces the armature up against the seal seat and the valve closes. The high-pressure and low-pressure sides are sealed off from each other and the rail pressure increases.

In the non-energized mode, the electromagnet no longer exerts force on the armature and the pressure-control valve opens so that some of the fuel from the rail can flow back to the tank through a collector line. The rail pressure drops.

It is possible to vary the pressure by pulsing (pwm) the triggering current. The degree to which the pressure-control valve is opened or closed depends on the pulse rate (duty cycle).

Glow control unit

The glow control unit is responsible for ensuring efficient cold starting. It also shortens the warm-up period, a fact which is highly relevant for exhaust emissions. The preheating time is a function of the coolant temperature. The further glow phases during engine start or when the engine is actually running are determined by a number of parameters which include engine speed and injected fuel quantity. Glow control utilises a power relay.

Electropneumatic transducer

The valves or flaps of the swirl controller, EGR positioner, and boost-pressure actuator are actuated mechanically using overpressure or negative pressure. Here, the engine ECU generates an electrical signal which is converted to overpressure or negative pressure by an electropneumatic transducer.

Boost-pressure actuator

Passenger-car engines with exhaust-gas turbocharging must develop high torques even at low engine speeds.

The turbocharger housing is therefore designed for a low exhaust-gas mass flow. But in order that the excessive charge-air pressure is not developed when larger exhaust-gas masses flow, part of this flow must be diverted past the turbine by means of a bypass valve ("wastegate") and into the exhaust system. To do so, the boost-pressure actuator (Fig. 9) changes the cross section at the wastegate as a function of engine speed and injected fuel quantity etc. Variable

Fig.9

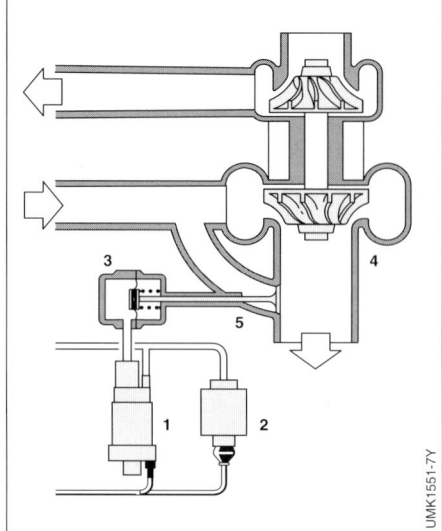

Boost-pressure actuator

1 Boost-pressure actuator,
2 Vacuum pump,
3 Pressure actuator,
4 Exhaust-gas turbocharger,
5 Wastegate.

UMK1551-7Y

turbine geometry (VTG) can be applied instead of the wastegate. This varies the approach angle at the turbine wheel and as a result influences the charge-air pressure.

Swirl controller

Swirl control serves to influence the swirl movement of the intake air. The swirl itself is usually generated by spiral-shaped inlet passages and determines the mixing of the fuel and the air in the combustion chamber. It therefore has considerable influence upon the combustion quality. As a rule, a pronounced swirl is generated at low engine speeds, and a weak swirl at high speeds. The swirl can be modified by means of the swirl controller (flap or slide valve) in the vicinity of the throttle valve.

EGR positioner

With exhaust-gas recirculation (EGR) a portion of the exhaust gas is led into the engine's intake tract. Up to a certain degree, an increasing portion of the residual exhaust gas content has a positive effect upon energy conversion and therefore upon the exhaust-gas emissions. Depending upon the engine's operating point, the air/gas mass drawn into the cylinders can be composed of up to 40% exhaust gas (Figs. 10 and 11).

For ECU control, the actual drawn-in fresh-air mass is measured and compared at each operating point with the air-mass setpoint value. Using the signal generated by the control circuit, the EGR positioner (a valve) opens so that exhaust gas can flow into the intake tract.

Throttle-valve control

The throttle valve in the diesel engine fulfils a completely different function to that in the gasoline engine. It serves to increase the exhaust-gas recirculation rate by reducing the overpressure in the intake manifold. Throttle-valve control is only operative in the lower speed range.

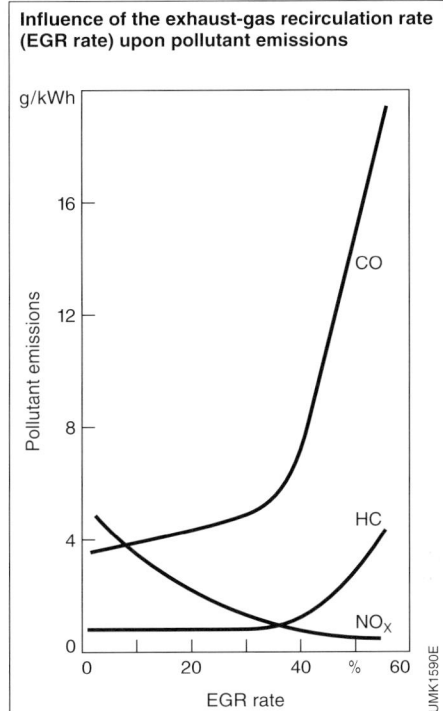

Influence of the exhaust-gas recirculation rate (EGR rate) upon pollutant emissions

Fig.10

Fig.11

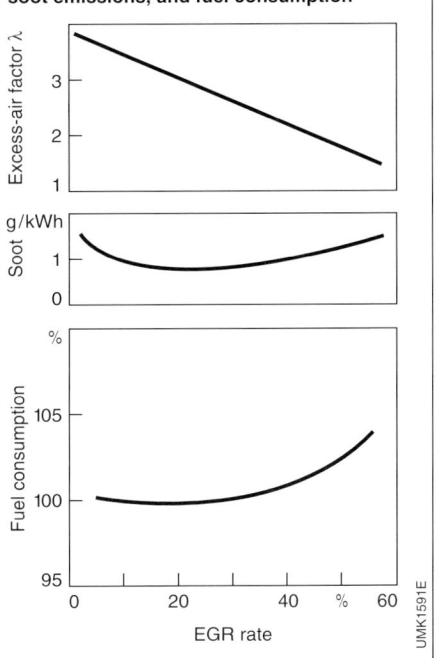

Influence of the exhaust-gas recirculation rate (EGR rate) upon the excess-air factor λ, soot emissions, and fuel consumption

Information exchange

Communication between the ECU's

Communication between the Common Rail ECU and the other ECU's takes place using the CAN-Bus (Controller Area Network). This is used to transmit the desired values, operating data, and status information needed for operation and for fault monitoring.

External intervention in injected fuel quantity

In this case, the injected fuel quantity is influenced not by the Common Rail ECU but by another ECU (for instance for ABS, TCS), which informs the Common Rail ECU that the engine's torque, and therefore the injected fuel quantity, is to be modified (and if so, by how much).

Electronic immobilizer

For theft-deterrence purposes, the immobilizer ECU prevents engine start. The driver can signal to this ECU, for instance with remote control, that he is authorized to use the vehicle. The immobilizer ECU then informs the Common Rail ECU that it is permissible for fuel to be injected so that the engine can be started and the vehicle driven.

Air conditioner

In order to maintain a pleasant temperature inside the vehicle when the temperatures outside are too high, the air conditioner cools down the air by means of a refrigeration compressor.
Depending upon the engine and the driving situation, its energy consumption is 1%...30% of the engine output power. The target therefore is not so much the improvement of the temperature control but rather the optimum use of engine torque. As soon as the driver accelerates strongly (and therefore needs maximum engine torque), the refrigeration compressor is switched off by the EDC.

Integrated diagnosis

Sensor monitoring

For sensor monitoring, the integrated diagnosis facility checks whether they are being supplied with power, and whether their O/P signals are plausible (within the permitted range, e.g. temperature between -40 und $150°C$). Where possible, the redundancy principle is applied for important signals. That is, in case of malfunction a switch is made to another similar signal.

Monitoring module

In addition to the microcontroller, the ECU also incorporates a monitoring module. The ECU and the monitoring module monitor each other. If a malfunction is detected, either of them can switch off the injection independent of the other.

Malfunction detection

Malfunction detection is only possible within the monitoring range of a given sensor. A signal path is classified as faulty when an error is present for longer than a predefined period. In such cases, the error is stored in the ECU's error memory together with details of the environmental conditions which prevailed when the error/malfunction occurred (e.g. coolant temperature, engine speed etc.). For a large number of errors/malfunctions, it is possible for the "OK again" status to be established. Here, the signal path must be identified as intact for a defined period of time.

Error procedure

If a sensor's permitted output-signal range is violated, a switch is made to a substitute value. This procedure is applied for the following input signals:
– Battery voltage,
– Coolant, air, and lube-oil temperature,
– Charge-air pressure,
– Atmospheric pressure and intake-air quantity.
In addition, in case of non-plausible signals from the accelerator-pedal sensor and/or the brakes, a substitute accelerator-pedal sensor signal is applied.

Fig.12

System overview of a Common Rail injection system and a variety of system components

1 High-pressure pump,
2 Element shutoff valve,
3 Pressure-control valve,
4 Fuel filter,
5 Fuel tank with preliminary filter and presupply pump,
6 ECU,
7 Glow control unit,
8 Battery,
9 High-pressure accumulator (rail),
10 Rail-pressure sensor,
11 Flow limiter,
12 Pressure limiter valve,
13 Fuel-temperature sensor,
14 Injector,
15 Sheathed-element glow plug,
16 Coolant-temperature sensor,
17 Crankshaft sensor,
18 Camshaft sensor,
19 Intake-air temperature sensor,
20 Boost-pressure sensor (BPS),
21 Air-mass meter,
22 Turbocharger,
23 EGR positioner,
24 Boost-pressure actuator,
25 Vacuum pump,
26 Instrument panel with display for
fuel consumption, engine speed etc.,
27 Accelerator-pedal sensor,
28 Brake contacts,
29 Clutch switch,
30 Road-speed sensor,
31 Operator unit for vehicle-speed controller
(Cruise Controller),
32 Air-conditioner compressor,
33 Air-conditioner operator unit,
34 Diagnosis display with connection
for diagnostic unit.

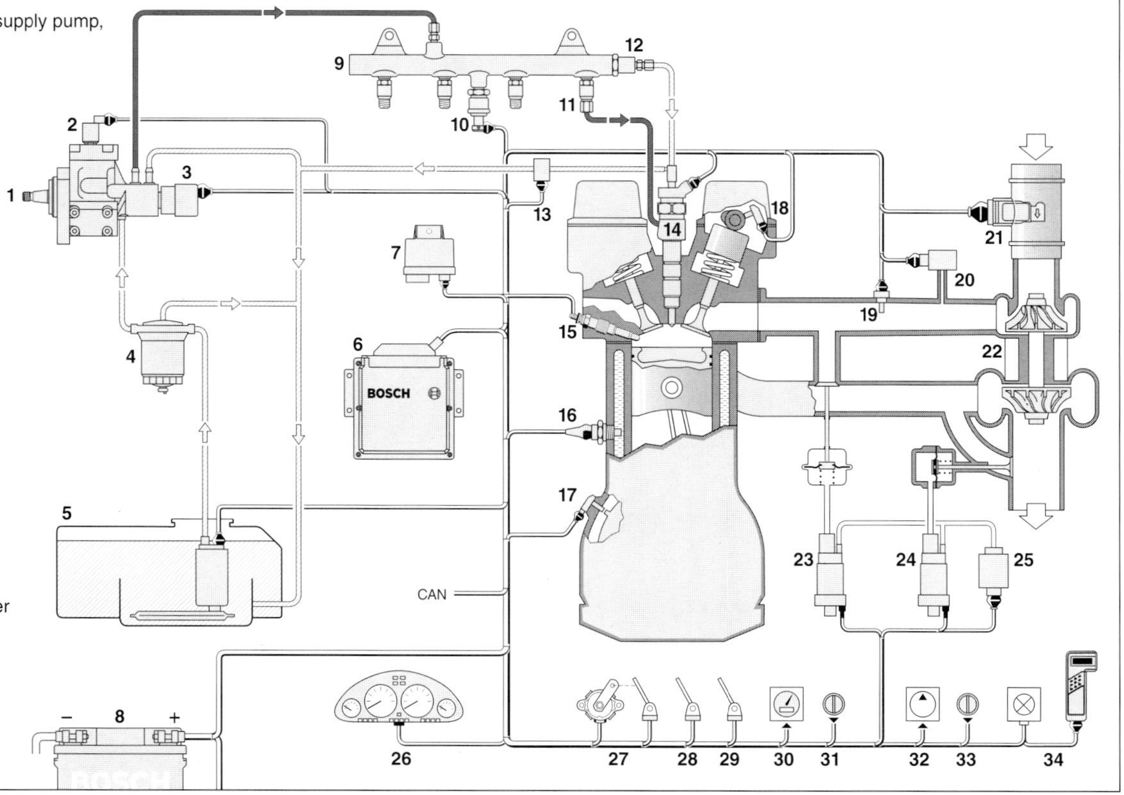

UMK1551Y

Single-plunger fuel-injection pumps

PF single-plunger fuel-injection pumps

Design and construction

The PF single-plunger injection pump has no integral camshaft of its own (the "F" in PF indicates external drive). The PF single-plunger injection pumps and the PE in-line injection pumps operate according to the same principles. The PF pumps are suitable for use with small, medium-size, and large engines (Fig. 1), to which they are flange-mounted.

Since each engine cylinder is allocated its own injection pump, the use of single-plunger injection pumps with multicylinder engines make it possible to use very short high-pressure delivery lines. This means that when single-cylinder pumps are used with multi-cylinder engines, only one type of pump and high-pressure line is employed. The PF injection pump is equipped with a locking device when it leaves the factory. This keeps the pump in its full-load delivery setting so that additional adjustment work is not needed when mounting it on the engine.

Engine-speed control

On large engines, the governor is fitted directly to the engine housing. The adjustment of injected fuel quantity needed for controlling engine speed is defined by the governor and transmitted to the

Fig. 1

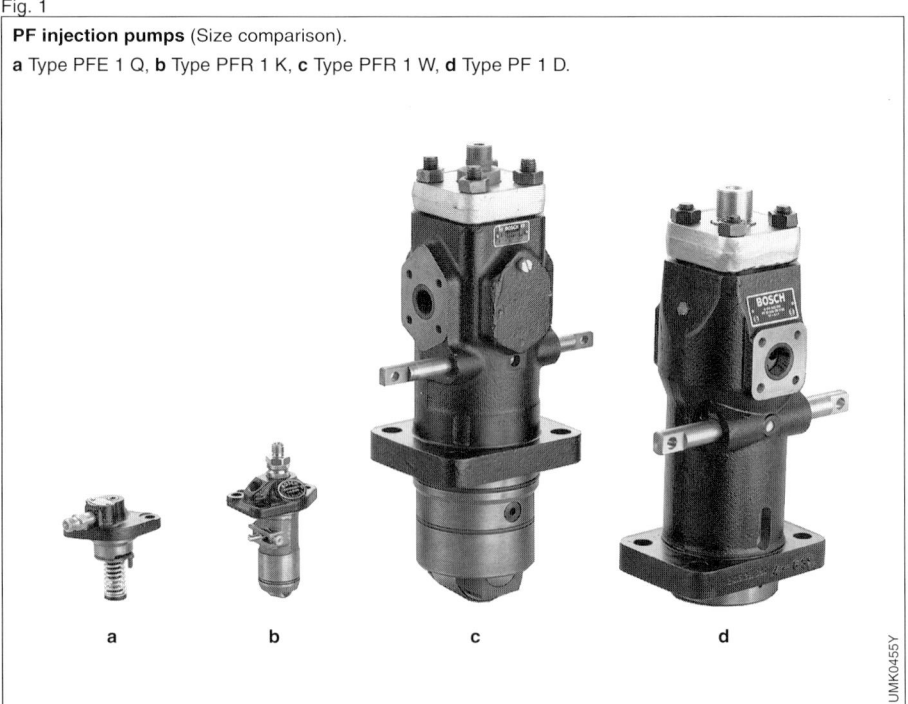

PF injection pumps (Size comparison).
a Type PFE 1 Q, **b** Type PFR 1 K, **c** Type PFR 1 W, **d** Type PF 1 D.

a b c d

UMK0455Y

individual pumps through a linkage integrated in the engine. Mechanical-hydraulic, electronic, and purely mechanical governors are in use, although the latter are rarely encountered. A sprung intermediate element in the transmission linkage to each pump ensures that governing can still take place even if the pump's adjustment mechanism should block.

Fuel supply and delivery

With regard to fuel supply and delivery, fuel filtering, and bleeding of the injection system, the same stipulations apply for the PF pump as they do for the PE in-line injection pump. The gear-type fuel-supply pump generates a pressure of 3…10 bar, and its delivered fuel quantity is 3…5 times the injected fuel quantity. Dirt is to be kept out of the injection system by using fine filtration with 5…30 μm aperture sizes. PF injection pumps developing powers of up to 100 kW/cylinder are used not only for diesel fuel injection, but also for injecting heavy-oil with viscosities up to 700 mm^2/s at 50 °C. In order to be able to deliver and inject this oil it must be heated to 150 °C before reaching the fuel-supply pump so that it has the necessary injection viscosity of approx. 10…20 mm^2/s.

Injection timing

The injection cams for the individual PF injection pumps are on the engine's timing-gear camshaft. This means that turning the camshaft relative to the timing gears is out of the question for injection-timing purposes. Instead, an intermediate member is used, for instance a rocker fitted between roller tappet and camshaft (Fig. 2), which provides for an advance angle of several degrees. This permits not only the optimization of fuel consumption and exhaust-gas emissions, but also the adaptation to the ignition quality of different fuels.

Fig. 2

Injection timing (At the eccentric rocker bearing)

1 Engine, **2** Timing-shaft contour, **3** Rocker bearing, **4** Injection cam, **5** Pump plunger,
6 Roller tappet, **7** Tappet roller, **8** Rocker, **9** Camshaft roller, **10** Engine camshaft.

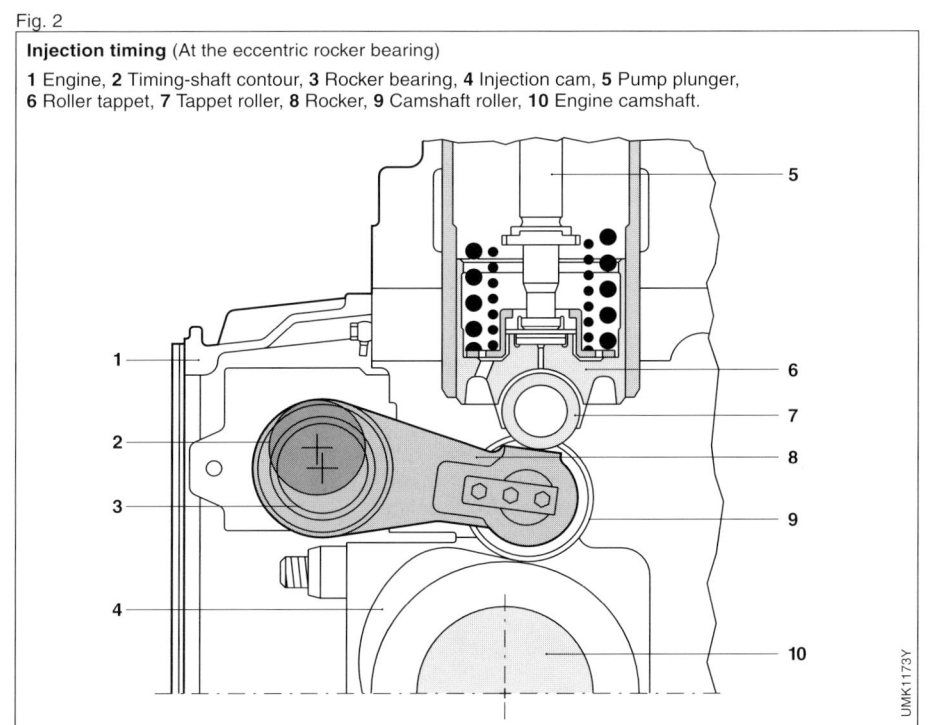

Sizes

PF pumps up to 50 kW/cylinder

These single-plunger injection pumps are used for instance in diesel engines for powering small construction machines, pumps, agricultural tractors, and engine-generator sets.

The pump types PFE 1 A.. and PFE 1 Q.. are of the single-plunger type without integral roller tappet (Fig. 4), which instead is integrated in the engine block. Control of fuel delivery on both pump ranges is by means of a control-sleeve lever which engages with the control rack running in the engine block. The PFR..K pump types with integrated roller tappet are fitted to 1-, 2-, 3-, or 4-cylinder engines. On all versions, the plunger is rotated by means of a toothed control sleeve which engages in a control rack mounted in the pump housing (Fig. 3).

The small PF pumps have a maximum pump speed of about 1800 min⁻¹.

Depending upon plunger diameter (5...9 mm), full-load injected fuel quantity is max. 95 mm^3 per plunger stroke, and maximum permissible peak injection pressure in the high-pressure lines (pump end) is 600 bar.

PF fuel-injection pumps are equipped with constant-volume valves (with or without return-flow restriction). Constant-pressure valves are used in applications characterized by high pump loading and increased demands on the stability of the injected fuel quantity.

PF pumps above 50 kW/cylinder

Single-plunger injection pumps (Fig. 5) are installed in diesel engines with output powers of up to 1000 kW per cylinder. They operate with diesel fuel and heavy oils of various viscosities. With peak injection pressures (pump end) of up to approx. 1200 bar, through-holes are drilled in the pump barrels. When peak pressures up to 1500 bar are concerned, blind-hole plunger-and-barrel assemblies are used to keep the deformation in the

Fig. 3

PFR 1 K injection pump

1 Delivery-valve holder, **2** Delivery valve,
3 Pump barrel, **4** Pump plunger,
5 Control rack, **6** Control sleeve,
7 Plunger control arm, **8** Roller tappet.

Fig. 4

PFE 1 Q injection pump

1 Delivery valve, **2** Delivery-valve holder,
3 Housing, **4** Pump barrel,
5 Pump plunger, **6** Control sleeve,
7 Plunger return spring.

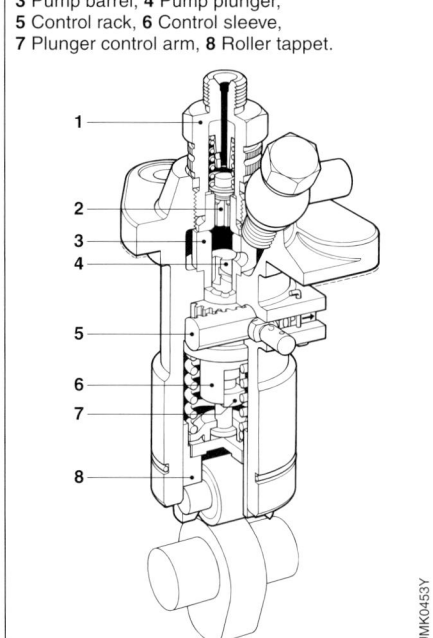

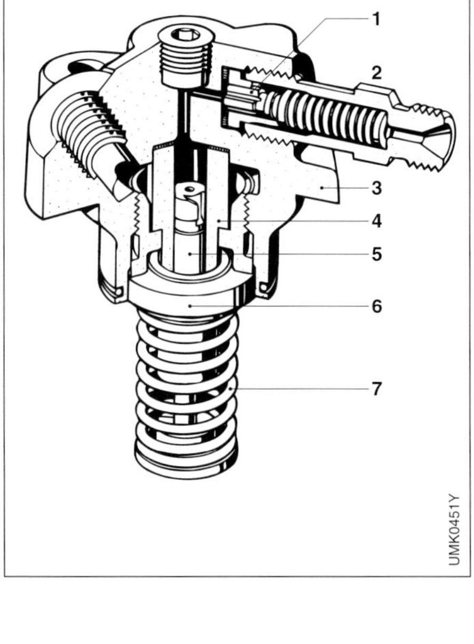

UMK0453Y

UMK0451Y

area of the plunger-and-barrel head to a minimum (Fig. 6).

To prevent pump-housing damage due to the high-energy jet of fuel generated at the end of delivery (port closing), baffle screws are fitted in the immediate vicinity of the pump-barrel spill ports.

In order to seal the delivery valve against very high pressures, the surfaces between it and the flange, as well as between it and the pump barrel are lapped and flat. Pressure compensation at the pump plunger prevents asymmetrical loads. The pump-plunger control helixes are arranged similarly to those on the in-line injection pump. The delivery quantity is adjusted by rotating the plunger via the control rack. Rack-travel indication is provided.

The plunger-and-barrel assembly is provided with leak-oil return and an extra oil block for leak fuel. A second ring-shaped groove together with a blocking-oil inlet passage is machined into the pump barrel. Filtered oil is forced into this groove at a pressure of 3…5 bar, a pressure which under normal circumstances exceeds that in the pump's fuel gallery, and suffices to form an oil block to prevent passage of the leak fuel. The extremely small leakage quantity concerned (a mixture of fuel and lube-oil) is drawn off through a separate outlet and led into a collecting tank.

In the case of heavy-oil operation, the roller tappet of the PFR pump, or the guide sleeve of the PF pump, are lubricated with engine lube-oil through a separate connection.

Fig. 5

PF 1 D injection pump

1 Delivery valve, **2** Vent screw,
3 Pump barrel, **4** Pump plunger,
5 Control rack, **6** Control sleeve,
7 Guide sleeve.

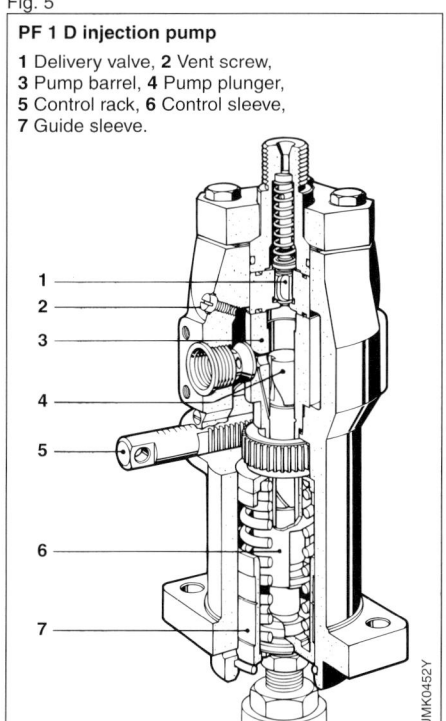

Fig. 6

PFR 1 CY injection pump

1 Flange, **2** Pilot-delivery valve, **3** Pump barrel, **4** Pump plunger, **5** Oil-mixture exit, **6** Control rack, **7** Plunger return spring, **8** Pump housing, **9** Roller tappet, **10** Pressure-holding valve, **11** Vent screw, **12** Baffle screw, **13** Leakage-oil return, **14** Control sleeve.

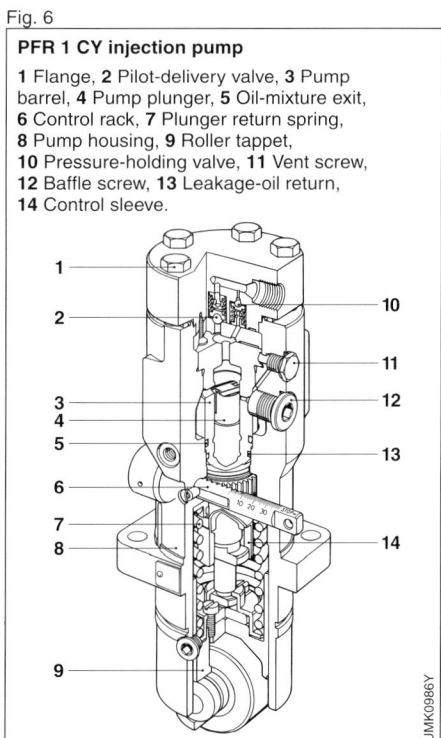

Solenoid-valve-controlled single-plunger fuel-injection pumps PF-MV

Applications and technical requirements

Single-plunger fuel-injection pumps Type PF are installed in large diesel engines for ships and diesel-powered locomotives etc.

Legislation on emission limits is already in place for certain applications and is planned for other areas of use. Such legislation influences engine development, and therefore also has a considerable effect upon the development of the fuel-injection systems used with these engines.

A freely selectable start-of-injection is one of the prerequisites which must be met by fuel-injection systems in complying with the requirements for very high power, low NO_X emissions and low fuel consumption.

Single-plunger injection pumps equipped with an electronically triggered solenoid valve for defining the start and end of fuel delivery comply with these demands for a freely selectable start-of-injection point.

Design and construction

By means of a roller tappet, the rotating injection cam on the engine's camshaft forces the pump plunger up and down. Contact between roller and cam is maintained by the pump spring.

Since it has no milled helixes, the pump plunger here is of far simpler design than the conventional PF version. There are no spill ports in the pump barrel, and the delivery valve is superseded by the solenoid valve.

A sleeve is fitted in the upper section of the pump barrel to hold and guide the cylindrical control plunger. At one end of this plunger there is a conical valve element which mates with a conical valve seat in the sleeve, and at the other there is an armature plate. When the solenoid is energized it pulls this armature plate in against the force of a spring so that the control plunger moves to its closed position and the valve plate is pushed up against its valve seat. In this closed position, there is a residual air gap between the armature plate and the solenoid body. As soon as the solenoid is de-energized, the spring pulls the armature plate, together with the control plunger, away from the solenoid and in doing so frees the conical valve seat and forces the control plunger against the stop plate. Compared to the mechanically governed PF fuel-injection pumps, on the electrically controlled system there is no longer a control rack, or control shaft with linkages to the pump's control rack, and the governor's actuator.

These elements are all superseded by control cables to the driver stages, from where they go to the solenoid valves of the single-plunger injection pumps.

This system's advantages are as follows:
- Less installation space required,
- Higher installation flexibility,
- Lower costs for material and application engineering,
- More precise possibilities of control,
- Precision fuel metering with the possibility of being able to control the injected fuel quantity and start of delivery individually for each cylinder, and
- Shut-off of individual engine cylinders (during part load operation, the remaining cylinders then operate at maximum efficiency).

It is even possible to apply different control data for the individual injection processes at each cylinder. This means that reports from the solenoid valves arising from specific operating conditions such as fuel temperature or wear can be taken into account.

Using two extra passages in the pump cylinder, it is possible to install a pressureless fuel-return line and/or forced-feed oil lubrication for the pump plunger.

Method of operation

During its downward movement, the pump plunger draws in fuel through the open solenoid valve, and during its upward movement forces fuel though the open valve and back into the suction chamber until an ECU pulse switches off the solenoid valve so that it closes. A detailed account of this process is given below (Fig. 1):

There are two suction chambers (2) in this pump. One with the armature plate (8), and the other with the stop plate (3). A low-pressure fuel pump forces fuel from the fuel inlet (1) into suction chamber (2) with armature plate (8). From here, the fuel flows through passages to the opposite suction chamber (2) with stop plate (3) from where a passage leads to the fuel exit (4) from where a line goes to the fuel tank. The suction chambers are designed so that the fresh fuel entering them from the fuel tank can permanently cool the solenoid-valve magnets (at full load, the flushing quantity is about 4 larger than the delivery quantity).

When the control plunger (11) is shifted to its open position by the spring, fuel is free to enter the pump chamber (12) at the conical valve seat. During its upward movement, the pump plunger (5) forces the fuel into the suction chamber (with stop plate) until the triggered solenoid closes the control plunger again. The fuel can now no longer enter the low-pressure section and is instead forced by the pump plunger through the high-pressure line to the injection nozzle (6) from where it is injected into the cylinder.

This instant in time is defined as the start of delivery. Filling the pump chamber through the control-plunger gap at the conical valve seat has the following advantage: In case the valve sticks in the open or closed position, or a malfunction occurs in the solenoid-valve triggering, dangerous engine operating states are impossible, and in particular the engine cannot overspeed due to too much fuel being injected. The valve spring opens the solenoid valve at the end of the closing pulse, so that the pressure collapses in the high-pressure system and fuel delivery stops.

As with the mechanically controlled single-plunger injection-pump system, the rate-of-discharge curve can be influenced in the solenoid-controlled system by the shape of the injection cam.

Fig. 1

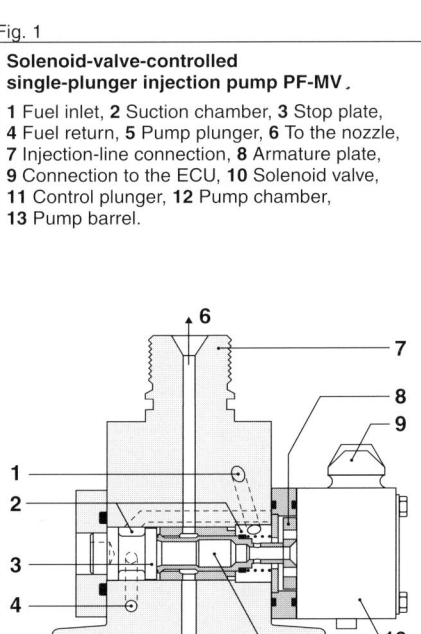

Solenoid-valve-controlled single-plunger injection pump PF-MV.

1 Fuel inlet, 2 Suction chamber, 3 Stop plate, 4 Fuel return, 5 Pump plunger, 6 To the nozzle, 7 Injection-line connection, 8 Armature plate, 9 Connection to the ECU, 10 Solenoid valve, 11 Control plunger, 12 Pump chamber, 13 Pump barrel.

UMK1562Y

Innovative fuel-injection systems

Unit-injector system (UIS)

The unit injector (Fig. 1) is screwed directly into the engine's cylinder head. This design combines the injection pump and the injection nozzle in a single unit which is driven by the engine camshaft. Each unit injector has its own high-speed solenoid valve which controls start and end of injection. With the solenoid valve open the unit injector forces fuel into the return line, and when the solenoid valve closes, into the engine cylinder. The start of injection is defined by the solenoid closing point, and the injected fuel quantity by the closing time (length of time the solenoid remains closed). The solenoid valve is triggered by an ECU with map-based control, which means that start and end of injection are programmable and therefore independent of piston position in the engine cylinder. Compared to the injection valve/injector in electronically controlled gasoline injection systems (Jetronic/EFI) the diesel solenoid valve must be able to control pressures which are 300...500 times higher, and must also be able to switch 10...20 times faster.

In today's conventional fuel-injection systems, the maximum injection pressure is limited by the physical characteristics of the high-pressure lines between injection pump and injection nozzle. The unit injector makes such lines superfluous, which means that injection pressures of up to 2000 bar are possible. Pressures of this magnitude, together with map-based control of start of injection and duration of injection (injected fuel quantity), lead to a considerable reduction in the diesel engine's pollutants emission. Using electronic control concepts, special functions such as temperature-controlled start of injection, engine smooth-running control, anti-buck damping, and pilot fuel injection for even further reductions in noise, be-

Fig. 1

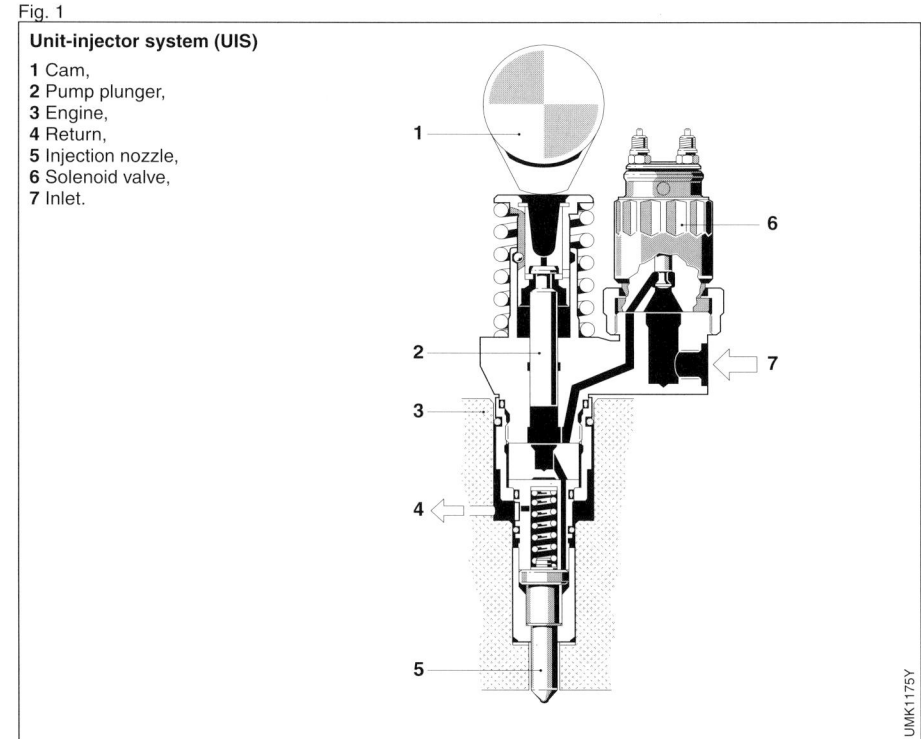

Unit-injector system (UIS)

1 Cam,
2 Pump plunger,
3 Engine,
4 Return,
5 Injection nozzle,
6 Solenoid valve,
7 Inlet.

UMK1175Y

come feasible. In addition, the use of unit injectors makes it possible to switch off individual engine cylinders during part-load operation (Fig. 1).

Unit-pump system (UPS)

The unit-pump system (Fig. 2) is a modular high-pressure injection system. From the control-engineering viewpoint, it is closely related to the unit injector.

Both systems use an individual injection pump for each engine cylinder which is driven by an extra cam (injection cam) on the engine's camshaft. The use of an electronically triggered high-speed solenoid valve enables the moment of injection and the injected fuel quantity to be precisely adjusted for each cylinder. This permits:
– Fuel delivery to the injection nozzle,
– Interruption of fuel delivery, and
– Return of excess fuel to the fuel tank.

Similar to the unit-injector system, the unit-pump system registers the most important engine and environmental parameters, and converts them into the optimal start of injection and optimal injected fuel quantity for the given operating conditions. The systems comprises the follow-ing modules:
– The high-pressure pump with attached solenoid valve,
– The short high-pressure delivery line, and
– The nozzle-and-holder assembly.
With this modular design, which contrasts with the compact design of the unit-injector system, the unit-pump system represents a directly controlled high-pressure injection system which is suitable for a wide range of different installation requirements.

This system is further characterized by a fault-recognition facility, the possibility of emergency operation and self-diagnosis, as well as the option of communicating with other control systems via already existing interfaces.

Fig. 2

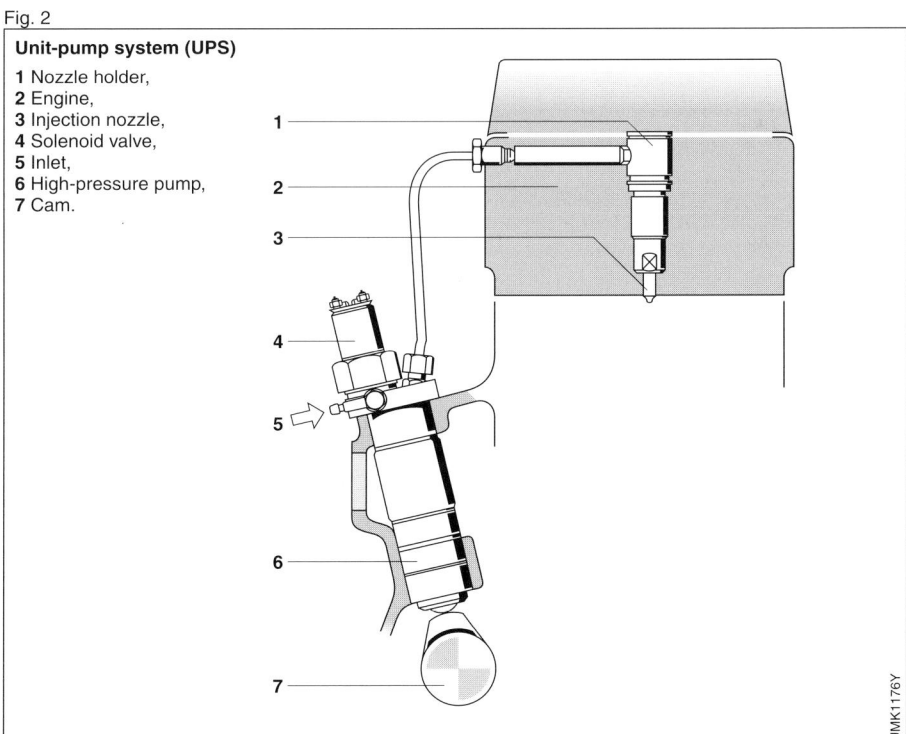

Unit-pump system (UPS)

1 Nozzle holder,
2 Engine,
3 Injection nozzle,
4 Solenoid valve,
5 Inlet,
6 High-pressure pump,
7 Cam.

UMK1176Y

Start-assist systems

Since leakage and heat losses reduce the pressure and the temperature of the A/F mixture at the end of the compression stroke, the cold diesel engine is more difficult to start and the mixture more difficult to ignite than it is when hot. These facts make it particularly important that start-assist systems are used. The minimum starting temperature depends upon the engine type. Pre-chamber and swirl-chamber engines are equipped with a sheathed-element glow plug (GSK) in the auxiliary combustion chamber which functions as a "hot spot". On small direct-injection (DI) engines, this "hot spot" is located on the combustion chamber's periphery. Large DI truck engines on the other hand have the alternative of using air preheating in the intake manifold (flame start) or special, easily ignitable fuel (Start Pilot) which is sprayed into the intake air. Today, the start-assist systems use sheathed-element glow plugs practically without exception.

Sheathed-element glow plug

The sheathed-element glow plug's tubular heating element is so firmly pressed into the glow-plug shell that a gas-tight seal is formed. The element is a metal tube which is resistant to both corrosion and hot gases, and which contains a heater (glow) element embedded in magnesium-oxide powder (Fig. 1). This heater element comprises two series-connected resistors: the heater filament in the glow-tube tip, and the control filament. Whereas the heater filament maintains virtually constant electrical resistance regardless of temperature, the control filament is made of material with a positive temperature coefficient (PTC). On newer-generation glow plugs (GSK2), its resistance increases even more rapidly with rising temperature than was the case with the conventional S-RSK glow plug. This means that the newer GSK2 glow plugs are characterized by reaching the temperature needed for ignition far more quickly (850 °C in 4s). They also feature a lower steady-state temperature (Fig. 2) which means that the glow plug's temperature is limited to a non-critical level. The result is that the GSK2 glow plug can remain on for up to 3 minutes following engine start. This post-glow feature improves both the warm-up and run-up phases with considerable improvements in noise and exhaust-gas emissions.

Fig. 1

Sheathed-element glow plug GSK2

1 Electrical connector terminal, **2** Insulating washer, **3** Double gasket, **4** Terminal pin, **5** Glow-plug shell, **6** Heater seal, **7** Heater and control filament, **8** Glow tube, **9** Filling powder.

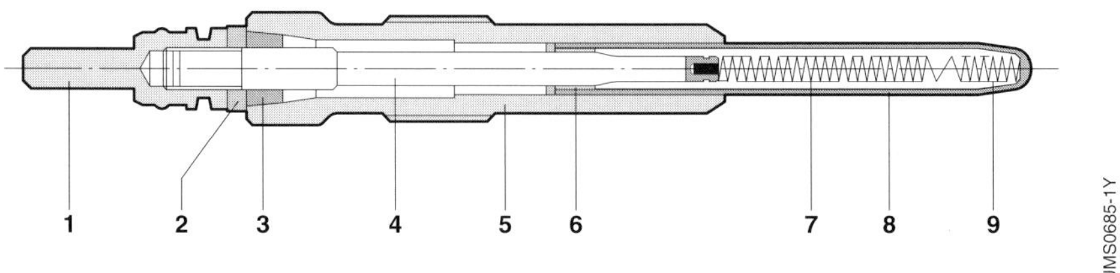

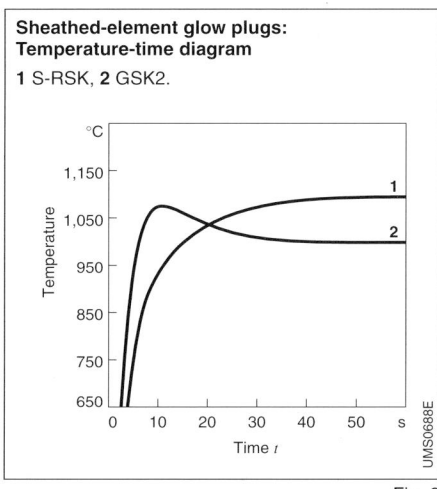

Sheathed-element glow plugs: Temperature-time diagram

1 S-RSK, **2** GSK2.

Fig. 2

Flame glow plug

The flame glow plug burns fuel to heat the intake air. Normally, the injection system's supply pump delivers fuel to the flame plug through a solenoid valve. The flame plug's connection fitting is provided with a filter, and a metering device which permits passage of precisely the correct amount of fuel appropriate to the particular engine. This fuel then evaporates in an evaporator tube surrounding the tubular heating element and mixes with the intake air. The resulting mixture ignites on the 1,000 °C heating element at the flame-plug tip.

Glow control unit

For triggering the glow plugs, the glow control unit (GZS) is provided with a power relay and a number of electronic switching blocks. These, for instance, control the glow duration of the glow plugs, or have safety and monitoring functions. Using their diagnosis functions, more sophisticated glow control units are also able to recognise the failure of individual glow plugs and inform the driver accordingly. Multiple plugs are used as the control inputs to the ECU. In order to avoid voltage drops, the power supply to the glow plugs is through suitable threaded pins or plugs.

Functional sequence

The diesel engine's glow plug and starter switch, which controls the preheat and starting sequence, functions in a similar manner to the ignition and starting switch on the spark-ignition (SI) engine. Switching to the "Ignition on" position starts the preheating process and the glow-plug indicator lamp lights up. This extinguishes to indicate that the glow plugs are hot enough for the engine to start, and cranking can begin. In the following starting phase, the droplets of injected fuel ignite in the hot, compressed air. The heat released as a result leads to the initiation of the combustion process (Fig. 3).

In the warm-up phase following a successful start, post-heating contributes to faultless engine running (no misfiring) and therefore to practically smokeless engine run-up and idle. At the same time, when the engine is cold, preheating reduces combustion noise. A glow-plug safety switchoff prevents battery discharge in case the engine cannot be started.

The glow-control unit can be coupled to the ECU of the Electronic Diesel Control (EDC) so that information available in the EDC control unit can be applied for optimum control of the glow plugs in accordance with the particular operating conditions. This is yet another possibility for reducing the levels of blue smoke and noise.

Fig. 3

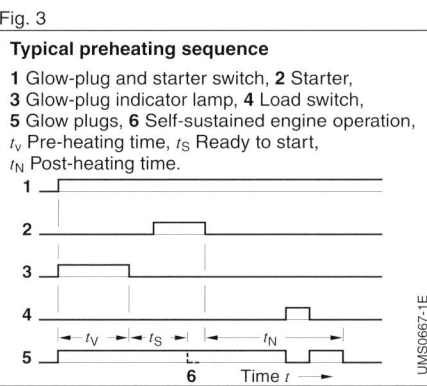

Typical preheating sequence

1 Glow-plug and starter switch, **2** Starter,
3 Glow-plug indicator lamp, **4** Load switch,
5 Glow plugs, **6** Self-sustained engine operation,
t_V Pre-heating time, t_S Ready to start,
t_N Post-heating time.

Index

A

A injection pump, 69
Accelerator-pedal sensor, 204, 240, 283
Accumulator injection system (CRS), 53, 256
Active surge-damping device, ARD, 143, 216, 245, 287
ADA, altitude-pressure compensator, 141, 184
Additives (diesel fuel), 19
Add-on modules, 134, 176
Advancing (start of injection), 15
Air filter, 20
Air filter, fender-mounted, 20
Air supply, 20
Air-conditioner, intervention in, 249, 290
Air-fuel mixture, 15
ALDA, manifold-pressure compensator, absolute metering, 142
Altitude-pressure compensator, ADA, 141, 184
Angle-of-rotation sensor, 218
ARD, active surge-damping device, 143, 216, 245, 287
Auto-ignition, 18, 28
Auxiliary combustion chamber, 6
Axial-piston distributor pump, electronically controlled, VE-EDC, 52, 212
Axial-piston distributor pump, mechanically governed, VE, 52, 152
Axial-piston distributor pump, solenoid-valve-controlled, VE-MV, 52, 218

B

Barrel (PE), 56
Black smoke, 26
Boiling range (diesel fuel), 19
Boost-pressure actuator, 246, 288
Bus arbitration (CAN), 203

C

Cam contours (PE), 58, 161
Cam plate (VE), 161
Cam shape (PE), 58
CAN, serial data transmission, 201
Carbon-dioxide analysis, 43
Carbon-monoxide analysis, 43
Catalytic converter, 31
Central air filter, 20
Cetane number, 18
Charge-air pressure sensor, 204, 240, 283
CLD measurement method (exhaust-gas test), 44
Cold behavior (diesel fuel), 18
Combination governor, 104
Combustion-pressure limit, 12
Common Rail, CR, 53, 256
Compression pressure, 14
Compression ratio, 28
Compression stroke, 4
Compression temperature, 16
Constant-pressure delivery-valve (PE), 61
Constant-volume delivery-valve (PE), 60
Content-based addressing (CAN), 202
Control loop (EDC), 207, 211, 215
Control-lever stops (PE), 134
Control-rack (PE), 102
Control-rack stops (PE), 135
Control-rack travel (PE), 101
Control-sleeve in-line fuel-injection pump, electronically controlled, PE-EDC, 210
Control-sleeve in-line fuel-injection pump, PE, 52, 73
Control-sleeve pump, PE, 52, 73
Control-sleeve, 73
CRT method (catalytic converter), 31
CVS test method (exhaust-gas test), 37
Cyclone prefilter, 21

D

Data transmission to other systems (EDC), 200

T

U

V

W

Comprehensive information made easy
Bosch Technical Books

Automotive electrics and electronics
Vehicle electrical systems, Symbols and circuit diagrams, EMC/interference-suppression, Batteries, Alternators, Starting systems, Lighting technology, Comfort and convenience systems, Diesel-engine management systems, Gasoline-engine management systems.

Hard cover,
Format: 17 x 24 cm,
3rd updated edition,
314 pages, bound,
with numerous illustrations.
ISBN 0-7680-0508-6

Gasoline-engine management
Combustion in the gasoline (SI) engine, Exhaust-gas control, Gasoline-engine management, Gasoline fuel-injection system (Jetronic), Ignition, Spark plugs, Engine-management systems (Motronic).

Hard cover,
Format: 17 x 24 cm,
1st edition,
370 pages, bound,
with numerous illustrations.
ISBN 0-7680-0510-8

Diesel-engine management
Combustion in the diesel engine, Mixture formation, Exhaust-gas control, In-line fuel-injection pumps, Axial piston and radial-piston distributor pumps, Distributor injection pumps, Common Rail (CR) accumulator injection systems, Single-plunger fuel-injection pumps, Start-assist systems.

Hard cover,
Format: 17 x 24 cm,
2nd updated and expanded edition,
306 pages, bound,
with numerous illustrations.
ISBN 0-7680-0509-4

Driving-safety systems
Driving safety in the vehicle, Basics of driving physics, Braking-system basics, Braking systems for passenger cars, ABS and TCS for passenger cars, Commercial vehicles – basic concepts, systems and schematic diagrams, Compressed-air equipment for commercial vehicles, ABS, TCS, EBS for commercial vehicles, Brake testing, Electronic stability program (ESP).

Hard cover,
Format: 17 x 24 cm,
2nd updated and expanded edition,
248 pages, bound,
with numerous illustrations.
ISBN 0-7680-0511-6

Automotive terminology
4,700 technical terms from automotive technology, in German, English and French, assembled from the above Bosch Technical Books: "Automotive Electrics and Electronics", "Diesel-engine management"; "Gasoline-engine management" and "Driving-safety systems".

Hard cover,
Format: 17 x 24 cm,
1st edition
378 pages, bound.
ISBN 0-7680-0338-5

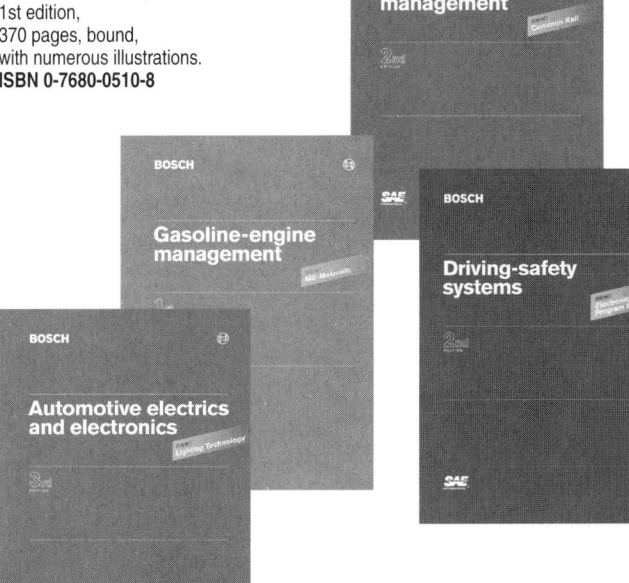
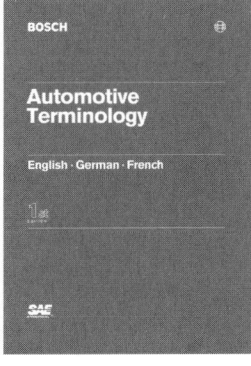